Recent Developments in Automatic Control and Systems Engineering

Recent Developments in Automatic Control and Systems Engineering

Guest Editors

John Anthony Rossiter
Maurício B. de Souza Jr.

MDPI

Basel • Beijing • Wuhan • Barcelona • Belgrade • Novi Sad • Cluj • Manchester

Guest Editors

John Anthony Rossiter
Department of Automatic Control and Systems Engineering
University of Sheffield
Sheffield
UK

Maurício B. de Souza Jr.
Department of Chemical Engineering
Universidade Federal do Rio de Janeiro (UFRJ)
Rio de Janeiro
Brazil

Editorial Office
MDPI AG
Grosspeteranlage 5
4052 Basel, Switzerland

This is a reprint of the Special Issue, published open access by the journal *Processes* (ISSN 2227-9717), freely accessible at: https://www.mdpi.com/journal/processes/special_issues/59O827D612.

For citation purposes, cite each article independently as indicated on the article page online and as indicated below:

Lastname, A.A.; Lastname, B.B. Article Title. *Journal Name* **Year**, *Volume Number*, Page Range.

ISBN 978-3-7258-4903-1 (Hbk)
ISBN 978-3-7258-4904-8 (PDF)
https://doi.org/10.3390/books978-3-7258-4904-8

Contents

About the Editors

John Anthony Rossiter

Anthony Rossiter currently holds a position at the University of Sheffield, United Kingdom. He received his M.A. and D.Phil. degrees from Oxford University in 1987 and 1990, respectively. Following a brief postdoctoral position, he obtained his first academic post at Loughborough University in 1992 and subsequently moved to the University of Sheffield in 2001.

He is best known for his contributions to predictive control and notably the early work on proposing and understanding dual-mode approaches, alongside being credited with providing the foundational ideas and publications behind tube model predictive control. He is also the author of several popular textbooks and online resources on predictive control, as well as numerous academic papers.

More recently, he has served as chair of both the IFAC and IEEE Technical Committees on Control Education and continues to hold leadership roles in the international control community, promoting the importance of both effective pedagogy and the potential for modern technology to enhance the student learning experience. His open-access teaching resources on the web "Modelling, dynamics and control" are widely used. His pedagogical work is based on the philosophy that what works effectively can also be implemented simply and cheaply for both staff and students.

Maurício B. de Souza Jr.

Maurício B. de Souza Jr. currently holds a post at the Department of Chemical Engineering, School of Chemistry, Federal University of Rio de Janeiro (UFRJ), Brazil.

His graduate studies at COPPE/UFRJ were pioneering in Brazil; his M.Sc. dissertation (1989) addressed model predictive control and his D.Sc. thesis (1993) focused on neural network-based modeling and control. Since then, his research has centered on the intersection of artificial intelligence and process control. As a professor for more than thirty years, he has devoted his career to forming professionals in process control and machine learning for chemical engineering applications, as well as disseminating knowledge in these areas. He has also collaborated extensively with industry, coordinating and participating in several P&ID projects across sectors ranging from oil and gas to pulp and paper.

Currently, his primary research interests lie in the development of hybrid models that combine AI techniques with phenomenological knowledge, and their application in the optimization and control of chemical and biochemical processes. In recent years, he has also led a series of academic seminars on process control, organized with colleagues from LADES/COPPE/UFRJ and GIMSCOP/UFRGS, with active participation from both academia and industry.

Preface

This Special Issue presents a rich panorama of approaches and applications, ranging from hydraulic drive pavers to oil and gas processes. Model Predictive Control (MPC) algorithms emerge as a widely used tool; however, the hardware limitations of many industries still constrain their application to simpler systems. While artificial intelligence tools hold great promise, their widespread adoption may require a transitional period, where greater trust must be established, before becoming fully realized. Nevertheless, the recent developments presented here demonstrate their potential in monitoring, control, and optimization applications. The authors hope that this snapshot into control will encourage more researchers to contribute to solving the challenges of the future.

John Anthony Rossiter and Maurício B. de Souza Jr.
Guest Editors

processes

MDPI

Editorial

Recent Developments in Automatic Control and Systems Engineering

Maurício B. de Souza Jr. [1,2,*] and John Anthony Rossiter [3,*]

1 Chemical Engineering Department, School of Chemistry, Universidade Federal do Rio de Janeiro (UFRJ), Rio de Janeiro 21941-914, RJ, Brazil
2 Chemical Engineering Program, COPPE, Universidade Federal do Rio de Janeiro (UFRJ), Rio de Janeiro 21941-914, RJ, Brazil
3 Department of Automatic Control and Systems Engineering, University of Sheffield, Sheffield S1 3JD, UK
* Correspondence: mbsj@eq.ufrj.br (M.B.d.S.J.); j.a.rossiter@sheffield.ac.uk (J.A.R.)

1. Introduction

The industry's pursuit of digitalization, with a focus on automating solutions, is presently being complemented by efforts to develop autonomous systems, where artificial intelligence (AI), advanced control, modeling and optimization play a significant role [1–3]. In this context, this Special Issue presents some recent developments in automatic control and systems engineering, including practical and experimental applications in various areas, ranging from autonomous vehicles to industrial processes. A core aim is to capture many of the important themes being pursued by researchers in this area, which in turn will allow new researchers to plan future research.

Computational methods and implementation challenges are two important sides of various works and include such topics as digital twins, hardware-in-the-loop (HIL) and software-in-the-loop (SIL) simulation techniques; real-time computing issues in model predictive control (MPC); edge computing for real-time control; and reinforcement learning (RL) for control and optimization. The continuing development of autonomous systems and their interactions with human expertise are also addressed, together with the use of machine learning and vision systems to support the real-time monitoring of industrial processes.

2. An Overview of the Published Papers

A common point of interest regarding automatic process control is the extent to which it is safe and efficient to allow machine autonomy, in comparison to human control and intervention. In the case of autonomous vehicles, as claimed by Yan and coworkers (contribution 1), human drivers and automated systems must still cooperate in a transition phase to autonomy. In this case, the question of "driver authority" is an import point of attention. The group of Xinyu Liu proposed a driving authority optimization framework using model predictive control for shared steering control. They presented simulations and experiments in order to evaluate their framework.

While the analogy with autonomous vehicles provides a relevant reference, the automatic control of industrial processes involves specific considerations. One core observation is the widespread benefits of model predictive controllers (MPC), which have led to their broad adoption and the need for computationally efficient algorithms for their implementation. Consequently, several papers in this Special Issue focus on MPC. For example, Mendes and coworkers (contribution 3) explored the use of a low-computational-cost

 https://doi.org/10.3390/pr13082406

MPC-modified Takagi–Sugeno–Kang-based MPC (MTSK-MPC) embedded on a PLC (programmable logic controller) to control a pressure swing adsorption process (PSA). PSA is a cyclic method adopted for complex gas-phase separation. The authors showed the feasibility of their implementation through software-in-the-loop and hardware-in-the-loop simulations. In another work, Trentini and collaborators (contribution 7) investigated computationally efficient MPC for the control of a biphasic oil separator in the presence of slug flow using unrestricted horizon linear model predictive control (UHMPC). They used this unrestricted version of MPC to reduce the computational cost of the controller, avoiding the demand for expensive microprocessors, and compared their results via simulation with a linear quadratic regulator (LQR), with the UHMPC leading to more stable system behavior. Finally, and again on the same conceptual lines, Santanna and coworkers (contribution 6) addressed the practical dependence on limited hardware for the MPC of oil production processes equipped with an electric submersible pump (ESP). Two control strategies, namely robust infinite-horizon model predictive control (RIHMPC) and nonlinear model predictive control (NMPC), embedded in a microcontroller were evaluated and compared in terms of robustness and computational performance.

The next group of papers demonstrate the need for and benefits of tailored solutions that exploit existing control techniques for the improved control of specific applications. Gouveia and coworkers (contribution 4) developed an edge-computing-based modular control system and Yin and collaborators' (contribution 8) study focused on the automation of hydraulic drive pavers. Edge computing brings computation closer to the source of data generation, minimizing delays and improving system autonomy and reliability. The first study used data vibration signals from a real cement ball mill and aimed to test the system in different industrial process environments. The second investigation aimed to achieve constant speed control and reduce wandering deviation during paving operations, by proposing a control scheme based on a Global Navigation Satellite System (GNSS). The automation enhanced the efficiency and quality of paving operations.

An area of increasing interest in the community is the use of images for control (Image-based control (IBC)), given that algorithms and our ability to efficiently extract useful information from images are continually improving. While uses for autonomous driving are well known and widely reported, many other possible applications are becoming increasingly evident. Professor Durand's group (contribution 5) developed a testbed for IBC and analysis, consisting of a framework that integrates an open-source visualization software and process physics. The framework was applied to three different cases, demonstrating its potential.

Of course, no Special Issue on control would be complete without some discussion of artificial intelligence (AI) approaches. In the last few years, interest in using AI techniques in the field of process system engineering has grown significantly. Several developments in the design, monitoring, control and optimization of processes have been investigated (e.g., [4,5]) and it is clear that there is huge potential within this general area [3]. Within the Special Issue, we include AI work with an industrial application (Queiroz and co-workers, contribution 2) that synergistically integrated knowledge-based phenomenological models with a purely data-based artificial intelligence approach (the random forest tree-based algorithm) to develop soft-sensors for a sour water treatment unit (SWTU) in an oil refinery. The results allow the prediction of contaminants in the effluents of the SWTU and also the analysis of the input variables affecting them. Along the same lines, in transfer learning (TL), a deep learning model built from a source dataset is used to generalize the representations observed in another dataset. Professor Hedengren's group (contribution 9) presented a study on transfer learning (TL) and model predictive control. In Ref. [6], TL is applied in the context of MPC by using a pre-trained deep learning (DL) model of the MPC, and then fine-tuning the MPC training for a new automation task. The authors showed an illustration of the TL operation in which the behavior of the linear MPC of a liquid-level system is learned by a dynamic neural network (LSTM, long short-term memory) and then used to control a thermal system.

While learning-based methods could be considered as part of the broader AI family, one could also argue that reinforcement learning (RL) is to some extent a distinct area, one that is greatly facilitated by modern computer hardware and tools. Faria and coworkers (contribution 10) presented a review article on the application of reinforcement learning (RL) to process control [7]. According to the authors, unlike other review articles, their work examines RL from the perspective of simulation-based offline training, process demonstrations, and policy deployment with transfer learning. Finally, the challenges of integrating RL into online process control, including hyperparameter optimization, are also discussed, with a batch control experiment used to illustrate several proposed guidelines.

The important topic of process safety is also addressed in this Special Issue. Process safety involves strategies to avoid and mitigate incidents [8]; here, the procedures are based on process systems engineering approaches. Queiroz and co-workers (contribution 2) discussed the operational challenges and environmental concerns associated with SWTUs, particularly the removal of contaminants, and emphasized the role of soft sensors in real-time monitoring and control to prevent faults and ensure compliance with environmental regulations. In the work of Professor Durand's group (contribution 5), the use of image-based control systems and virtual testbeds to simulate and test monitoring and control strategies for chemical processes can help identify potential challenges and optimize control systems and cybersecurity strategies before implementation in real-world scenarios. Mendes and coworkers (contribution 3) and Santana and coworkers (contribution 6) ensured the safe implementation of MPCs in a PSA process and ESP-lifted oil wells, aiming to operate their systems within safe regions. Finally, Trentini and coworkers (contribution 7) discussed safety-critical issues posed by slug flow in offshore oil production control and developed a control strategy aiming to stabilize the process and mitigate disturbances, thereby enhancing its safety and reliability.

Finally, it is important to remember the future, and determine how the students of today might be encouraged to study the control challenges facing society [3] and take action. Consequently, the Special Issue also includes contribution 11, which addresses education and how, especially within the control arena, practice is changing to reflect current and future needs [9].

3. Conclusions

This Special Issue presents a rich panorama of approaches and applications, ranging from hydraulic drive pavers to oil and gas processes. Model Predictive Control (MPC) algorithms emerge as a widely used tool; however, the hardware limitations of many industries still constrain their application to simpler versions. While artificial intelligence tools hold great promise, their widespread adoption may require a transitional period where greater trust must be established before becoming fully realized. Nevertheless, the recent developments presented here demonstrate their potential in monitoring, control, and optimization. The authors hope this snapshot into control will promote more researchers to contribute to solving the challenges of the future.

Author Contributions: Writing—original draft preparation, M.B.d.S.J.; writing—review and editing, J.A.R. All authors have read and agreed to the published version of the manuscript.

Funding: Professor Maurício B. de Souza Jr. is grateful for financial support from CNPq (Grant No. 311153/2021-6) and Fundação Carlos Chagas Filho de Amparo à Pesquisa do Estado do Rio de Janeiro (FAPERJ) (Grant No. E-26/201.148/2022).

Conflicts of Interest: The author declares no conflicts of interest.

Abbreviations

The following abbreviations are used in this manuscript:

AI	Artificial intelligence
DL	Deep learning
ESP	Electric submersible pump
GNSS	Global Navigation Satellite System
HIL	Hardware-in-the-loop
IBC	Image-based control
LSTM	Long short-term memory
LQR	Linear quadratic regulator
MPC	Model predictive control
MTSK-MPC	Modified Takagi–Sugeno–Kang-based model predictive control
NMPC	Nonlinear model predictive control
PLC	Programmable logic controller
PSA	Pressure swing adsorption
RIHMPC	Robust infinite-horizon model predictive control
RL	Reinforcement learning
SIL	Software-in-the-loop
SWTU	Sour water treatment unit
TL	Transfer learning
UHMPC	Unrestricted horizon linear model predictive control

List of Contributions:

1. Yan, S.; Wei, Q.; Xie, X.; Zhao, D.; Liu, X. Driver–Automated Cooperation Driving Authority Optimization Framework for Shared Steering Control. *Processes* **2024**, *12*, 2313. https://doi.org/10.3390/pr12112313.
2. Queiroz, D.G.; Rodrigues, F.D.B.; Nogueira, J.N.P.; Melo, P.A.; Souza, M.B., Jr. Synergizing Phenomenological and AI-Based Models with Industrial Data to Develop Soft Sensors for a Sour Water Treatment Unit. *Processes* **2024**, *12*, 1900. https://doi.org/10.3390/pr12091900.
3. Mendes, T.P.G.; Ribeiro, A.M.; Schnitman, L.; Nogueira, I.B.R. A PLC-Embedded Implementation of a Modified Takagi–Sugeno–Kang-Based MPC to Control a Pressure Swing Adsorption Process. *Processes* **2024**, *12*, 1738. https://doi.org/10.3390/pr12081738.
4. Gouveia, G.; Alves, J.; Sousa, P.; Araújo, R.; Mendes, J. Edge Computing-Based Modular Control System for Industrial Environments. *Processes* **2024**, *12*, 1165. https://doi.org/10.3390/pr12061165.
5. Akkarakaran, L.A.; Gjonaj, G.; Rahman, M.; Durand, H.E. Virtual Test Beds for Image-Based Control Simulations Using Blender, *Processes* **2024**, *12*, 279. https://doi.org/10.3390/pr12020279.
6. Santana, B.A.; Matos, V.S.; Santana, D.D.; Martins, M.A.F. Embedded MPC Strategies for ESP-Lifted Oil Wells: Hardware-in-the-Loop Performance Analysis of Nonlinear and Robust Techniques. *Processes* **2023**, *11*, 1354. https://doi.org/10.3390/pr11051354.
7. Trentini, R.; Campos, A.; Salvador, M.A.; Scheuer, Y.M.; Santos, C.H.F. Unrestricted Horizon Predictive Controller Applied in a Biphasic Oil Separator under Periodic Slug Disturbances. *Processes* **2023**, *11*, 928. https://doi.org/10.3390/pr11030928.
8. Yin, X.; Yin, S.; Zhu, H.; Zhan, Z. Experimental Study on Constant Speed Control Technology of Hydraulic Drive Pavers. *Processes* **2023**, *11*, 477. https://doi.org/10.3390/pr11020477.
9. Munoz, S.A.; Park, J.; Stewart, C.M.; Martin, A.M.; Hedengren, J.D. Deep Transfer Learning for Approximate Model Predictive Control. *Processes* **2023**, *11*, 197. https://doi.org/10.3390/pr11010197.

10. Faria, R.R.; Capron, B.D.O.; Secchi, A.R.; Souza, M.B., Jr. Where Reinforcement Learning Meets Process Control: Review and Guidelines. *Processes* **2022**, *10*, 2311. https://doi.org/10.3390/pr10112311.
11. Rossiter, J.A. Modernising the control curriculum and delivery to meet 21st century needs. *Processes* **2025**, *13*, 531. https://doi.org/10.3390/pr13020531.

References

1. Daoutidis, P.; Megan, L.; Tang, W. The Future of Control of Process Systems. *Comput. Chem. Eng.* **2023**, *178*, 108365. [CrossRef]
2. Baldea, M.; Georgiou, A.T.; Gopaluni, B.; Mercangöz, M.; Pantelides, C.C.; Sheth, K.; Zavala, V.M.; Georgakis, C. From Automated to Autonomous Process Operations. *Comput. Chem. Eng.* **2025**, *196*, 109064. [CrossRef]
3. Annaswamy, A.M.; Johansson, K.H.; Pappas, G.J.; Alleyne, A.; Allgower, F.; Ames, A.D.; Amin, S.; Anderson, J.; Antsaklis, P.J.; Bagheri, N.; et al. Control for Societal-Scale Challenges: Roadmap 2030. IEEE Control Systems Society Publication. 2023. Available online: https://ieeecss.org/control-societal-scale-challenges-roadmap-2030 (accessed on 11 February 2025).
4. Ren, Q. AI in Chemical Engineering: A New Chapter of Innovation. *Engineering* **2024**, *39*, 1–2. [CrossRef]
5. Raven, D.B.; Chikkula, Y.; Patel, K.M.; Al Ghazal, A.H.; Salloum, H.S.; Bakhurji, A.S.; Patwardhan, R.S. Machine Learning & Conventional Approaches to Process Control & Optimization: Industrial Applications & Perspectives. *Comput. Chem. Eng.* **2024**, *189*, 108789. [CrossRef]
6. Munoz, S.A.; Park, J.; Stewart, C.M.; Martin, A.M.; Hedengren, J.D. Deep Transfer Learning for Approximate Model Predictive Control. *Processes* **2023**, *11*, 197. [CrossRef]
7. Faria, R.R.; Capron, B.D.O.; Secchi, A.R.; Souza, M.B., Jr. Where Reinforcement Learning Meets Process Control: Review and Guidelines. *Processes* **2022**, *10*, 2311. [CrossRef]
8. Mannan, S. (Ed.) *Lees' Loss Prevention in the Process Industries: Hazard Identification, Assessment and Control*, 4th ed.; reprinted through 2020; Butterworth-Heinemann: Oxford, UK, 2014.
9. Rossiter, J.A. Modernising the control curriculum and delivery to meet 21st century needs. *Processes* **2025**, *13*, 531. [CrossRef]

processes

MDPI

Review

Where Reinforcement Learning Meets Process Control: Review and Guidelines

Ruan de Rezende Faria [1,*], Bruno Didier Olivier Capron [1], Argimiro Resende Secchi [2] and Maurício B. de Souza, Jr. [1,2]

[1] Escola de Química, EPQB, Universidade Federal do Rio de Janeiro, Rio de Janeiro 21941-909, Brazil
[2] Programa de Engenharia Química, PEQ/COPPE, Universidade Federal do Rio de Janeiro, Rio de Janeiro 21941-972, Brazil
* Correspondence: rrfaria@eq.ufrj.br

Abstract: This paper presents a literature review of reinforcement learning (RL) and its applications to process control and optimization. These applications were evaluated from a new perspective on simulation-based offline training and process demonstrations, policy deployment with transfer learning (TL) and the challenges of integrating it by proposing a feasible approach to online process control. The study elucidates how learning from demonstrations can be accomplished through imitation learning (IL) and reinforcement learning, and presents a hyperparameter-optimization framework to obtain a feasible algorithm and deep neural network (DNN). The study details a batch process control experiment using the deep-deterministic-policy-gradient (DDPG) algorithm modified with adversarial imitation learning.

Keywords: Markov decision process; imitation learning; transfer learning; process optimization

1. Introduction

Reinforcement learning is a machine-learning (ML) technique characterized by an agent capable of self-learning in an environment guided only by numerical rewards [1]. Historically, its evolution was mainly related to artificial intelligence (AI), but it cannot be ignored that RL is directly derived from optimal control theory [2,3]. Nevertheless, it was a long time before the process-control community realized the potential of this technique to address control, as reported in Hoskins and Himmelblau [4]. There was a reduction in research incentives and state-of-the-art development at that time, as the results showed a technique with inferior performance to the proportional–integral–derivative controller (PID) that was also algorithmically complex, data-driven, and a black-box model.

This situation began to change in 2012 due to the consolidation of deep-learning (DL) theory, in which deep neural networks (DNN) began to be used as feature extractors [5,6]. This allowed RL in problems with high-dimensional state space (i.e., deep reinforcement learning (DRL)) such as, for example, cyber-physical systems (e.g., robot control; see Wulfmeier et al. [7], Peng et al. [8]), fixed, strict and complex environments (e.g., AlphaGO; see Silver et al. [9]; and AI in gaming; see Mnih et al. [10]), and large-scale environments (e.g., Vinyals et al. [11]).

For these reasons, as reported in [12,13], the number of publications on RL applied to process control began to grow again. The majority of the reported works are related to batch operation (e.g., [14–17]), although there are also applications for cases of continuous operation (e.g., [18–20]).

This paper presents a literature review on RL and addresses the online implementation and maintenance of RL techniques to control and optimize batch and continuous processes. At this point, model-based predictive control (MPC) theory and real-time optimization (RTO) are well-established technologies for application in chemical processes considering a decision-making scale of minutes and hours, respectively [21]. However, they depend on complex models with periodic recalibration. Decisions are based on open-control-loop simulations, with the process and economic constraints encompassing an entire unit. Under these conditions, such integration may be unfeasible when the approach does not consider the model mismatch, presents a long waiting time to reach the steady state and demands high computational costs associated with the computation of the control actions [22–24].

Likewise, developing an RL agent (control policy) for application and maintenance in an online process faces challenges similar to those discussed above. However, RL can result in adaptive and low-computational-cost control (i.e., after training). Theoretically, it seems to be an ideal technology for application in process control. However, it depends on the evolution of theory to incorporate process constraints and to ensure a level of control stability. It also needs extensive offline learning and requires adjustment of the deep neural network and algorithms' hyperparameters [1,25].

To take into account process constraints, three state-of-the-art developments can be mentioned: (1) modifying the Markov decision process (MDP) framework to incorporate process constraints implicitly (e.g., Pan et al. [26]); (2) combining imitation learning and reinforcement learning to learn from demonstrations (e.g., Mowbray et al. [27] used inverse reinforcement learning); (3) combining MPC and RL to obtain a control policy with the mutual benefits of the synergistic integration of both methodologies (e.g., Shah and Gopal [28], Alhazmi et al. [29], Kim et al. [30]).

Some review papers have addressed RL techniques applied to process control. Buşoniu et al. [13] established a detailed review of the methodology. Badgwell et al. [31] reviewed recent progress and implications for process control. Nian et al. [12] covered applications to the process industry. Görges [32] analyzed the relationships between model-based nonlinear predictive control (NMPC) and RL through tutorials. The present article reviews RL from a new perspective on simulation-based offline training and process demonstrations, policy deployment with transfer learning and the challenges of integrating it, by proposing a feasible approach to online process control. A control experiment elucidates how learning from demonstrations can be accomplished through IL and RL. A hyperparameter-optimization framework is proposed to obtain a feasible algorithm and DNN.

This review starts in Section 2 with RL theory, DRL and the state-of-the-art algorithms for DRL. Section 3 details simulation-based offline training and process demonstrations, and policy deployment with transfer learning. Section 4 presents the challenges of implementing reinforcement learning for process control. Section 5 presents an approach based on IL and RL for a batch-process-control experiment. The review is concluded in Section 6.

2. Reinforcement Learning

Four distinct but complementary phases mark the historical evolution of the RL methodology; in chronological order, they are [1,33]:

1. Definition of the term from animal psychology;
2. Analysis for optimal control theory and machine learning;
3. Evolution of training procedures and pattern recognition;
4. Development of DNN, powerful hardware, data availability and more stable algorithms.

References that are important for the evolution of RL theory or have influenced the current literature are listed in Table 1. Difficulties in defining the learning elements, as well as creating and implementing algorithms, were the main challenges during Phase 2. Howard [34] was the first to propose a viable RL algorithm based on dynamic programming (DP). In Phase 3, there was a return of interest in the field of AI in general [1], where developments for RL remain current, specifically with Q-learning and REINFORCE algorithms and stochastic RL theory. Phase 4 is when RL algorithms incorporated DNN, in addition to methodologies to store information in memory (i.e., buffer replay), in order to stabilize training and improve convergence.

Table 1. Several historical references and application contexts.

Author	Main Topic	Approach
	(Phase 1)	
[35]	Animal psychology	Definition of RL
	(Phase 2)	
[2,3]	Optimal control	MDP, DP and Bellman's equation
[36]	AI	Discussed RL models
[34]	AI	DP algorithm (policy iteration)
[37]	AI	Pointed out directions for the evolution of RL
	(Phase 3)	
[38]	AI	Neuro-like network
[39]	AI	Temporal difference (TD)
[40]	AI	Q-learning algorithm
[41]	AI	Stochastic RL algorithm
[42]	AI	REINFORCE algorithm
	(Phase 4)	
[10]	AI	RL in games (Atari)
[9]	AI	AlphaGo

2.1. Basics of RL

ML technologies are divided into three core classes: supervised learning, unsupervised learning and reinforcement learning [25,43]. In supervised learning, knowledge about the case study results from labeling its input and output features. An example of this is classification and regression tasks where only cause (X) and effect (Y) are known; their relationship is approximated ($Y = f(X)$) using some data-driven technique (e.g., neural networks) that has adequate generalization and accuracy when applied to the case study. On the other hand, unsupervised learning obtains information about the case study only through knowledge of the cause (X), and it is not necessary to include knowledge in the form of a "teacher" to label its effect (Y). A typical application is clustering [44].

RL differs significantly from those technologies discussed above. According to Sutton and Barto [1], this is because RL does not require external interference in the form of a "teacher" to explore its environment, and is also not limited to learning how input data is distributed ($p(X)$). Instead, it learns the best way to keep up with a given task through repeated interactions with its environment. This makes the task complex, as more elements have to be studied to characterize the problem. However, it can make machine learning automated.

Defining an RL problem depends on understanding its essential elements: agent, environment and reward. A widely studied example is the bandit problem. This is a problem in which a fixed limited set of resources must be allocated between competing choices (i.e., actions (a_t)) in a way that maximizes their expected gain. In other words, the agent's objective is to maximize the rewards sum along the trajectory $h = (a_0, a_1, \cdots, a_T)$ [45].

Although the definition of the RL problem seems to be simple, selecting which actions are the best without any a priori knowledge about the environment is a challenging task. Sutton and Barto [1] detailed that it is necessary to estimate the value of the actions (A_t) so that it is possible to predict the most valuable actions in the long term. However, through repeated executions of the bandit problem (episode), just selecting the k actions with the highest value (exploitation) will likely lead to a sub-optimal expected sum of rewards (Equation (1)), since RL does not explore enough combinations of actions throughout the episodes, and neither approaches the expectation of Equation (1). The exploration–exploitation trade-off dilemma is a classic problem in the RL literature. An alternative would be to select some actions randomly at a specific rate, thus preventing the selection of greedy actions and furthering exploration of the environment.

$$q^*(a) = E[R(h) | A_t = a_t] \quad (1)$$

2.2. Mathematical Background

2.2.1. Definition

Mathematically, the RL problem is formulated as an MDP (Figure 1): a stochastic sequential decision-making problem through the interaction of an agent with the environment [3]. That is, the agent, faced with the current representation of the state of the environment $s_t \in S$, executes strategies (*policies*) and implements the best action available at that moment $a_t \in A$, which will result in the transition of the environment to a new state $s_{t+1} \in S'$, evaluated employing a reinforcement sign (Equation (2)).

$$r_t = r(s_t, a_t, s_{t+1}) \in R_t \tag{2}$$

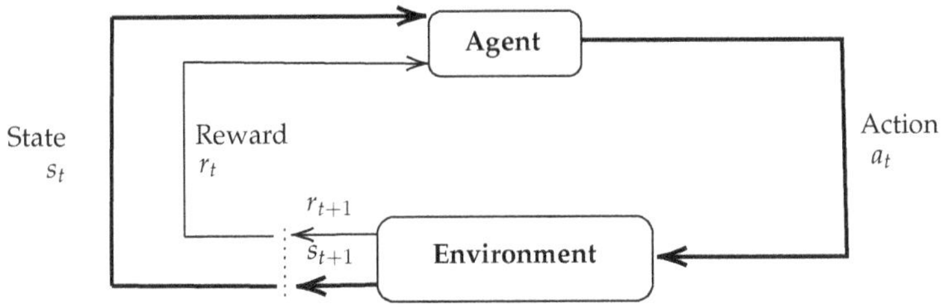

Figure 1. Simplified outline of the RL process.

For the case of continuous variables, the stochastic elements of this process include the initial condition of the state s, in the form of Equation (3). In this equation, the probability density function returns probability values for all possible states belonging to the set S.

$$\begin{gathered} p(s) \geq 0, \forall s \in S_t \\ \int_{s \in S_t} p(s) ds = 1 \end{gathered} \tag{3}$$

The probability transition from state s to state s' when action a is taken (state transition) defines the conditional probability density function $p(s'|s,a)$ (Equation (4)).

$$\begin{gathered} p(s'|s,a) \geq 0, \forall s, s' \in S_{t+1}, \forall a \in A_t \\ \int_{s' \in S_{t+1}} p(s'|s,a) ds' = 1, \forall s \in S_t, \forall a \in A_t \end{gathered} \tag{4}$$

The decision made by the agent is determined by a policy (π), which aims to map states to actions. In other words, it is a rule for deciding what to do given the current state of the environment, and is robust enough to specify what to do in any situation and for the entire state space. When using a deterministic policy, the action to be taken in each state is unique, as shown in Equation (5).

$$\pi(s) \in A_t, \forall s \in S_t \tag{5}$$

On the other hand, when the action to be taken in each state is stochastic, such a policy is a function of the conditional probability density of taking action a in-state s, as shown in Equation (6) [1,33,40].

$$\begin{gathered} \pi(a|s) \geq 0, \forall s \in S_t, \forall a \in A_t \\ \int_{a \in A_t} \pi(a|s) da = 1, \forall s \in S_t \end{gathered} \tag{6}$$

2.2.2. Optimization Objective

The sampling of the trajectory (h) of the MDP is illustrated in Figure 2. This sampling is an extension of Markov chain processes, in which the current state of the system (s_t) must contain all the information needed to decide which action (a_t) to take, resulting in the subsequent transition to the new state s_{t+1}, making it unnecessary to know about actions, states and rewards selected in the past. Hence, the agent's objective is to maximize the reward sum along the trajectory h, known as return [40]. Generally, the choice is between three alternatives known as total reward, average reward and discounted reward, the latter being the most common choice according to Sutton and Barto [1]. The corresponding return associated to a trajectory h is defined by Equation (7).

$$R(h) = \sum_{t=1}^{T} \gamma^{t-1} r(s_t, a_t, s_{t+1}) \tag{7}$$

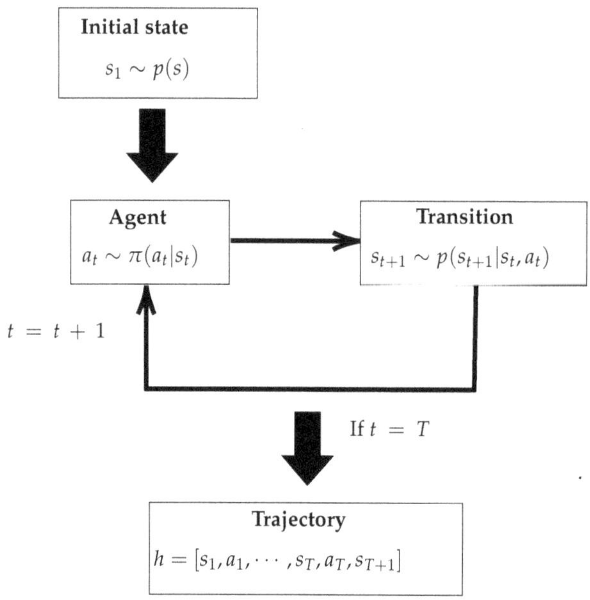

Figure 2. Trajectory sample generation.

In this equation, γ is a parameter known as the discount factor (ranging between 0 and 1), which determines the present value of future rewards. When $\gamma = 0$, the agent acts "greedily", maximizing the reward obtained in the next step, whereas if $\gamma = 1$, the agent will emphasize all future rewards equally, corresponding to the total reward alternative, or to the average reward when this return value is divided by T. Additionally, values of γ much less than 1 can make future reward values eventually negligible [31,40].

According to the logical sequence of Figure 2, the objective of RL-based methodologies is to learn an optimal policy that, regardless of the chosen initial state s_1, will produce as much of the return value as possible starting from that state (Equation (8)).

$$\pi^* = \underset{\pi}{argmax} E_{p^{\pi}(h)}[R(h)] \tag{8}$$

In this equation, $E_{p^{\pi}(h)}$ denotes the expectation about the trajectory h extracted from $p^{\pi}(h)$, and $p^{\pi}(h)$ denotes the probability density of observing the trajectory h under policy π (Equation (9)).

$$p^{\pi}(h) = p(s_1)\prod_{t=1}^{T} p(s_{t+1}|s_t, a_t)\pi(a_t|s_t) \tag{9}$$

The procedure for computing the optimal policy π^* is not an obvious task due to the problem of sequentially defining actions that can result in delayed rewards. Because of this difficulty in determining which actions are right or wrong, it can be challenging to decide which changes have to be made to improve the suboptimal policy. Therefore, it is necessary to have efficient ways of discovering changes to the employed policy so that it is improved [40,42,46].

2.2.3. Algorithms

Approaches employing value function are a traditional way to learn optimal policy. The overall objective of value functions is to approximate the return value for all possible trajectories (h) to improve the employed policy (π). When only the state value is taken for the computation of the value function, it results in Equation (10), while if the action value is also included, Equation (11) is obtained.

$$V^{\pi}(s) = E_{p^{\pi}(h)}[R(h)|s_1 = s] \tag{10}$$

$$Q^{\pi}(s,a) = E_{p^{\pi}(h)}[R(h)|s_1 = s, a_1 = a] \tag{11}$$

Howard [34] proposed the first viable algorithm to approximate the RL problem. However, it considered $E_{p^{\pi}(h)}[R(h)]$ as known and was specific to low-dimensional and completely specified problems. First, it calculates the optimal value function (*policy evaluation*) and then improves the optimal policy (*policy improvement*). When the "value" of the state (s) is used for this purpose, the steps of the algorithm are in the form of Equations (12) and (13), respectively, with δ representing the Dirac delta function.

$$V^{\pi}(s) = E_{p^{\pi}(h)}[r(s,a,s') + \gamma V^{\pi}(s')] \tag{12}$$

$$\begin{gathered} \pi^*(a|s) = \delta(a - a^{\pi}(s)) \\ a^{\pi}(s) = \underset{a\in A_t}{argmax} E_{p^{\pi}(h)}[r(s,a,s') + \gamma V^{\pi}(s')] \end{gathered} \tag{13}$$

The other option is to use the value function of the state–action pair, such that the optimal policy depends on Equations (14) and (15), respectively.

$$Q^{\pi}(s,a) = r(s,a) + \gamma E_{\pi(a'|s')p(s'|a,s)}[Q^{\pi}(s',a')] \tag{14}$$

$$\pi^*(a|s) = \delta(a - \underset{a'\in A_t}{argmax}\, Q^{\pi}(s,a')) \tag{15}$$

Considering a model-free RL approach (i.e., when a model of $E_{p^{\pi}(h)}[R(h)]$ is not available), it is not easy to learn from sampled data, especially when the state and action space is continuous [1]. One alternative is to approximate the value function of the state–action pair in the form of Equation (16) (i.e., approximate RL) to improve generalization and deal with the dimensionality problem. In this equation, $\phi(s,a)$ is the base function vector and θ is the parameter to be adjusted.

$$Q^{\pi}(s,a,\theta) = \sum_{t=1}^{T} \phi_i(s,a)\theta_i = \phi^{\mathrm{T}}(s,a)\theta \tag{16}$$

The problem now is to estimate θ to minimize Equation (17), using Monte Carlo (MC) or temporal difference estimates for $Q^{\pi}(s,a)$ [31].

$$\theta^{*} = \underset{\theta}{argmin}[Q^{\pi}(s,a,\theta) - Q^{\pi}(s,a)] \quad (17)$$

Approximating $Q^{\pi}(s,a)$ by temporal difference is a good choice because it is a combination of Monte Carlo ideas and dynamic programming (Equation (18)). As in Monte Carlo methods, temporal difference methods can learn directly from current experiences without an environment model. As in dynamic programming, temporal difference methods update estimates based in part on other learned estimates without waiting to complete their search for the entire state and action space (i.e., using *bootstrapping*) [1].

$$Q^{\pi}(s,a) = r(s,a) + \gamma Q^{\pi}(s',a',\theta) \quad (18)$$

A policy gradient is an alternative to deal with some limitations of value-based methods, mainly when it is crucial to consider the physical system's stability criteria. An example of this is control tasks in which the agent must compute continuous actions [1]. Unlike what is seen in value-based approaches, the optimal auto-parameterized policy (θ^{*}) should maximize the expected return $J(\theta)$ (Equations (19)–(21)). However, directly maximizing $J(\theta)$ can be difficult due to the high nonlinearities for θ; thus, it is necessary to improve and adapt solutions based on methods of the gradient, so that at least a feasible local optimum is found.

$$\theta^{*} = \underset{\theta}{argmax} J(\theta) \quad (19)$$

$$J(\theta) = E_{p^{\pi}(h|\theta)}[R(h)] = \int p(h|\theta)R(h)dh \quad (20)$$

$$p(h|\theta) = p(s_1)\prod_{t=1}^{T} p(s_{t+1}|s_t,a_t)\pi(a_t|s_t,\theta) \quad (21)$$

2.3. Deep Reinforcement Learning

Two achievements were essential for DL and RL theory development. First, Hinton et al. [6] demonstrated that it is possible to learn a model capable of classifying image classes without any pre-processing, mainly in the form of feature extractors to deal with the invariance dilemma [44]. Afterwards, Mnih et al. [10] developed a DRL algorithm for the Atari 2600 game platform, which in some cases surpassed the scores of human players and could address new situations distinct from those previously used for training. Specifically speaking, some general advances are listed below:

- Using elements of the mathematical theory of communication in terms of encoding information based on entropy [47];
- Updating DNNs to operate on different dynamic forms of the data distribution in space and time [25];
- Updating gradient descent (GD) and backpropagation algorithms;
- Since 2012, the DRL methodology has moved towards more stable, adaptable and robust optimization algorithms, thus avoiding obtaining time-correlated samples with a high variance which, in effect, could result in sub-optimal and divergent policies [1,48];
- Advances in parallel and distributed computing;
- Open-source software (e.g., Python code).

2.3.1. Deep Q-Learning

Approximating $E_{p^{\pi}(h)}[R(h)]$ with Q-learning depends on the number of combinations of valid actions and states required for MDP sampling. The use of tabular methods is indicated by the limit on the number of updates of all states and actions that can be efficiently performed individually. On the other hand, with parametric approximators, the update of the value function for each state–action pair individually also influences the estimate of the value function for the other state–action pairs, thus giving it a capacity for generalization that makes the learning process more powerful, but potentially more challenging to manage and understand [1,25,46].

Within this perspective, in DNNs, updating the value function of each state–action pair is conducted following Equations (22) and (23). The TD technique is used for the value function evaluation (i.e., $TD(0)$) and defines the loss function (L_t), with backpropagation and stochastic descending-gradient algorithms to update the neural-network parameters (θ). As a result, Sutton and Barto [1] did not expect to find a value function with zero error for all states. But an approximation that balances the errors in different states.

$$L_t = \left\{[r_t + \gamma\, max_{a'} Q(s_{t+1}, a', \theta_t)] - Q(s_t, a_t, \theta_t)\right\}^2 \tag{22}$$

$$\theta_{t+1} = \theta_t + \alpha\left\{[r_t + \gamma\, max_{a'} Q(s_{t+1}, a', \theta_t)] - Q(s_t, a_t, \theta_t)\right\}\frac{\delta Q(s_t, a_t, \theta_t)}{\delta \theta_t} \tag{23}$$

Figure 3 presents a simplified scheme describing the procedure for training the critic network (i.e., until each episode of size T ends). It is an iterative procedure in which the sampled initial state is restarted after the completion of each episode of size T, depending on how the environment was constructed and the rewards were formulated (e.g., setting a threshold a priori for accumulated rewards, or declaring state transitions considered impossible or unfeasible) [25]. In addition, in the terminal state, the target value $\hat{y}_t = r_t$.

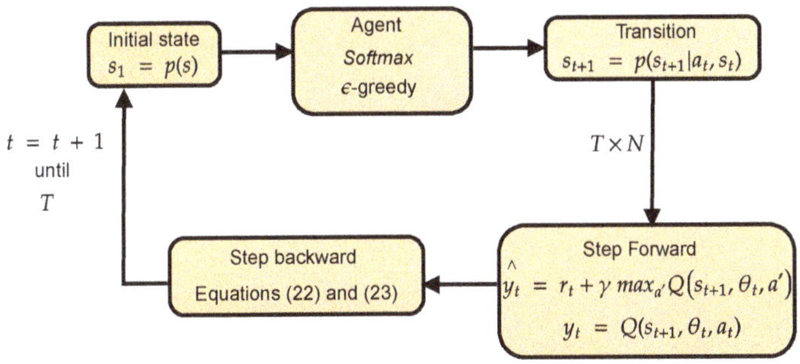

Figure 3. Simplified diagram describing the training procedure of a critic (*Q-learning*) parameterized by a DNN.

In its original form, the procedure described above may result in inadequate learning because the target $\hat{y}_t$ is never exactly approximated in the forward step. It can result in a critic with divergent policy for overestimating $\hat{y}_t$, leaving some amount of residual TD-error ($\delta_t = \hat{y}_t - y_t$). In Equation (24), the dispersion of the critic estimates depends on its complexity, variance of future rewards and the TD-error, accentuated by the value of $\gamma \sim 1$ [48–50].

$$\begin{aligned} Q(s_t, a_t, \theta_t) &= r_t + \gamma E[Q(s_{t+1}, a_{t+1}, \theta_{t+1})] - \delta_t \\ &= r_t + \gamma E[[r_{t+1} + Q(s_{t+2}, a_{t+2}, \theta_{t+2})] - \delta_{t+1}] - \delta_t \\ &= E_{p^{\pi}(h)}\left[\sum_{i=t}^{T} \gamma^{i-t}(r_i - \delta_i)\right] \end{aligned} \tag{24}$$

The alternatives proposed in the literature to alleviate the mentioned problems are:

- Controlling the exploratory component of the critic model used by the agent (e.g., *softmax* or *ϵ-greedy* [46])) will contribute to adequately exploring enough transitions of state, avoiding obtaining sub-optimal policies;
- Using experience replay to reduce the effect of temporal correlations between transitions uniformly sampled at random from the buffer, which allows estimating θ with important dynamic information [10];
- Updating from the target value $\hat{y}_t$ with delayed (or filtered) copies of the original DNN (i.e., $\theta_{t+1} = \kappa\theta_t + (1-\kappa)\theta_{t+1}$) [48].

2.3.2. Deep Policy Gradient

During the development of RL theory, new complementary methodologies emerged to overcome the main limitations of methods based on value function, which are a consequence of using a deterministic model to select the actions. Thus, it was difficult to obtain a stable model since small changes made in the estimation of the value function led to significant changes in the action performed by the agent. Therefore, an under-optimized and inadequate critic for exploiting states and actions was obtained [51–53].

Historically, policy-gradient methods were the first option developed to deal with such disadvantages. They are stable compared to value-based methods, as they use an auto-parameterized policy known as an actor and implement the actions without employing a critic, allowing them to approach stochastic dynamics and learn in a more stable way from the interaction between agent and environment.

Mathematically, they inherit concepts from Equations (17)–(19), in which the challenge is to carry out the optimization of $J(\theta)$ simultaneously with the exploration of more rewarding actions and states for $R(h)$. For that purpose, the most-used technique is the ascending-gradient-optimization algorithm (Equation (25)), which depends on a new approximation for Equation (19), aiming to compute the gradient with relation to the chosen parameterized model (Equation (26)) [33]. According to Sutton and Barto [1] and Grondman et al. [54], the solution of this last equation results from the policy-gradient theorem, which explains that the unknown effect of changes in $\pi(a_t|s_t,\theta)$ over the state distribution $p(h|\theta)$ does not involve its derivative (Equation (27)), so its expectation can be approximated by the empirical average in the form of Equation (28), where $h_n = [s_{1,n}, a_{1,n}, \cdots, s_{T,n}, s_{T+1,n}]$.

$$\theta_{t+1} = \theta_t + \alpha\nabla_\theta J(\theta_t) \tag{25}$$

$$\begin{aligned} \nabla_\theta J(\theta) &= \int \nabla_\theta p(h|\theta)R(h)dh \\ &= \int p(h|\theta)\nabla_\theta \log p(h|\theta)R(h)dh \\ &= \int p(h|\theta)\sum_{t=1}^{T}\nabla_\theta \log \pi(a_t|s_t,\theta)R(h)dh \end{aligned} \tag{26}$$

$$\nabla_\theta J(\theta) = E_{p^\pi(h|\theta)}\left[\sum_{t=1}^{T}\nabla_\theta \log \pi(a_t|s_t,\theta)R(h)\right] \tag{27}$$

$$\nabla_\theta J(\theta) = \frac{1}{N}\sum_{n=1}^{N}\sum_{t=1}^{T}\nabla_\theta \log \pi(a_{t,n}|s_{t,n},\theta)R(h_n) \tag{28}$$

The actor categorizes actions instead of their value $\pi(a_t|s_t,\theta)$, and there is no way to compute ∇J directly using gradient-based optimization techniques. Because of this, the Monte Carlo simulation approximates the value of Equation (28) (i.e., random sampling N episodes of size T). As a result, the actor is updated by Equation (25), where $R(h_n)$ is directly proportional to $\nabla_\theta J$, and $\log \pi(a_{t,n}|s_{t,n},\theta)$ is inversely proportional to the policy adopted (i.e., it follows from the identity $\nabla \ln x = \frac{\nabla x}{x}$), as actions sampled with frequency are chosen even if they do not produce the highest expected return [1,13,25,33,54].

For example, training with the REINFORCE [42] algorithm requires the execution of a very large number of cycles, as represented in the diagram shown in Figure 4 (i.e., when $\dim(T) \sim \inf$), which is computationally unfeasible for some online applications [1,25]. Therefore, it uses a baseline to reduce variance, similar to the temporal-difference-learning methodology. For example, Williams [42] used a constant to represent it (e.g., mean reward). However, it is more appropriate to employ a state-specific option, with the value taken immediately before the actor samples the action.

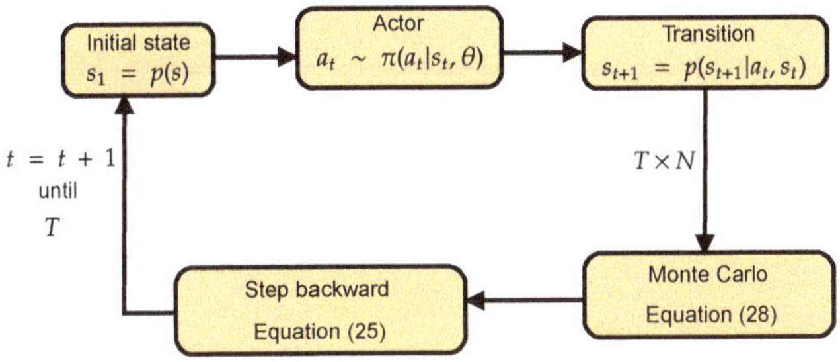

Figure 4. Simplified diagram describing the REINFORCE algorithm.

2.3.3. Deep Actor–Critic

The actor–critic methodology evolved due to the individual limitations of the methods based on the policy gradient and value function (Figure 5). Historically, it first addressed adaptive elements consisting of an associative search element and an adaptive critic element [38]. However, they were not yet structured in an actor combined with a critic. This only happened after formulating the policy-gradient theorem [51], which guaranteed convergence for both methodologies. Currently, its methodology stands out as a complete approach to RL (e.g., see algorithms developed in the works of Silver et al. [48], Fujimoto et al. [50], Ramicic and Bonarini [55] and Haarnoja et al. [56]), as they add the advantages of methods based on the value function for accelerating training and reducing variance while maintaining stability and convergence properties resulting from policy-gradient methods during training [1,54].

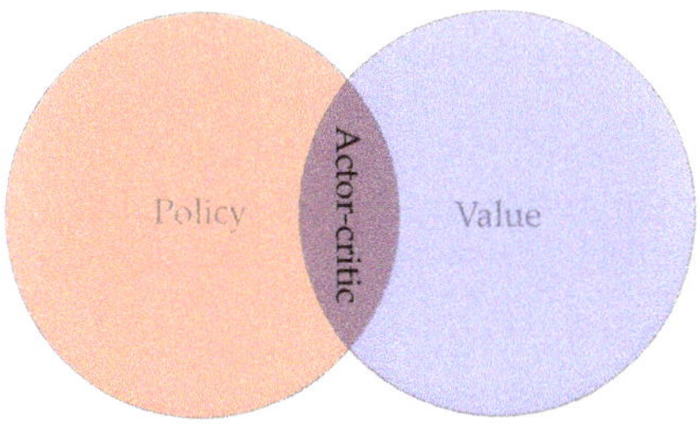

Figure 5. Venn diagram of the methods that constitute the actor–critic methodology.

Mathematically, the main difference concerning policy-gradient methodology is the replacement of the complete return (e.g., obtained with Equation (28)) by the return in one time step. According to Benhamou [57], the algorithms must remain unbiased concerning the approximation of Equation (28) and follow the principle of the control variant, where $R(h)$ is replaced by TD methods such as TD(0), Sarsa(0) and Q-learning; or by methods based on advantage, according to Equation (29).

$$A(s,a) = E_{\pi_\theta}\left[r_{t+1} + \gamma V(s_{t+1}|s_t = s, a_t = a)\right] - V(s_t) \tag{29}$$

This paradigm for DNN is shown in Figure 6, with the actor selecting the action a_t. After this step, the critic weighs the descending gradient in the algorithm for updating $\theta_a(t+1)$ (e.g., Q-AC, TD-AC, A-AC). In the backward step, Equation (30) characterizes the updating of the actor's parameters. Additionally, the critic updates the values of such actions (Equation (31)). At the end, both parameterized models must achieve good generalization so that the actor does not becomes stuck around sub-optimal solutions and the critic minimizes the residual TD-error, as seen in the form of Equation (24) [1,13,25].

$$\theta_a(t+1) = \theta_a(t) + \alpha_a Q(s_t, a_t, \theta_c) \nabla\theta_a \log(\pi(a_t|s_t, \theta_a)) \tag{30}$$

$$\begin{aligned} \delta_t &= r_t + \gamma Q(s_{t+1}, a_{t+1}, \theta_c) - Q(s_t, a_t, \theta_c) \\ \theta_c(t+1) &= \theta_c(t) + \alpha_c \delta_t \nabla\theta_c Q(s_t, a_t, \theta_c) \end{aligned} \tag{31}$$

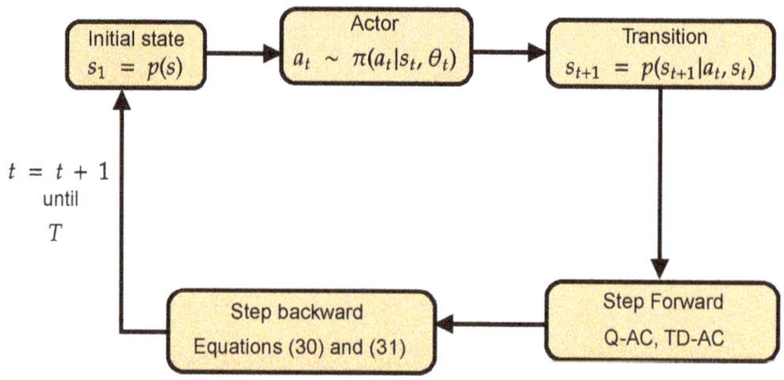

Figure 6. Simplified diagram describing the training procedure of a critic and actor parameterized by a DNN.

2.3.4. State-of-the-Art Algorithms

Dating from 2012, the created algorithms are mainly categorized as off-policy or on-policy. The former defines a learning process in which the policy update and prediction are disassociated. Such a paradigm is shown in Figure 7, where π_{k+1} is updated from samples of policy rollout data up to π_k (i.e., $\pi_1, \cdots, \pi_k$), which are contained in a large buffer. On the other hand, the latter defines a process where prediction and update are associated so that the rollout data to update the policy π_{k+1} is exclusively taken from the previous policy (i.e., π_k), meaning that it does not depend on a buffer [1,49]. Some key examples are detailed below:

- Deterministic policy gradient (DPG) (i.e., actor–critic and off-policy). According to Silver et al. [48], this is an adaptation to the policy-gradient and Q-learning algorithms, since the stochastic component intrinsic to the policy-gradient algorithms ($\pi(s_t, a_t)$) has a parameterized function in the form of $\pi(s_t, \mu_\theta(s_t))$, while depending on the computation of gradients to approximate the optimal values of a^* and $Q(s,a)^*$, which is guaranteed according to the deterministic-policy-gradient theorem, as shown in Equation (32):

$$\nabla J(\mu_\theta) = E_{p^\pi(h|\mu_\theta)}\left[\nabla_\theta \mu_\theta(s)\, \nabla_a Q^\mu(s,a)|_{a=\mu_\theta(s)}\right]; \quad (32)$$

- Deep deterministic policy gradient (DDPG) (i.e., actor–critic and off-policy) [58]. This is an updated version of the DPG algorithm regarding the use of DNN, replay buffer, target networks and batch normalization, in addition to the possibility of handling the exploration problem independent of the learning algorithm used;
- Proximal policy optimization (PPO) [59]. Contrary to the algorithms above (i.e., off-policy), PPO is an algorithm that learns while interacting with the environment over different episodes (i.e., on-policy). Methodologically, this property comes from another similar algorithm considered more complex (trust region policy optimization (TRPO)), addressing the Kullback–Leibler (KL) divergence effect and surrogate objective functions;
- Soft actor–critic (SAC) (i.e., actor–critic and off-policy) [56]. This algorithm is composed of an actor and a critic, and includes a smooth value function, which is responsible for stabilizing the training of the actor and the critic. In addition, it also has similar properties to the DDPG algorithm; however, it adds an entropy value to compose the buffer.

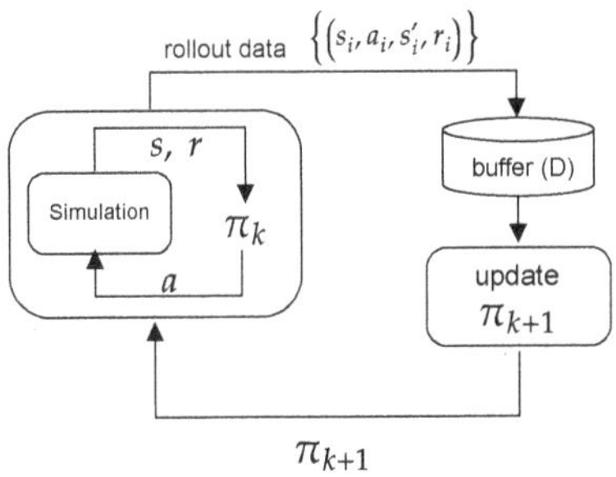

Figure 7. Off-policy reinforcement learning.

3. Reinforcement Learning for Process Control

For the definition of the MDP, it is crucial to understand the dynamics of the process rather than just focusing on the RL agent itself. This article uses batch and continuous operating regimes as case studies. The main difference between them lies in considering a single operating point in continuous operation. From a process-control point of view, for both operating regimes, it is difficult to establish a robust methodology to account for the effects of the process uncertainty, with model-based (MPC) or data-based approaches [12,13,32]. At this point, the analyzed RL paradigm comprises methodologies that rely on extensive offline training, both simulation-based and data-driven, and how to implement the control policy with transfer learning. It is illustrated in Figure 8, where Modules (1) and (2) characterize the offline training phase. In contrast, Module (3) characterizes the deployment phase of the policy obtained in the preceding phase ($\pi_S + \pi_E$). Explicitly speaking, Module (1) defines a type of training known as off-policy, demanding a reliable model from the process. The obtained policy $\pi_s = \pi_{k+1}$ is updated through samples stored in a large buffer, which is appended from samples (i.e., rollout data) taken from other policies $\pi_0, \pi_1, \cdots, \pi_k$. On the other hand, Module (2) learning depends on a fixed policy sample (π_β), which can be either known or unknown, to fill the buffer. Thus, the final policy π_E will equal the base policy π_β, characterizing a learning process similar to supervised learning. However, information from the process can be helpful to delimit the exploration and exploitation of agents trained from it.

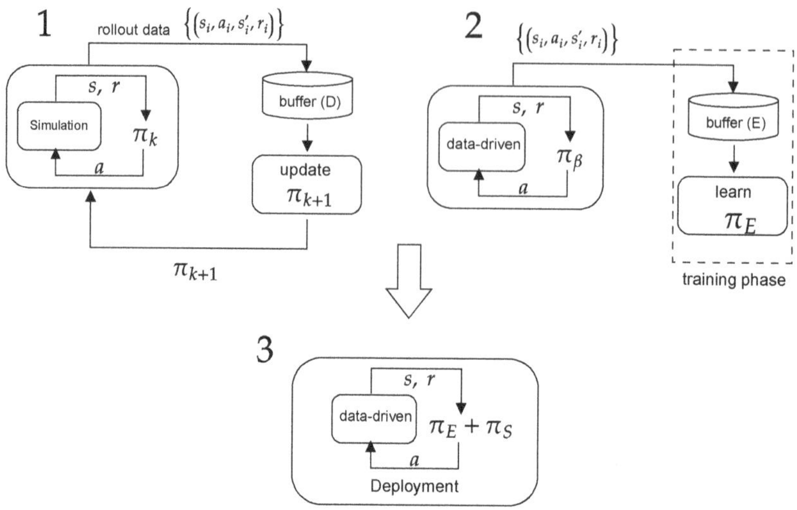

Figure 8. Diagram describing the modules for offline agent training (Modules (1) e (2)) and process-line deployment (Module (3)).

3.1. *Defining Elements of RL*

Whether the operating regime is continuous or batch, the RL elements compose a non-stationary MDP with an agent acting under uncertainty. According to RL theory, the state considered in the design of the MDP must be Markov, and implicitly contains all the relevant information of the past, which may not be the case for a batch or continuous process, as the prediction of the future will also depend on a few past states. For example, Yoo et al. [16] used a phase-segmentation strategy for a polymerization process, in which the MDP consists of RL elements for each phase of the process. Furthermore, they used historical information by adopting the DDPG algorithm with an MC estimator instead of TD(0). Nian et al. [12] defined the RL elements for an experiment to set-point tracking control, and they considered it crucial to modify the reward function to avoid its saturation due to the size of the time horizon (i.e., $T = 2000$). They also increased the dimension of the state–action pair to contain historical information (i.e., with a delay of one or two steps in time) and stabilized the training with the DDPG algorithm. Petsagkourakis et al. [14] addressed a process with batch-to-batch dynamics. To that end, they included historical information from the state–action pair to compose the input vector, which is fed to the REINFORCE algorithm (i.e., $s_k, s_{k-1}, a_k, a_{k-1}, t$). It is parameterized by a vanilla recurrent neural network (VRNN). It uses an MC estimator, storing historical information by the recurrence of VRNN hidden states and approximating the return considering the entire process trajectory.

The examples cited above assume that the current state of the system is observable, which is desirable to approximate the MDP without relying on historical information to predict the state transition. An extension of this is called partially observable MDP (POMDP) [60]. It is still not a critical problem for process control with RL, as there is an offline training step in which an attempt is made to reduce the interference of hidden states, adding to the acquired data experience, to obtain the control policy to be implemented in the actual process.

Under all the conditions mentioned above, there may still be incompatibilities between offline and online MDP elements, especially given the stochastic nature of the online process. An alternative is to combine RL techniques with IL and TL. The former can augment the MDP with data-driven information, thereby implicitly including factual process information about control objectives and constraints. The latter helps compose the online MDP by transferring information from the offline MDP and then starting to learn from a condition that is at least sub-optimal.

3.2. Batch Process

Batch processes are intrinsically dynamic, as the operation point constantly changes between the start-up and the end of the batch. This operation regime is specific to units in which the processed load volume is relatively small compared to units operating in a continuous regime, in addition to being multi-objective regarding the processing steps in order to reach the desired end product. Therefore, understanding this dynamic is not simple, as there may be several challenges to adopting an adequate control structure in this context, especially considering technical problems related to [61–63]:

- Modeling: processes in batches normally exhibit nonlinear dynamics;
- Measurements: these are typically only available at the end of the batch;
- Uncertainty: present in practically all batch processes, whether resulting from reactant quality, modeling errors, process disturbances and measurement errors;
- Constraints: there is usually the added effect of terminal constraints and the existing security and operational constraints.

Optimizing a control structure can be challenging, as it depends on managing all the issues discussed above. In this context, Bonvin [61] and Arpornwichanop et al. [63] commented that the most appropriate option for dynamic optimization of such processes is based on uncertainty, measurement and model-free optimization.

Therefore, the RL methodology can be a viable option for the dynamic optimization of these processes. Specifically speaking, the control of robots by RL agents is conceptually similar to the control of batch processes. Hence, the number of applications aimed at batch process control is increasing due to the structure of the optimization problem being easily adaptable.

For example, Petsagkourakis et al. [14,15] analyzed the optimization of batch bio-processes using the REINFORCE algorithm, adapting it to handle continuous domains by using an agent with control outputs represented by a multivariate normal distribution. At the end, the control structure only includes Modules (1) and (3), with a batch-to-batch configuration, in which the policy initially obtained in (1) (i.e., in the simulated environment π_S), is updated in (3) according to each new batch made $(\pi_S^1, \cdots, \pi_S^N)$. For this, the policy-transfer technique is used so that the actor approximated by DL in π_S^1 as the batch progresses results in a new agent with updated weights π_S^2 until the batches are completed, π_S^N.

Mowbray et al. [64] also analyzed batch bio-processes. They included process constraints to define the MDP (Module (1)) and proposed the entropy-regularized proximal-policy-optimization algorithm instead of the REINFORCE algorithm to stabilize training in the deployment phase (Module(3)).

Oh et al. [65] proposed an integrated algorithm of a double-deep Q-network and model predictive control. The proposed method learns the action–value function in an off-policy fashion (i.e., Module (1)) and solves the model-based optimal control problem where the terminal cost is assigned by the action–value function (Module (3)).

Yoo et al. [16] analyzed the control problem of a batch process using an RL agent with the DDPG algorithm and Monte Carlo sampling instead of TD (i.e., polymerization process). Likewise, the controller design depends only on Modules 1 and 3. The policy is obtained in the simulated environment given different process conditions, and is segmented into phases to facilitate the definition of learning elements and interaction between agent and environment. Then, the trained policy was applied online and proven efficient under new process conditions.

Although applications for batch processes with RL can be used to replace standard approaches (e.g., with MPC methodologies), the most significant difficulty for its implementation is in the proper design of the offline training phase, followed by the deployment of the obtained policy. Additionally, it is currently common to have measurements of the dynamics of the process and control actions, which enables the inclusion of Module (2) applied in the offline training of the RL agent. This allows obtaining an RL agent with information about the actual process, including process constraints and control objectives.

3.3. Continuous Process

On the other hand, for continuous process control, reaching a steady state is essential to meet the plant optimum and evaluate the process trajectory rather than just considering the final optimization objective. Contrary to the batch process, measurements and good models are prerequisites for an effective approach to the problem, given that the operation is practically uninterrupted. Currently, the problem may be that the benefits achieved by following an optimal steady-state trajectory may not be economically the most viable path, as can be evidenced in the chemical industry's hierarchical structure for process optimization and control at the descending decision-making scale level: RTO, supervisory control, regulatory control and process. Namely, the RTO layer depends on a rigorous process model with hour-scale control decisions. When the new optimal values for other layers are not transmitted, they may operate in a sub-optimal stationary condition, causing economic losses [12,13,21,66].

Based on this, Ellis et al. [66] pointed out the advantages and disadvantages of an alternative methodology called economic model-based predictive control (EMPC), which implicitly accounts for the effect of economic cost based on a less rigorous model of the process, but with minute-by-minute control decisions, in such a way that it dispenses with the optimization layer of higher-level processes (e.g., RTO). However, they found that the path for its consolidation, similar to MPC, is still open, due to the difficulty of meeting the economic objective of optimization while addressing restrictions, uncertainties, measurement noise and process disturbances. These challenges are also problems for RL-based methodologies, as it is intrinsically related to the field of AI, in which the objective is to achieve the greatest possible return regardless of the trajectory followed.

Table 2 summarizes the main state-of-the-art developments regarding RL applied to continuous process control, where the implementation and training details of the agent (i.e., control algorithm, learning method, value function evaluation method) applied in the process are shown for each reference cited.

Table 2. Several references related to the control of continuous processes.

Author	Control Algorithm	Learning	Estimator	Process
[18]	Deep actor–critic	Off-policy (1)	TD(0)	CSTR
[67]	Deep Q-learning	Off-policy (1)	TD(0)	Conical tank systems
[68]	Deep Q-learning	Off-policy (1)	TD(0)	Liquid–liquid separation
[30]	MPC plus DRL	Online	MC	Nonlinear control-affine system
[28]	MPC plus DRL	Online	—	Nonlinear process
[69]	MADDPG	Off-policy (1)	TD(0) and TD(λ)	Waste treatment
[70]	A2C	Off-policy (1)	TD(0)	Hydrocracking
[20]	A3C	Off-policy (1)	TD(0)	Hybrid tank system

In general, all control algorithms are parameterized by DNNs for the actor and critic, as they have already proven to be efficient in handling large and complex data [10]. Furthermore, only Ramanathan et al. [67] and Hwangbo and Sin [68] have employed value-based methods, while the other authors used algorithms derived from actor–critic methods. The explanation for this is the benefit of integrating an actor to decide the control actions weighted by a critic, accelerating the learning process while reducing the variance of actions selected by the actor resulting from TD methods.

Except for Shah and Gopal [28] and Kim et al. [30], other authors cited employed off-policy learning, with the index (1) referring to training carried out in a simulated environment. The practicality of the offline training step depends on a simulated environment that is trustworthy, allowing the testing of various process conditions. Furthermore, TD methods are preferable to MC methods precisely because they combine MC and DP ideas to obtain process estimates and can be improved with eligibility traces (λ), which is a way of weighting between TD(0) "targets" and Monte-Carlo "returns".

Another important consideration is the agents used by Dogru et al. [20], Chen et al. [69] and Oh et al. [70], who developed control structures with agents learning asynchronously. The A2C and A3C algorithms are variations of the actor–critic algorithm with agents learning asynchronously, with two or three agents in parallel [71]. In multi-agent DDPG (i.e., MADDPG with two agents), the training is decentralized regarding the control actions taken by each actor, while it is centralized by only one critic to evaluate the actions taken by each actor. Thus, such approaches allow agents with competitive and cooperative control objectives to improve offline learning and facilitate deployment in the entire process, as seen in Chen et al. [69].

Powell et al. [18] innovated by proposing the first algorithm for RTO. They employed an agent with a deep actor–critic algorithm and replaced the standard descending gradient optimization algorithm with the particle swarm method, justifying it as a global optimization method.

3.4. Policy Deployment with Transfer Learning

The transfer-learning methodology is about the process of reusing and transmitting information obtained by a specific agent (e.g., animals, robots or human beings), which is trained in a particular environment (i.e., source task), in another unexplored environment (target task). According to DL theory, TL (i.e., through a DNN) relaxes the assumption that training data should be independent and identically distributed with test data, which motivates using it against the problem of insufficient training data. As a result, what would be the benefits of using this approach for RL (e.g., see Tan et al. [72] for more details)? Briefly, in principle, it could result in an improvement in the sampling efficiency of RL methodologies applied to an online problem, in which there are usually few samples available to train the agent and obtain a suitable policy [72,73].

The definition of transfer-learning methodology depends on the similarity between the analyzed tasks, in other words, the similarity between the offline MDP (i.e., source task, Γ_S) and the online MDP (i.e., target task, Γ_T). In the literature, this is called inductive transfer learning when the tasks and rewards are different (i.e., $\Gamma_S \neq \Gamma_T$ and $r_S \neq r_T$), despite having the same state and action space ($D_S = D_T$). It is called transductive transfer learning when $\Gamma_S = \Gamma_T$, $r_S = r_T$ and $D_S \neq D_T$ [74]. For both scenarios, TL must improve initial performance, learning performance, asymptotic performance and convergence [73–75].

TL methods aim to improve the learning and sampling rate in an online MDP, as it is impractical to learn while exploring the real environment in process-control applications. In Table 3, several references on TL for RL are summarized, giving a view of the techniques and leading control-oriented applications that may be promising for process control. Zhu et al. [76] analyzed the different ways the TL process can take place in an RL context. In general, they categorized the approaches by distinguishing them according to what is learned, namely: through demonstrations (learning from demonstrations (LD)); assigning external knowledge to reconstruct target domain reward distributions to guide agent learning (reward shaping (RS)); using external knowledge of pre-trained policies from one or multiple source domains (policy transfer (PT)); using inter-member transfer learning (inter-task mapping (IM)).

Table 3. Several references on TL for RL to robot control-oriented.

Author	TL Methodology	Environment
Wulfmeier et al. [7]	RS and LD	Robot control-oriented
Peng et al. [8]	PT	Robot control-oriented
Yan et al. [77]	LD with IL	Robot control-oriented
Christiano et al. [78]	LD with IRL	Robot control-oriented
Joshi and Chowdhary [75]	RS and PT	OpenAIGym
Kostrikov et al. [79]	LD with AIL	Robot control-oriented

The most employed TL methodology for the RL-based control of batch and continuous processes is the policy-transfer technique, which is similar to the approach used in Peng et al. [8] (e.g., see [16–18,80]). It depends on a reliable process approximation for extensive offline learning before application to the real process.

For example, Petsagkourakis et al. [14] implemented this TL methodology to develop an RL-based controller for a biochemical batch process. First, they obtained the optimal policy for the simulated environment. Then, they applied it online by allowing the output layer of the DNN to learn from consecutive batches while freezing the remaining layers. Mowbray et al. [27] developed a TL framework employing an inverse RL technique, which allowed them to analyze historical process data for synchronous identification of a reward function and then obtain the control policy in a step before its application in the online process. According to the work's objective, this was proven necessary, as the proposed RL agent started the learning process following the policy already obtained in the previous step.

The other forms of TL in RL (e.g., RS, LD, IM) have still not been significantly explored because they combine all forms of learning and obtaining a stable policy therefore becomes challenging. An example of these difficulties is the inclusion of the process constraints. However, this is the way forward in chemical processes in the future, as it will result in a more efficient and faster control agent to adapt to changes in the process dynamics [12,32].

3.5. Conclusions about RL for Process Control

In general, offline learning methods take advantage of both simulated and process data. Pan et al. [26] and Mowbray et al. [27] developed methodologies enabling Modules (1), (2) and (3) shown in Figure 8, which is the way forward for the design of RL agents suitable for control in online processes. Although not contemplated in this review article, there are many adaptations of these algorithms whose function is to improve the sampling efficiency of the MDP. For example, the actor approximated by DL does not have saturated neurons (i.e., inverting gradient [81]); rather, the actor selects samples with information prioritized by the critic (i.e., prioritized buffer replay [82]).

4. Challenges for the Implementation of RL to Process Control

In this section, the challenges to implementing a general framework for optimal process control are considered:

1. The MDP design based on the process dynamics;
2. The offline training step;
3. Policy transfer to the process line;
4. Keeping the policy stable against new process changes.

4.1. Overview

First, implementing a new methodology for optimal process control depends on designing an MDP based on the process dynamics. Through the sections already described in this review, the available knowledge about the process dynamics will define the stages of offline training and TL. In other words, the objective is to extract useful information from already established policies in the process and simulation data.

The focus has been on batch and continuous processes, in which the inclusion of the transfer-learning process is complex. At this point, state-of-the-art development is still being consolidated, as the inclusion of process constraints remains to be addressed. Some exceptions are the works of Pan et al. [26] and Mowbray et al. [27,64].

As long as the process states can be observed, the configuration of the offline training step is directly related to the training of the learning agent itself, namely:

1. The choice of the algorithm;
2. The exploration–exploitation trade-off dilemma;
3. Hyperparameter optimization.

The choice of algorithm depends on the complexity of the process. There is a preference for off-policy algorithms for batch and continuous processes, as they allow the storage of information from other policies. However, the exploration–exploitation trade-off dilemma analysis indicates the most appropriate policy to approximate the MDP. Being stochastic, it deals better with the uncertainties inherent in the process. Finally, hyperparameter optimization is essential to obtain a feasible algorithm and DNN.

The transferring of knowledge obtained in the previous phases is also necessary to implement the agent for optimal process control. In Section 3.4, common approaches to robot control were detailed. However, for chemical processes, a conservative approach must consider process constraints. Thus, learning from demonstrations, transferring actor and critic policies, adapting rewards (i.e., reward shaping) and observing the correspondence between offline and online environments are essential to fulfilling this purpose.

Lastly, process-line maintenance is another critical issue to consider. At this point, the control dynamics and complexity of the process indicate how often measurements are available. This allows establishing when to retrain the obtained policy. Generally, the preference is to directly apply an alternative in the online environment without performing a backup. However, the current state of the art is still embryonic. For example, Wang and Ye [83] proposed consciousness-driven reinforcement learning, which showed superior results to standard algorithms (i.e., DDPG, PPO) for simplified examples only (i.e., OpenAIGym). Hence, the current state of the art in chemical processes is exploring RL techniques that include information from demonstration to discover unknown a priori information and limit the search space of the chosen algorithm.

4.2. Proposed Learning Structure

The proposed learning structure is derived from the one shown in Figure 8. Figure 9 characterizes the modules to design the proposed learning structure. Again, Module (1) characterizes off-policy training. The simulated environment must approximate the actual process that allows for evaluating the conditions to which the control agent may be subject. The policy is continually updated from samples stored in a large buffer in this configuration. Thus, π_{k+1} is updated with transitions obtained through the implementation of policies up to π_k, which provides information about various dynamic conditions of the process and facilitates the gradient approximation for policies based on a DNN. Feise and Schaer [84] described the importance of simulation for chemical engineering, highlighting its importance in the context of Industry 4.0 and big data. In addition, they comment that it is crucial to introduce new methodologies that integrate AI with process control to add value to new businesses in the chemical industry.

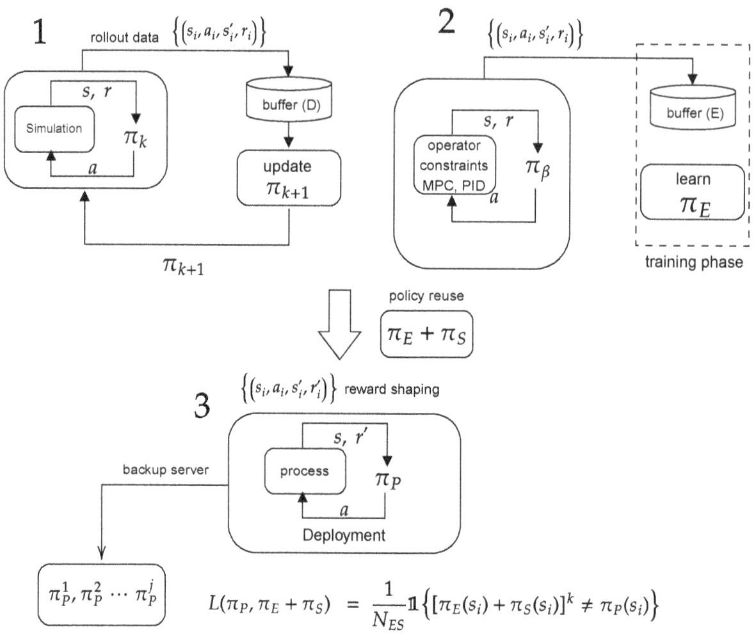

Figure 9. Diagram containing the modules for offline agent training (Modules (1) and (2)) and process-line deployment (Module (3)) adapted for chemical processes.

In Module (2), all available knowledge from demonstrations, such as sub-optimal policy information (e.g., MPC), from plant operators and from process constraints can also be included to help off-policy training. It is essential to extract useful information from the buffer E, containing implicit knowledge about control objectives and process constraints. However, the policy π_β will at most be equal to that of the expert, π_E.

Despite this, simply reusing policies previously obtained directly over the online process can result in sub-optimal policies, as the process undergoes unknown changes to its dynamics (e.g., resulting from the degradation of a catalyst). For this reason, an alternative is to choose a function for the reward regularized by the knowledge about the offline process (i.e., reward shaping) so that $\pi_P = [\pi_E + \pi_S]^{k+1}$ does not result in erroneous rewards, as well as taking advantage of the new samples of r' from the actual process.

Furthermore, for training to be effective in the online process, an ideal loss function $L(\pi_P, \pi_E + \pi_S)$ should minimize the discrepancy between π_P and $\pi_E + \pi_S$ while adding new process information. This dilemma is a problem for RL in general. Therefore, new works are investing in its solution (e.g., see [64]). Finally, a backup server could store policies updated across episodes, which then, depending on the dynamics of the process (i.e., batch or continuous), could reused again in an offline training phase, since the current policy would continue to act in the process.

5. MDP Design and Agent Training through Imitation Learning

This section highlights two points of Section 4, elucidating how learning from demonstrations can be accomplished through imitation learning and proposing a hyperparameter-optimization framework to obtain a feasible algorithm and DNN. For that purpose, a complete approach to imitation learning (IL) and reinforcement learning is detailed in a batch-process-control experiment using the deep-deterministic-policy-gradient algorithm modified with adversarial imitation learning.

5.1. Imitation Learning Techniques

Imitation learning for AI means learning from demonstrations [85]. The basic technique directly uses these demonstrations without adding experience to the learning process (i.e., direct learning). The policy trained on these principles will be equal to that used to generate demonstration samples, similar to supervised learning techniques (i.e., behavior cloning). The problem with this is generalization to unseen scenarios, which is often observed in robot control (e.g., see [8,54]). This results from an approach that does not consider different learning scenarios, thus generating sub-optimal and divergent policies.

There is currently a preference for so-called indirect-imitation-learning techniques (e.g., inverse-reinforcement learning and adversarial-imitation learning), intending to learn from experience when refining the policy obtained a priori through RL techniques and TL, among others. These techniques allows the combination of information available from the actual process and simulation, dealing with the difficulties of establishing an efficient offline training procedure, such as [86]:

- Exploration that is outside the scope of the algorithm (Module (2));
- Learning a policy that does something different from the pattern of behavior observed in the dataset E (Module (2));
- Scaling up to complex high-dimensional function approximators, such as DNN, high-dimensional state or observation spaces and temporally extended tasks.

DDPG Algorithm with Adversarial Imitation Learning

The IL approach is based on the algorithm proposed by Kostrikov et al. [79], which uses an off-policy discriminator and an off-policy actor–critic reinforcement learning algorithm. The discriminator is derived from the generative-adversary-network (GANs) theory, whose objective is to train two architectures of "adversary" neural networks, known as the generative and the discriminate [87]. Based on this, we propose here a DDPG algorithm modified by a discriminator. In Figure 10, the generator represents the DDPG algorithm learning in the simulated environment and the expert learns from the samples obtained from demonstrations. Hence, the discriminator (Ds) learns from samples of D and E until the moment it can no longer distinguish them. For this, a new reward weighs the information of the expert and the generator (r_s), as shown in Equation (33).

$$r_t = \beta r_s - (1-\beta)\log(1 - Ds(s_t, a_t)) \tag{33}$$

The generator employs a zero-mean Ornstein–Uhlenbeck process to generate temporally correlated exploration samples (i.e., η_t). In Equation (34), this is used in the actor-network μ considering time horizon T and N episodes. Subsequently, the new rollout data is stored in the buffer D (s_t, a_t, s_{t+1}, r_t). After filling the buffer of size D, random samples of size K are selected for the actor. Then, the error-TD is approximated using Equations (35) and (36) to update the critic network, with the loss function shown in Equation (37). Equation (38) defines the actor update following the deterministic policy theorem. In addition, the target networks for the critic and actor have delayed (or filtered) copies of the original network, i.e., $\theta_c(t+1) = \kappa\theta_c(t) + (1-\kappa)\theta_c(t+1)$ and $\theta_a(t+1) = \kappa\theta_a(t) + (1-\kappa)\theta_a(t)$. At the same time, the discriminator learns to discriminate between expert and generator samples by minimizing the loss function shown in Equation (39), where the expert and generator samples (i.e., $y_d = Ds(E)$ and $x_d = Ds(D)$) must be equal to one and zero (i.e., binary cross-entropy loss).

$$a_t = \mu(a_t, \theta_a) + \eta_t \tag{34}$$

$$y_t = r_t + \gamma Q(s_t, \mu(s_{t+1}, \theta_a'), \theta_c) \tag{35}$$

$$\delta_t = y_t - Q(s_t, a_t, \theta_c) \tag{36}$$

$$L_c = \frac{1}{K}\sum \delta_t^2 \tag{37}$$

$$\nabla J(\mu_\theta) = \frac{1}{K}\sum\left[\nabla_\theta \mu(s_t, \mu(s_{t+1}, \theta'_a)\, \nabla_a Q(s_t, a_t, \theta_a)|_{a=\mu(s_t,\theta_a)}\right] \tag{38}$$

$$L_d = \frac{1}{2K}\sum y_d \log(x_d) + (1 - y_d)\log(1 - x_d) \tag{39}$$

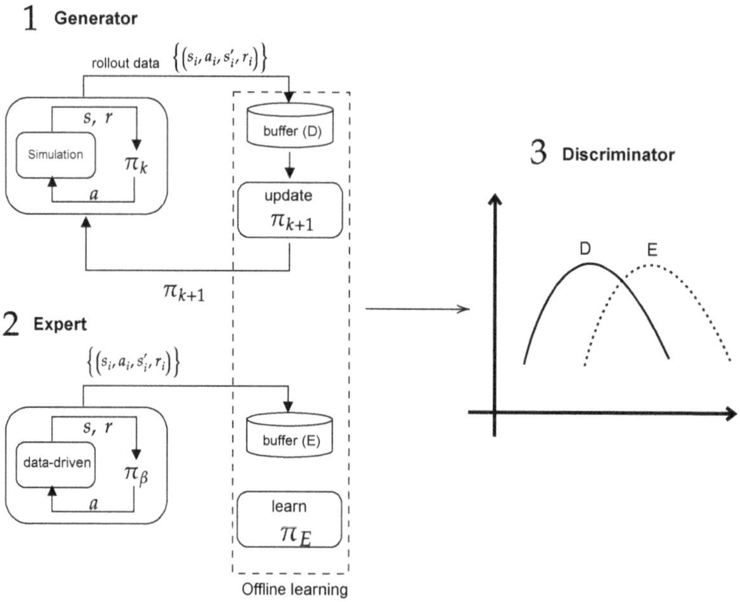

Figure 10. Diagram describing the modules for offline training of a generator and expert with a discriminator.

5.2. Hyperparameter Optimization

In general, hyperparameter optimization is recommended for complex data-driven algorithms. At the beginning, it was directed to problems with parameters characterized exclusively numerically and of low dimension. This approach became unfeasible when applied to algorithms depending on the adjustment of a large number of parameters, which may be, in addition to numeric, ordinal, and categorical, possibly dependent on each other (e.g., the CPLEX optimizer is based on a mixed-integer programming algorithm and has 80 hyperparameters to be automatically configured) [88,89]. To overcome this, Hutter et al. [88] proposed the creation of an automatic algorithm configuration structure, known as ParamILS (Hutter [89] considers this algorithm the first viable alternative to configure hyperparameters efficiently, automatically and robustly), applying it to adjust the CPLEX algorithm and return the best set of hyperparameters to meet the objective sought by the user. At that point, the value of the cost function must be efficient concerning the computational time required to complete the task, and robust so that it does not deviate significantly from the best possible solution.

Since then, this theory has been adapted for DL to optimize its complex structures (e.g., DNN), which are deep and whose training also depends on other algorithms (i.e., backpropagation and descending stochastic gradient). At this point, Coates and Ng [90] and Coates et al. [91] were the first to realize how important it can be to focus on the DL structure itself rather than just improving the learning algorithms. They tested several configurations of their main parameters, concluding that previously unfeasible alternatives can also reach results comparable to other more tested methodologies.

5.2.1. Hyperparameter Optimization Software

A good hyperparameter-optimization software framework must improve the steps seen in Figure 11. First, it is necessary to define the limit for the domain containing the hyperparameters and the starting values to carry out the other steps. This is done by categorizing all selected hyperparameters as integer, real (i.e., *float*), ordinal, or fixed value. Then, the software in charge of this will run some trials, resulting in different optimization scenarios, which are evaluated on different test samples (e.g., offline training) and returned as a solution of the employed cost function. In the end, this makes it possible to update the information about the values of the hyperparameters, directing them to produce better future optimization scenarios. A promising example is Optuna [92] due to the following features:

- Simplicity and practicality for writing the source code, an improvement over Nevergrad [92,93], Ray tune [94] and Hyperopt [95]);
- Open-source software (i.e., Python).

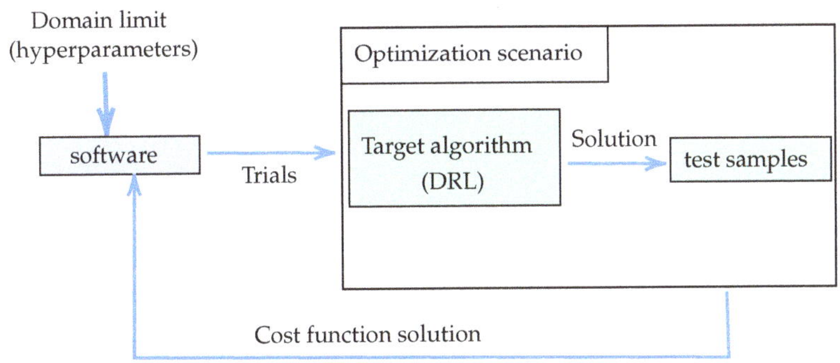

Figure 11. Simplified hyperparameter-optimization scheme.

Besides, Optuna has two optimization approaches:

- Sequential optimization relies on multiple runs of such a cycle (trials) to obtain credible information about how the parameters influence the value of the cost function. The Bayesian-optimization algorithms use an approximate model for the unknown cost function a priori, being Gaussian processes that maximize expectancy of the cost function concerning the current best solution (e.g., the expected improvement algorithms, such as the tree-structured Parzen estimator (TPE)) [95,96];
- Parallel optimization algorithms are complementary alternatives to sequential algorithms, as they update the information of hyperparameter values based on distributed executions, in contrast to the single-cycle approach used in sequential optimization.

The best of both approaches are being exploited to guarantee robust results (i.e., resulting from the plausibility of Bayes models to explore hyperparameters) and are computationally efficient (i.e., a result of distributed optimization). Successful examples of this are algorithms known as ASHA (asynchronous successive halving algorithm) [97], HYPERBAND [98] and those in which training is based on populations [99].

5.3. An Offline RL Control Experiment

For the elaboration of the offline RL control experiment, the control of batch process using DDPG algorithm with adversarial imitation learning will be investigated. The characteristics of the algorithm are detailed below:

- The algorithm is specific to MDP where the sampled state–action pairs are continuous;
- The simplicity of the algorithm, which makes writing the source code easier, the proposed updates to the algorithm (e.g., prioritized buffer replay, inverting gradient) and distributed optimization;
- There are applications for optimal process control (e.g., for DDPG, see Ma et al. [17] and Spielberg et al. [80]);
- The algorithm combines RL and adversarial imitation learning to learn from demonstrations.

5.3.1. Case Study: Batch Process

The case study is a simplified version of a batch process (Petsagkourakis et al. [14]). Following to Figure 10, the expert uses sub-optimal demonstrations from Equations (40) and (41) to fill the buffer E, representing the actual process. At the same time, the generator depends on a simulation based on a simplified version of the actual process to fill the buffer D (Equations (42) and (43)).

$$\frac{ds_1}{dt} = -(a_1 + 0.5a_1^2)s_1 + 0.5\frac{a_2 s_2}{s_1 + s_2} \tag{40}$$

$$\frac{ds_2}{dt} = a_1 s_1 - 0.7a_2 s_1 \tag{41}$$

$$\frac{ds_1}{dt} = -(a_1 + 0.5a_1^2)s_1 + a_2 \tag{42}$$

$$\frac{ds_2}{dt} = a_1 s_1 - a_2 s_1 \tag{43}$$

The optimization objective is to maximize $E[s_2(T)]$ given the limits of the control actions (i.e., $a_1 \in (0,5)$ and $a_2 \in (0,5)$) for batches of ten time intervals, with the reward for each time interval taken from the simulation and discriminated with the expert reward, as shown in Equation (44). In this equation, a standard DDPG is derived from $\beta = 1$ and an indirect-imitation-learning approach from $\beta = 0$. In addition, the initial condition is fixed at the $s_1 = 1$ and $s_2 = 0$ ($s_0 = (1,0)$).

$$r = \beta s_2(t) - (1-\beta)\log(1 - Ds(s,a)) \tag{44}$$

5.3.2. Tree-Structured Parzen Estimator

As described in Section 5.2.1, the Optuna software was chosen to perform hyperparameter optimization applied to the control experiment. Thus, it remains to determine the optimization algorithm. Specifically, the sequential optimization algorithm (TPE) is suitable to serve this purpose [92,95,100]:

1. It is computationally efficient compared to standard approaches;
2. Hyperparameters can belong to any set: integer, ordinal, real (i.e., float) and categorical.
3. It optimizes the expected improvement with a Gaussian process and Parzen estimator;
4. The default algorithm implemented in Optuna offers an alternative to overcome its main disadvantage, that is, it also models the interactions between hyperparameters, which improves the efficiency and robustness of the algorithm.

5.3.3. Validation of the Control Experiment

For validation, the policies obtained through the different trials were tested on the actual process, and the objective function defined by the amount of s_2 obtained at the end of the batch was used for comparison (Figure 12).

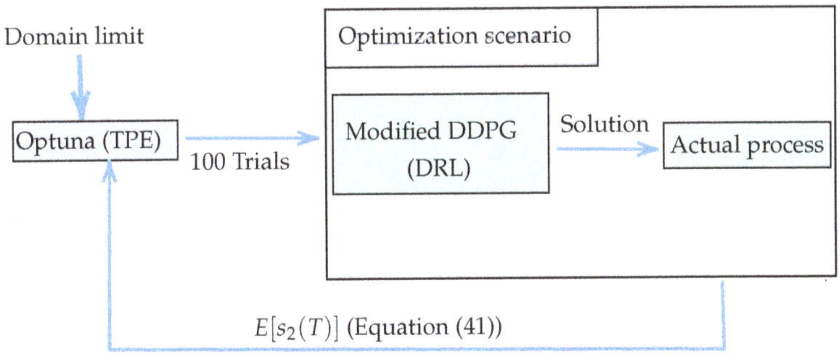

Figure 12. Optimization scheme with Optuna to validate the online control experiment with domain limit according to Table 4.

Table 4 summarizes the essential hyperparameters for optimization with the TPE algorithm and the search space for the domain containing the hyperparameters. β is the main hyperparameter to validate the proposed algorithm against the standard DDPG algorithm [58] and sub-optimal demonstrations from the actual process (E), which were obtained with REINFORCE algorithm [14]. In addition, the hyperparameters N, D, the activation function and the optimizer (i.e., Adam [101]) do not influence the optimization of $E[s_2(T)]$. Then, a limit value considered sufficient for the control experiment was defined.

Table 4. Hyperparameters for modified DDPG algorithm.

Hyperparameters	Search Space
RL	
γ	(0.80, 0.99)
β	(0, 1)
Batch size (K)	(10, 150)
Buffer (D)	500
Expert buffer (E)	500
Episodes (N)	2000
κ	(0.005, 0.01)
Actor Network	
Activation function	ReLU, Tanh
Layers (La_a)	(1, 5)
Neurons (Na_i)	(4, 250)
Critic Network	
Activation function	ReLU, Tanh
Layers (La_c)	(1, 5)
Neurons (Nc_i)	(4, 250)
Discriminator	
Activation function	Linear, Sigmoid
Layers (La_{ds})	(1, 5)
Neurons (Nds_i)	(4, 250)
NN training algorithm	
Optimizer	Adam
Actor learning rate (α_a)	$(1 \times e^{-5}, 1 \times e^{-2})$
Critic learning rate (α_c)	$(1 \times e^{-5}, 1 \times e^{-2})$
Discriminator learning rate (α_{ds})	$(1 \times e^{-5}, 1 \times e^{-2})$

When running the online control experiment according to the algorithmic implementation details described above, Figure 13 shows that the TPE algorithm took 18 trials to obtain policies with an objective function value comparable to that obtained by Petsagkourakis et al. [15] (i.e., 0.583) against 0.575 e 0.55 with the DDPG algorithm and

MPC, respectively. The TPE algorithm obtained some erroneous policies until trial 33, where the best value was obtained (i.e., 0.595). At the end, the TPE algorithm showed dispersion of the objective function values precisely because it was not able to find a solution that exceeded the best value, as detailed in Bergstra et al. [95].

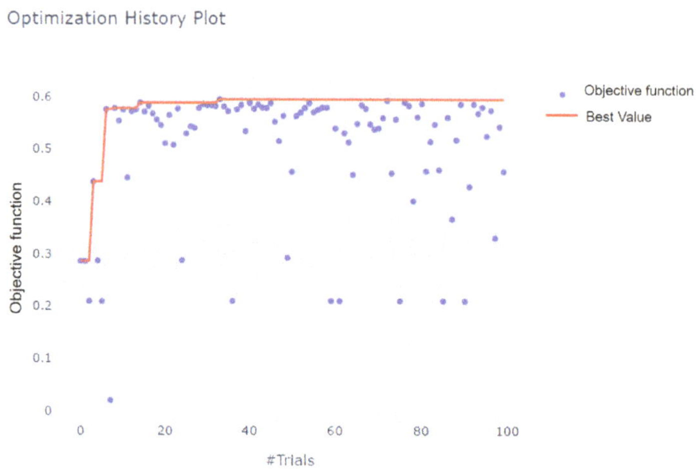

Figure 13. Optimization history plot for modified DDPG considering 100 trials with the TPE algorithm.

Figure 14 shows the hyperparameters that had more influence on the objective function. The highlight here is the hyperparameter β, which weighs the importance of the discriminator for the objective value; according to [79] it is crucial to properly know the dynamics of the environment to specify this hyperparameter correctly. This can be seen in Figure 15, where values of β important to the objective function are dispersed around its optimal value (i.e., $\beta = 0.8$), indicating that including demonstrations in the process of offline learning results in appropriate online control policies.

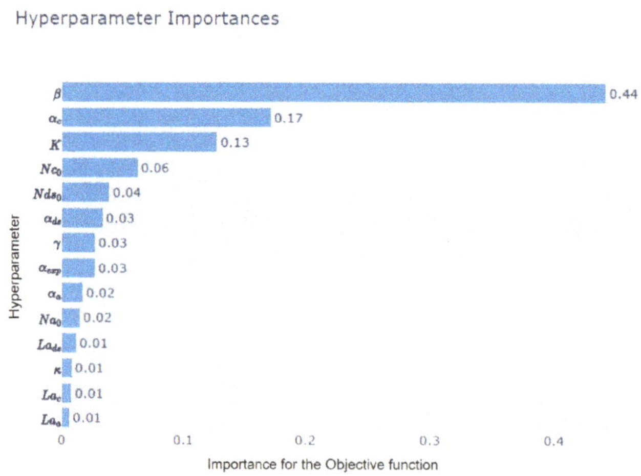

Figure 14. Hyperparameter importance plot for modified DDPG considering 100 trials with the TPE algorithm.

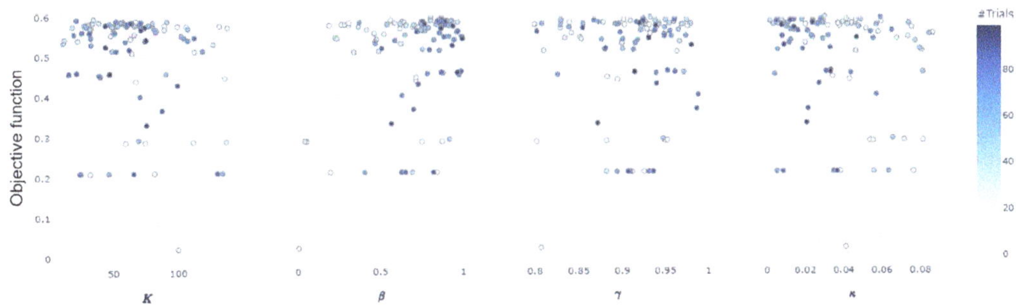

Figure 15. Hyperparameter importance plot considering 100 trials of the TPE algorithm for data distribution and sampling.

Regarding the structures of the actor, critic and discriminator networks regarding the number of layers and neurons and the learning rate, the critic network has the most significant contribution to the objective function (e.g., see α_c and Nc_0). More details can be found in Figure 16, where a shallow architecture (i.e., $La_c = 2$) is sufficient to approximate the value function, which depends on a significant number of neurons only in the first layer for local approximation of the input features (i.e., state–action pair), as explained in Das et al. [102]. Furthermore, training with the gradient-descent algorithm is less complex compared to the actor and discriminator. The former depends on a deeper architecture in general (i.e., $L_a = 4$) and training with a conservative GD optimization algorithm due to its greater complexity and nonlinearity to approximate the deterministic gradient (Figure 17). The latter also depends on a deep architecture and a conservative GD optimization (Figure 18), which is due to the complexity of the objective function resulting from generative and discriminative networks (i.e., by employing binary cross-entropy loss) [87].

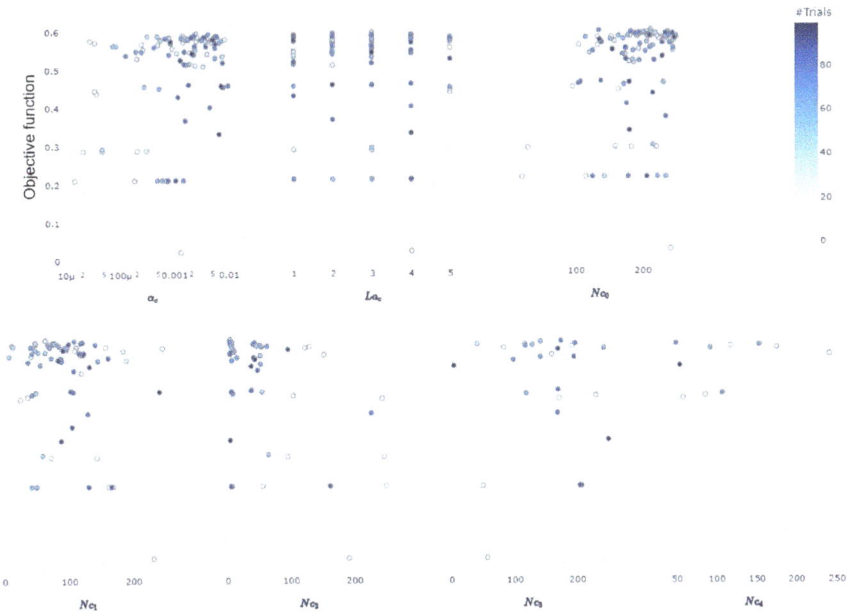

Figure 16. Hyperparameter importance slice plot for modified DDPG considering 100 trials of the TPE algorithm for the critic network.

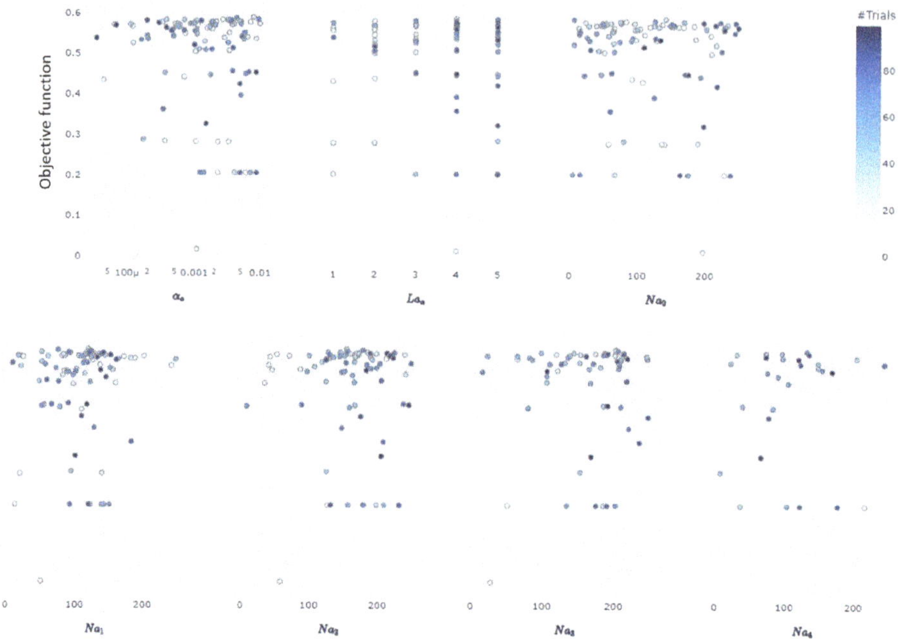

Figure 17. Hyperparameter importance slice plot for modified DDPG considering 100 trials of the TPE algorithm for the actor network.

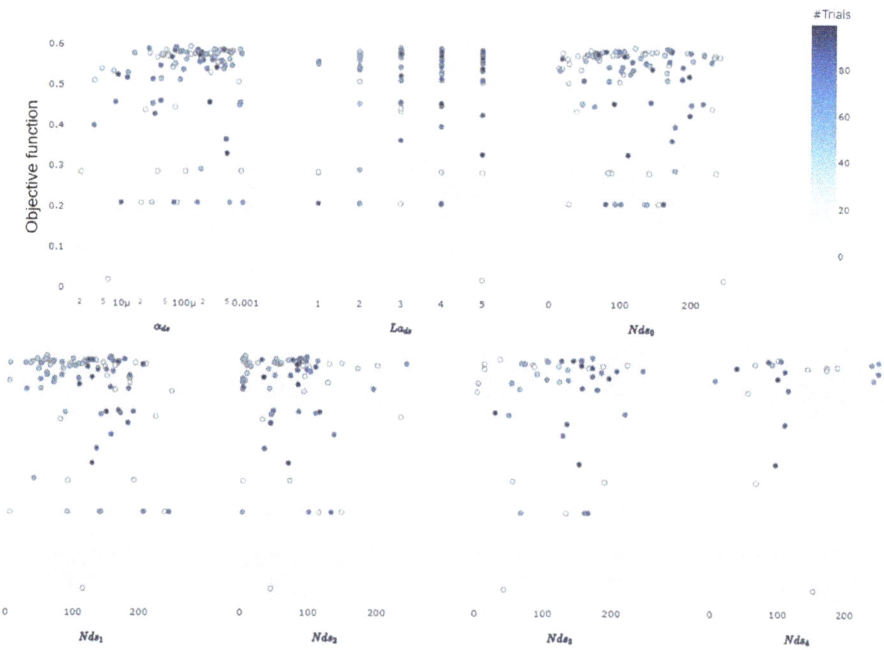

Figure 18. Hyperparameter importance slice plot for modified DDPG considering 100 trials of the TPE algorithm for the discriminative network.

6. Conclusions

These final considerations result from an in-depth state-of-the-art review of artificial intelligence and optimal process control. The objective was to develop a complete guide for hyperparameter optimization, imitation learning and transfer learning, since there are no articles covering such subjects together. Thus, the conclusions made answer several claims about the integration of reinforcement learning and process control:

- State-of-the-art technologies are still embryonic;
- Batch and continuous processes require different learning structures;
- Developing state-of-the-art offline training technologies is essential;
- Transfer learning has a broad meaning in RL, since it can encompass learning from demonstration, reward shaping, policy transfer and inter-task mapping;
- The proposed modified DDPG algorithm with an off-policy discriminator confirmed the hypothesis that information from process demonstrations improves the performance of the standard DDPG algorithm, as detailed in Section 5.

Finally, the control experiment carried out in Section 5 provides some guidelines for extending this approach to more complex systems (batch and continuous). These results can support the study of hyperparameter optimization for other case studies. Furthermore, the main challenge is developing new control structures appropriate to process constraints. To achieve this, it remains necessary to improve Modules 1, 2 and 3, shown in Figure 9.

Author Contributions: R.d.R.F. participated in all steps of the research method: conceptualization, methodology, writing—original draft preparation. For review and editing, all authors participated. Conceptualization and supervision, B.D.O.C., A.R.S. and M.B.d.S.J. All authors have read and agreed to the published version of the manuscript.

Funding: This study was financed in part by the Coordenação de Aperfeiçoamento de Pessoal de Nível Superior—Brasil (CAPES)—Finance Code 001. Maurício B. de Souza Jr. is grateful for financial support from CNPq (Grant No. 311153/2021-6) and Fundação Carlos Chagas Filho de Amparo à Pesquisa do Estado do Rio de Janeiro (FAPERJ) (Grant No. E-26/201.148/2022).

Data Availability Statement: Not applicable.

Conflicts of Interest: The authors declare no conflict of interest.

References

1. Sutton, R.S.; Barto, A.G. *Reinforcement Learning: An Introduction*; MIT Press: Cambridge, MA, USA, 2018.
2. Bellman, R. *Dynamic Programming*; Princeton University Press: Princeton, NJ, USA, 1957; Volume 95.
3. Bellman, R. A Markovian decision process. *J. Math. Mech.* **1957**, *6*, 679–684. [CrossRef]
4. Hoskins, J.; Himmelblau, D. Process control via artificial neural networks and reinforcement learning. *Comput. Chem. Eng.* **1992**, *16*, 241–251. [CrossRef]
5. Hinton, G.; Srivastava, N.; Swersky, K. Neural networks for machine learning lecture 6a overview of mini-batch gradient descent. *Cited On* **2012**, *14*, 2.
6. Hinton, G.E.; Srivastava, N.; Krizhevsky, A.; Sutskever, I.; Salakhutdinov, R.R. Improving neural networks by preventing co-adaptation of feature detectors. *arXiv* **2012**, arXiv:1207.0580.
7. Wulfmeier, M.; Posner, I.; Abbeel, P. Mutual alignment transfer learning. In Proceedings of the Conference on Robot Learning (PMLR), Mountain View, CA, USA, 13–15 November 2017; pp. 281–290.
8. Peng, X.B.; Andrychowicz, M.; Zaremba, W.; Abbeel, P. Sim-to-real transfer of robotic control with dynamics randomization. In Proceedings of the 2018 IEEE International Conference on Robotics and Automation (ICRA), Brisbane, Australia, 21–25 May 2018; pp. 3803–3810.
9. Silver, D.; Schrittwieser, J.; Simonyan, K.; Antonoglou, I.; Huang, A.; Guez, A.; Hubert, T.; Baker, L.; Lai, M.; Bolton, A.; et al. Mastering the game of go without human knowledge. *Nature* **2017**, *550*, 354–359. [CrossRef]
10. Mnih, V.; Kavukcuoglu, K.; Silver, D.; Rusu, A.A.; Veness, J.; Bellemare, M.G.; 38 Graves, A.; Riedmiller, M.; Fidjeland, A.K.; Ostrovski, G.; et al. Human-level control through deep reinforcement learning. *Nature* **2015**, *518*, 529–533. [CrossRef]
11. Vinyals, O.; Babuschkin, I.; Czarnecki, W.M.; Mathieu, M.; Dudzik, A.; Chung, J.; Choi, D.H.; Powell, R.; Ewalds, T.; Georgiev, P.; et al. Grandmaster level in StarCraft II using multi-agent reinforcement learning. *Nature* **2019**, *575*, 350–354. [CrossRef]
12. Nian, R.; Liu, J.; Huang, B. A review on reinforcement learning: Introduction and applications in industrial process control. *Comput. Chem. Eng.* **2020**, *139*, 106886. [CrossRef]

13. Buşoniu, L.; de Bruin, T.; Tolić, D.; Kober, J.; Palunko, I. Reinforcement learning for control: Performance, stability, and deep approximators. *Annu. Rev. Control* **2018**, *46*, 8–28. [CrossRef]
14. Petsagkourakis, P.; Sandoval, I.O.; Bradford, E.; Zhang, D.; del Rio-Chanona, E.A. Reinforcement learning for batch bioprocess optimization. *Comput. Chem. Eng.* **2020**, *133*, 106649. [CrossRef]
15. Petsagkourakis, P.; Sandoval, I.O.; Bradford, E.; Zhang, D.; del Rio-Chanona, E.A. Reinforcement learning for batch-to-batch bioprocess optimisation. In *Computer Aided Chemical Engineering*; Elsevier: Amsterdam, The Netherlands, 2019; Volume 46, pp. 919–924.
16. Yoo, H.; Kim, B.; Kim, J.W.; Lee, J.H. Reinforcement learning based optimal control of batch processes using Monte-Carlo deep deterministic policy gradient with phase segmentation. *Comput. Chem. Eng.* **2021**, *144*, 107133. [CrossRef]
17. Ma, Y.; Zhu, W.; Benton, M.G.; Romagnoli, J. Continuous control of a polymerization system with deep reinforcement learning. *J. Process Control* **2019**, *75*, 40–47. [CrossRef]
18. Powell, K.M.; Machalek, D.; Quah, T. Real-time optimization using reinforcement learning. *Comput. Chem. Eng.* **2020**, *143*, 107077. [CrossRef]
19. Nikita, S.; Tiwari, A.; Sonawat, D.; Kodamana, H.; Rathore, A.S. Reinforcement learning based optimization of process chromatography for continuous processing of biopharmaceuticals. *Chem. Eng. Sci.* **2021**, *230*, 116171. [CrossRef]
20. Dogru, O.; Wieczorek, N.; Velswamy, K.; Ibrahim, F.; Huang, B. Online reinforcement learning for a continuous space system with experimental validation. *J. Process Control* **2021**, *104*, 86–100. [CrossRef]
21. Ławryńczuk, M.; Marusak, P.M.; Tatjewski, P. Cooperation of model predictive control with steady-state economic optimisation. *Control Cybern.* **2008**, *37*, 133–158.
22. Skogestad, S. Control structure design for complete chemical plants. *Comput. Chem. Eng.* **2004**, *28*, 219–234. [CrossRef]
23. Backx, T.; Bosgra, O.; Marquardt, W. Integration of model predictive control and optimization of processes: Enabling technology for market driven process operation. *IFAC Proc. Vol.* **2000**, *33*, 249–260. [CrossRef]
24. Adetola, V.; Guay, M. Integration of real-time optimization and model predictive control. *J. Process Control* **2010**, *20*, 125–133. [CrossRef]
25. Aggarwal, C.C. *Neural Networks and Deep Learning*; Springer: Berlin/Heidelberg, Germany, 2018; Volume 10, pp. 978–983.
26. Pan, E.; Petsagkourakis, P.; Mowbray, M.; Zhang, D.; del Rio-Chanona, E.A. Constrained model-free reinforcement learning for process optimization. *Comput. Chem. Eng.* **2021**, *154*, 107462. [CrossRef]
27. Mowbray, M.; Smith, R.; Del Rio-Chanona, E.A.; Zhang, D. Using process data to generate an optimal control policy via apprenticeship and reinforcement learning. *AIChE J.* **2021**, *67*, e17306. [CrossRef]
28. Shah, H.; Gopal, M. Model-free predictive control of nonlinear processes based on reinforcement learning. *IFAC-PapersOnLine* **2016**, *49*, 89–94. [CrossRef]
29. Alhazmi, K.; Albalawi, F.; Sarathy, S.M. A reinforcement learning-based economic model predictive control framework for autonomous operation of chemical reactors. *Chem. Eng. J.* **2022**, *428*, 130993. [CrossRef]
30. Kim, J.W.; Park, B.J.; Yoo, H.; Oh, T.H.; Lee, J.H.; Lee, J.M. A model-based deep reinforcement learning method applied to finite-horizon optimal control of nonlinear control-affine system. *J. Process Control* **2020**, *87*, 166–178. [CrossRef]
31. Badgwell, T.A.; Lee, J.H.; Liu, K.H. Reinforcement learning–overview of recent progress and implications for process control. In *Computer Aided Chemical Engineering*; Elsevier: Amsterdam, The Netherlands, 2018; Volume 44, pp. 71–85.
32. Görges, D. Relations between model predictive control and reinforcement learning. *IFAC-PapersOnLine* **2017**, *50*, 4920–4928. [CrossRef]
33. Sugiyama, M. *Statistical Reinforcement Learning: Modern Machine Learning Approaches*; CRC Press: Boca Raton, FL, USA, 2015.
34. Howard, R.A. *Dynamic Programming and Markov Processes*; MITPL: Cambridge, MA, USA, 1960.
35. Thorndike, E.L. Animal intelligence: An experimental study of the associative processes in animals. *Psychol. Rev. Monogr. Suppl.* **1898**, *2*, 1.
36. Minsky, M. Neural Nets and the Brain-Model Problem. Doctoral Dissertation, Princeton University, Princeton, NJ, USA, 1954; *Unpublished*.
37. Minsky, M. Steps toward artificial intelligence. *Proc. IRE* **1961**, *49*, 8–30. [CrossRef]
38. Barto, A.G.; Sutton, R.S.; Anderson, C.W. Neuronlike adaptive elements that can solve difficult learning control problems. *IEEE Trans. Syst. Man Cybern.* **1983**, *SMC-13*, 834–846. [CrossRef]
39. Sutton, R.S. Learning to predict by the methods of temporal differences. *Mach. Learn.* **1988**, *3*, 9–44. [CrossRef]
40. Watkins, C.J.C.H. *Learning from Delayed Rewards*; University of Cambridge: Cambridge, UK, 1989.
41. Gullapalli, V. A stochastic reinforcement learning algorithm for learning real-valued functions. *Neural Netw.* **1990**, *3*, 671–692. [CrossRef]
42. Williams, R.J. Simple statistical gradient-following algorithms for connectionist reinforcement learning. *Mach. Learn.* **1992**, *8*, 229–256. [CrossRef]
43. Bishop, C.M. *Pattern Recognition and Machine Learning*; Springer: Berlin/Heidelberg, Germany, 2006.
44. LeCun, Y.; Bengio, Y.; Hinton, G. Deep learning. *Nature* **2015**, *521*, 436–444. [CrossRef] [PubMed]
45. Berry, D.A.; Fristedt, B. *Bandit Problems: Sequential Allocation of Experiments (Monographs on Statistics and Applied Probability)*; Chapman and Hall: London, UK, 1985; Volume 5, pp. 71–87.
46. Sutton, R.S.; Barto, A.G. *Introduction to Reinforcement Learning*; MIT Press Cambridge: Cambridge, MA, USA, 1998; Volume 135.

47. Shannon, C.E. A mathematical theory of communication. *ACM SIGMOBILE Mob. Comput. Commun. Rev.* **2001**, *5*, 3–55. [CrossRef]
48. Silver, D.; Lever, G.; Heess, N.; Degris, T.; Wierstra, D.; Riedmiller, M. Deterministic policy gradient algorithms. In Proceedings of the International Conference on Machine Learning (PMLR), Bejing, China, 22–24 June 2014.
49. Thrun, S.; Schwartz, A. Issues in using function approximation for reinforcement learning. In *Proceedings of the 1993 Connectionist Models Summer School*; Lawrence Erlbaum: Hillsdale, NJ, USA, 1993.
50. Fujimoto, S.; Van Hoof, H.; Meger, D. Addressing function approximation error in actor-critic methods. *arXiv* **2018**, arXiv:1802.09477.
51. Sutton, R.S.; McAllester, D.A.; Singh, S.P.; Mansour, Y. Policy gradient methods for reinforcement learning with function approximation. In Proceedings of the Advances in Neural Information Processing Systems, Denver, CO, USA, 29 November–4 December 2000; pp. 1057–1063.
52. Gordon, G.J. Stable function approximation in dynamic programming. In *Machine Learning Proceedings 1995*; Elsevier: Amsterdam, The Netherlands, 1995; pp. 261–268.
53. Tsitsiklis, J.N.; Van Roy, B. Feature-based methods for large scale dynamic programming. *Mach. Learn.* **1996**, *22*, 59–94. [CrossRef]
54. Grondman, I.; Busoniu, L.; Lopes, G.A.; Babuska, R. A survey of actor-critic reinforcement learning: Standard and natural policy gradients. *IEEE Trans. Syst. Man Cybern. Part C (Appl. Rev.)* **2012**, *42*, 1291–1307. [CrossRef]
55. Ramicic, M.; Bonarini, A. Augmented Replay Memory in Reinforcement Learning With Continuous Control. *arXiv* **2019**, arXiv:1912.12719.
56. Haarnoja, T.; Zhou, A.; Abbeel, P.; Levine, S. Soft actor-critic: Off-policy maximum entropy deep reinforcement learning with a stochastic actor. In Proceedings of the International Conference on Machine Learning (PMLR), Stockholm, Sweden, 10–15 July 2018; pp. 1861–1870.
57. Benhamou, E. Variance Reduction in Actor Critic Methods (ACM). *arXiv* **2019**, arXiv:1907.09765.
58. Lillicrap, T.P.; Hunt, J.J.; Pritzel, A.; Heess, N.; Erez, T.; Tassa, Y.; Silver, D.; Wierstra, D. Continuous control with deep reinforcement learning. *arXiv* **2015**, arXiv:1509.02971.
59. Schulman, J.; Wolski, F.; Dhariwal, P.; Radford, A.; Klimov, O. Proximal policy optimization algorithms. *arXiv* **2017**, arXiv:1707.06347.
60. Kaelbling, L.P.; Littman, M.L.; Cassandra, A.R. Planning and acting in partially observable stochastic domains. *Artif. Intell.* **1998**, *101*, 99–134. [CrossRef]
61. Bonvin, D. Optimal operation of batch reactors—A personal view. *J. Process Control* **1998**, *8*, 355–368. [CrossRef]
62. Bonvin, D.; Srinivasan, B.; Ruppen, D. *Dynamic Optimization in the Batch Chemical Industry*; Technical Report; NTNU: Trondheim, Norway, 2001.
63. Arpornwichanop, A.; Kittisupakorn, P.; Mujtaba, I. On-line dynamic optimization and control strategy for improving the performance of batch reactors. *Chem. Eng. Process. Process. Intensif.* **2005**, *44*, 101–114. [CrossRef]
64. Mowbray, M.; Petsagkourakis, P.; Chanona, E.A.d.R.; Smith, R.; Zhang, D. Safe Chance Constrained Reinforcement Learning for Batch Process Control. *arXiv* **2021**, arXiv:2104.11706.
65. Oh, T.H.; Park, H.M.; Kim, J.W.; Lee, J.M. Integration of reinforcement learning and model predictive control to optimize semi-batch bioreactor. *AIChE J.* **2022**, *68*, e17658. [CrossRef]
66. Ellis, M.; Durand, H.; Christofides, P.D. A tutorial review of economic model predictive control methods. *J. Process Control* **2014**, *24*, 1156–1178. [CrossRef]
67. Ramanathan, P.; Mangla, K.K.; Satpathy, S. Smart controller for conical tank system using reinforcement learning algorithm. *Measurement* **2018**, *116*, 422–428. [CrossRef]
68. Hwangbo, S.; Sin, G. Design of control framework based on deep reinforcement learning and Monte-Carlo sampling in downstream separation. *Comput. Chem. Eng.* **2020**, *140*, 106910. [CrossRef]
69. Chen, K.; Wang, H.; Valverde-Pérez, B.; Zhai, S.; Vezzaro, L.; Wang, A. Optimal control towards sustainable wastewater treatment plants based on multi-agent reinforcement learning. *Chemosphere* **2021**, *279*, 130498. [CrossRef]
70. Oh, D.H.; Adams, D.; Vo, N.D.; Gbadago, D.Q.; Lee, C.H.; Oh, M. Actor-critic reinforcement learning to estimate the optimal operating conditions of the hydrocracking process. *Comput. Chem. Eng.* **2021**, *149*, 107280. [CrossRef]
71. Mnih, V.; Badia, A.P.; Mirza, M.; Graves, A.; Lillicrap, T.; Harley, T.; Silver, D.; Kavukcuoglu, K. Asynchronous methods for deep reinforcement learning. In Proceedings of the International Conference on Machine Learning, New York, NY, USA, 19–24 June 2016; pp. 1928–1937.
72. Tan, C.; Sun, F.; Kong, T.; Zhang, W.; Yang, C.; Liu, C. A survey on deep transfer learning. In Proceedings of the International Conference on Artificial Neural Networks, Rhodes, Greece, 4–7 October 2018; Springer: Berlin/Heidelberg, Germany, 2018; pp. 270–279.
73. Taylor, M.E.; Stone, P. Transfer learning for reinforcement learning domains: A survey. *J. Mach. Learn. Res.* **2009**, *10*, 1633–1685.
74. Peirelinck, T.; Kazmi, H.; Mbuwir, B.V.; Hermans, C.; Spiessens, F.; Suykens, J.; Deconinck, G. Transfer learning in demand response: A review of algorithms for data-efficient modelling and control. *Energy AI* **2022**, *7*, 100126. [CrossRef]
75. Joshi, G.; Chowdhary, G. Cross-domain transfer in reinforcement learning using target apprentice. In Proceedings of the 2018 IEEE International Conference on Robotics and Automation (ICRA), Brisbane, Australia, 21–25 May 2018; pp. 7525–7532.
76. Zhu, Z.; Lin, K.; Dai, B.; Zhou, J. Learning sparse rewarded tasks from sub-optimal demonstrations. *arXiv* **2020**, arXiv:2004.00530.

77. Yan, M.; Frosio, I.; Tyree, S.; Kautz, J. Sim-to-real transfer of accurate grasping with eye-in-hand observations and continuous control. *arXiv* **2017**, arXiv:1712.03303.
78. Christiano, P.; Shah, Z.; Mordatch, I.; Schneider, J.; Blackwell, T.; Tobin, J.; Abbeel, P.; Zaremba, W. Transfer from simulation to real world through learning deep inverse dynamics model. *arXiv* **2016**, arXiv:1610.03518.
79. Kostrikov, I.; Agrawal, K.K.; Dwibedi, D.; Levine, S.; Tompson, J. Discriminator-actor-critic: Addressing sample inefficiency and reward bias in adversarial imitation learning. *arXiv* **2018**, arXiv:1809.02925.
80. Spielberg, S.; Tulsyan, A.; Lawrence, N.P.; Loewen, P.D.; Bhushan Gopaluni, R. Toward self-driving processes: A deep reinforcement learning approach to control. *AIChE J.* **2019**, *65*, e16689. [CrossRef]
81. Hausknecht, M.; Stone, P. Deep reinforcement learning in parameterized action space. *arXiv* **2015**, arXiv:1511.04143.
82. Hou, Y.; Liu, L.; Wei, Q.; Xu, X.; Chen, C. A novel ddpg method with prioritized experience replay. In Proceedings of the 2017 IEEE international conference on systems, man, and cybernetics (SMC), Banff, AB, Canada, 5–8 October 2017; pp. 316–321.
83. Wang, X.; Ye, X. Consciousness-driven reinforcement learning: An online learning control framework. *Int. J. Intell. Syst.* **2022**, *37*, 770–798. [CrossRef]
84. Feise, H.J.; Schaer, E. Mastering digitized chemical engineering. *Educ. Chem. Eng.* **2021**, *34*, 78–86. [CrossRef]
85. Hua, J.; Zeng, L.; Li, G.; Ju, Z. Learning for a robot: Deep reinforcement learning, imitation learning, transfer learning. *Sensors* **2021**, *21*, 1278. [CrossRef]
86. Hussein, A.; Gaber, M.M.; Elyan, E.; Jayne, C. Imitation learning: A survey of learning methods. *ACM Comput. Surv. (CSUR)* **2017**, *50*, 1–35. [CrossRef]
87. Goodfellow, I.; Pouget-Abadie, J.; Mirza, M.; Xu, B.; Warde-Farley, D.; Ozair, S.; Courville, A.; Bengio, Y. Generative adversarial nets. *Adv. Neural Inf. Process. Syst.* **2014**, *27*.
88. Hutter, F.; Hoos, H.H.; Leyton-Brown, K.; Stützle, T. ParamILS: An automatic algorithm configuration framework. *J. Artif. Intell. Res.* **2009**, *36*, 267–306. [CrossRef]
89. Hutter, F. Automated Configuration of Algorithms for Solving Hard Computational Problems. Ph.D. Thesis, University of British Columbia, Vancouver, BC, Canada, 2009.
90. Coates, A.; Ng, A.Y. The importance of encoding versus training with sparse coding and vector quantization. In Proceedings of the 28th International Conference on Machine Learning (ICML), Washington, DC, USA, 28 June–2 July 2011.
91. Coates, A.; Ng, A.; Lee, H. An analysis of single-layer networks in unsupervised feature learning. In Proceedings of the Fourteenth International Conference on Artificial Intelligence and Statistics, Fort Lauderdale, FL, USA, 11–13 April 2011; pp. 215–223.
92. Akiba, T.; Sano, S.; Yanase, T.; Ohta, T.; Koyama, M. Optuna: A next-generation hyperparameter optimization framework. In Proceedings of the 25th ACM SIGKDD International Conference on Knowledge Discovery & Data Mining, Anchorage, AK, USA, 4–8 August 2019; pp. 2623–2631.
93. Rapin, J.; Teytaud, O. Nevergrad—A Gradient-Free Optimization Platform. 2018. Available online: https://GitHub.com/FacebookResearch/Nevergrad (accessed on 10 September 2022).
94. Liaw, R.; Liang, E.; Nishihara, R.; Moritz, P.; Gonzalez, J.E.; Stoica, I. Tune: A Research Platform for Distributed Model Selection and Training. *arXiv* **2018**, arXiv:1807.05118.
95. Bergstra, J.S.; Bardenet, R.; Bengio, Y.; Kégl, B. Algorithms for hyper-parameter optimization. In Proceedings of the Advances in Neural Information Processing Systems, Granada, Spain, 12–15 December 2011; pp. 2546–2554.
96. Snoek, J.; Larochelle, H.; Adams, R.P. Practical bayesian optimization of machine learning algorithms. In Proceedings of the Advances in Neural Information Processing Systems, Lake Tahoe, NA, USA, 3–6 December 2012; pp. 2951–2959.
97. Li, L.; Jamieson, K.; Rostamizadeh, A.; Gonina, E.; Hardt, M.; Recht, B.; Talwalkar, A. Massively parallel hyperparameter tuning. *arXiv* **2018**, arXiv:1810.05934.
98. Li, L.; Jamieson, K.; DeSalvo, G.; Rostamizadeh, A.; Talwalkar, A. Hyperband: A novel bandit-based approach to hyperparameter optimization. *J. Mach. Learn. Res.* **2017**, *18*, 6765–6816.
99. Jaderberg, M.; Dalibard, V.; Osindero, S.; Czarnecki, W.M.; Donahue, J.; Razavi, A.; Vinyals, O.; Green, T.; Dunning, I.; Simonyan, K.; et al. Population based training of neural networks. *arXiv* **2017**, arXiv:1711.09846.
100. Bergstra, J.; Bardenet, R.; Kégl, B.; Bengio, Y. Implementations of algorithms for hyper-parameter optimization. In Proceedings of the NIPS Workshop on Bayesian Optimization, Sierra Nevada, Spain, 16–17 December 2011; p. 29.
101. Kingma, D.P.; Ba, J. Adam: A method for stochastic optimization. *arXiv* **2014**, arXiv:1412.6980.
102. Das, L.; Sivaram, A.; Venkatasubramanian, V. Hidden representations in deep neural networks: Part 2. Regression problems. *Comput. Chem. Eng.* **2020**, *139*, 106895. [CrossRef]

processes

MDPI

Article

Driver–Automated Cooperation Driving Authority Optimization Framework for Shared Steering Control

Shuting Yan [1,2], Qingsong Wei [3], Xianyi Xie [3,*], Dingxuan Zhao [1] and Xinyu Liu [3]

1 School of Mechanical Engineering, Yanshan University, Qinhuangdao 066000, China; styan@ysu.edu.cn (S.Y.); zdx@ysu.edu.cn (D.Z.)

2 Institute of Science and Technology, Yanshan University, Qinhuangdao 066000, China

3 School of Vehicle and Energy, Yanshan University, Qinhuangdao 066000, China; weiqs1999@stumail.ysu.edu.cn (Q.W.); liuxinyu@stumail.ysu.edu.cn (X.L.)

* Correspondence: xiexianyi@ysu.edu.cn

Abstract: In this paper, we introduce the preview-follower theory for modeling the trajectory-tracking controller of an automated system using model predictive control (MPC). The primary contribution of this research lies in enhancing tracking accuracy and driving safety when the driver and automated system share similar driving intentions, while also enabling a rapid transfer of driving authority to the human driver in cases of differing intentions. To verify the effectiveness of the proposed driving authority optimization framework, both simulation and driver-in-the-loop experiments were conducted under conditions of consistent and inconsistent driving intentions between the human driver and the autonomous driving system. The results of both experiments demonstrated that the proposed shared steering cooperative control and driving authority optimization framework not only significantly improves vehicle tracking accuracy but also promptly aligns with the driver's intentions.

Keywords: driver–automated cooperation; driving authority; driver model; model predictive control; driving intention; shared steering control

1. Introduction

In recent years, with advancements in artificial intelligence and technological progress, autonomous driving (AD) vehicles have emerged as a promising solution to alleviate traffic congestion and reduce traffic accidents [1], leading to their gradual commercialization [2,3]. However, the prospect of fully automated driving systems becoming available in the near future remains unlikely [4,5]. This is due to various challenges, including but not limited to, technical requirements, safety concerns, legal restrictions, and ethical dilemmas [6]. Although automated driving systems are becoming increasingly sophisticated, they cannot yet fully replace human drivers [7]. It is widely acknowledged that human drivers, with their superior ability to handle complex situations, should remain in the control loop until fully automated systems are realized. Until that point, human drivers and automated systems must cooperate to prevent vehicle collisions and mitigate potential dangers [8].

Under such circumstances, steering control shared between human drivers and machines is worth considering as a transitional approach from human-driven vehicles to fully automated driving [9]. According to the SAE J3016 standard [10], human–machine shared steering control can be categorized into three types: single driving with double control, double driving with single control, and double driving with double control [11]. In general, modern vehicles equipped with ABS (anti-lock braking system) represent the single driving with double control type [12]. This is akin to a low-level ADAS, which does not interfere with the driver's subjective intentions [13]. The double driving with single control type involves switching driving control between the driver and the machine under different conditions, with both the driver and the intelligent system taking turns to control the vehicle. However, in this setup, the intelligent system must alert the driver and transfer control when the vehicle encounters conditions beyond the machine's capabilities. Thus, selecting an appropriate strategy to switch driving authority between the driver and the controller at the right time is a challenging task for this control type. The double driving with double control type involves both the driver and the automated system sharing driving authority. Here, driving control is distributed between the two based on a specific mechanism, although either the driver or the intelligent system can take full control under certain conditions when necessary [14]. This control type considers the driver's intentions while benefiting from the precise control and quick response capabilities of the automated system. In most instances, both the driver and the intelligent system remain in the control loop. By employing this control type, vehicle safety and stability can be enhanced, particularly for drivers with lower skill levels. Simultaneously, this control type increases autonomous vehicle trajectory-tracking accuracy while reducing driver workload [15].

Additionally, the dual driving with dual control type encounters a significant and critical challenge: determining an appropriate and effective strategy for shared steering control. A substantial body of research has been dedicated to this issue, particularly concerning decision-making for driving authority [16–18]. Despite these efforts, it remains a topic that merits further in-depth investigation.

In recent years, steering-by-wire technology has advanced vehicle steering systems by decoupling the driver's commands from direct vehicle execution. Instead, controllers modify the driver's inputs to enhance vehicle handling stability, thus enabling shared steering control. Prior to this, shared steering control primarily relied on haptic interaction, where co-driving was achieved through torque interaction between the driver and the automated system [19]. Reference [20] introduced a driver–automation cooperative control approach that effectively distributes driving control authority between human drivers and the lane-keeping assist (LKA) system. This approach addresses vehicle stability and trajectory tracking by implementing damping compensation and conflict minimization strategies to manage control authority between the two driving entities. Additionally, a novel cooperative planning and tracking system for safe lane changes in automated vehicles was proposed [21]. This system predicts vehicle driving trajectories based on historical data and assesses the risk level associated with human driver behavior. Adaptive human–machine cooperation is then activated during trajectory planning to resolve conflicts. Hang [22] proposed a cooperative control framework incorporating both human drivers and an active rear steering (ARS) system. Performance metrics focused on active safety, including lateral stability, handling performance, rollover prevention, and path tracking. Constraints for the ARS controller were established, and a linear time-varying (LTV) model predictive control (MPC) system was employed for ARS controller design, considering multiple constraints. This cooperative control framework aims to enhance active vehicle safety by assisting drivers with varying driving characteristics.

Chen J [23] proposed a human–machine cooperative scheme for car-following control in connected and automated vehicles (CAVs), which ensures that drivers remain consistently engaged and enhances car-following performance. In this scheme, the automated system handles velocity tracking, leveraging the precise perception and communication capabilities of CAVs, while the human driver is responsible for headway adjustment, given their accurate perception. Stability analysis based on the Lyapunov function demonstrated that a simple proportional feedback control by the driver can maintain system stability when cooperating with automated velocity tracking.

The research in this area focuses on trajectory planning, shared steering cooperation, and driving assistance, among other aspects, aiming to optimize and improve human–machine systems from various perspectives.

However, two critical issues remain in the domain of shared steering control amidst extensive research on driving authority distribution. The first issue is accurately and effectively discerning the driver's driving intentions. The second is timely and effectively resolving conflicts between the driver and the automated system. In the context of shared steering, model predictive control (MPC) is considered an effective method for modeling driver behavior and characteristics [24–27]. The primary advantage of MPC in this context is its predictive capability, which aligns with driver behavior and adheres to the preview-follower theory. For addressing conflicts between drivers and machines, game theory [28] offers an effective framework to describe and manage non-cooperative conflict situations and to develop shared steering control strategies under such conditions. However, while game theory can effectively address conflicts between two players, the resolution often involves adhering to the desired lane of one party or following traffic rules, rather than finding a middle ground. This approach may not always align with the original intent of resolving conflicts. Therefore, addressing these conflicts requires solutions that meet actual needs and conditions.

For Level-2 autonomy according to the SAE J3016 standard, the driver–automated cooperation strategy in shared steering control should prioritize the driver's demands. Reference [11] proposed an adaptive steering torque coupling framework to address human–machine conflicts. This framework utilizes long short-term memory (LSTM) networks to identify the driver's potential intentions and adjusts the Q and R matrices to achieve dynamic load allocation between the human and machine at a Nash equilibrium. Li [29] introduced an indirect shared control method to model driver–automated system conflicts and enhance driving performance. The Nash equilibrium strategy guides the automated system to respond in a manner that ensures safety and comfort, with approximate dynamic programming used to find Nash equilibria for practical applications.

Essentially, the two key issues can be consolidated into one: maintaining vehicle driving safety by keeping drivers with lower-level skills in the control loop, with the controller handling most of the workload in the absence of conflict. Conversely, skilled drivers should be assigned driving authority when conflicts arise between the driver and automated system. Machine learning methods can be used to identify driver intentions [11], though this issue is beyond the scope of this paper.

To address conflicts between the driver and the automated system, this paper proposes a driving authority optimization framework for shared steering control. This framework models driver–automated system interactions using model predictive control (MPC) and outlines a strategy for distributing driving authority. The driver model, which describes the driver's lateral handling behavior, is also based on MPC. The automated system for trajectory tracking is established using single-point preview-follow theory. Conflicts between the driver and the automated system are formulated and resolved using MPC. The solution for driving authority distribution is framed as a nonlinear programming problem, which is solved using the "fmincon" function in MATLAB. Simulations under two typical scenarios—where the driver and automated system have either consistent or inconsistent driving intentions—are conducted to evaluate the effectiveness of the proposed framework.

The purpose of this paper is to present an optimal framework for driver–automated driving authority in shared steering control that enhances trajectory-tracking accuracy and resolves conflicts between the driver and the automated system. The main contributions of this paper are summarized as follows:

(1) Development of an optimal framework. An optimal framework for driver–automated driving authority in shared steering control is proposed, which adjusts the driving authority between the driver and the automated system.
(2) Introduction of a novel adjustment method. A novel method for adjusting driving authority, based on model predictive control (MPC), is established to resolve conflicts between the driver and the automated system.
(3) Enhancement of trajectory tracking and conflict resolution. The proposed framework improves the trajectory-tracking accuracy of the autonomous system when the driver and automated system have consistent driving intentions. It also effectively resolves conflicts between the driver and the automated system, prioritizing the driver's desired trajectory.

The remainder of this paper is organized as follows: Section 2 outlines the driver modeling, automated controller, and the structure of the driver–automated driving authority optimal framework. Section 3 details the design of the driver–automated shared steering controller based on MPC. Section 4 presents simulation results and discussions, validating the proposed optimal strategy under conditions of both consistent and inconsistent driving intentions. Section 5 includes validation of the proposed framework through driver-in-the-loop experiments. Finally, Section 6 concludes the paper.

2. System Modeling and Driver–Automated Driving Authority Optimal Framework

2.1. Driver-Automation Sharing Steering Control Framework

The structure of the driver–automated shared control framework is shown in Figure 1, where r_d and r_c denote the target trajectories or driving intention of the driver and automated system, respectively. δ_d, δ_c represent the steering wheel angle of the human driver and automated controller. The δ_f is the actual steering wheel angle executed to the autonomous vehicle, which consists of δ_d and δ_c, which is influenced by driving authority α. How to adjust the driving authority α will be discussed in detail later.

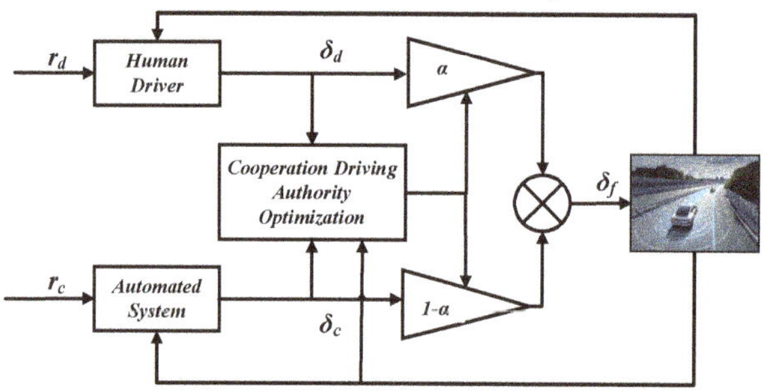

Figure 1. The structure of driver–automated shared control framework.

2.2. Vehicle Modeling

In this section, we describe the vehicle model employed for the trajectory-tracking controller. For the shared steering control system of the autonomous vehicle, we focus on the vehicle's lateral performance while assuming the longitudinal speed is constant. Consequently, a two-degree-of-freedom vehicle dynamics model is established, as depicted in Figure 2 and detailed in the following Formulation (1).

$$\begin{cases} \dot{\beta} = \frac{C_f+C_r}{mv_x}\beta + (\frac{aC_f-bC_r}{mv_x^2} - 1)\omega_r - \frac{C_f}{mv_x}\delta_f \\ \dot{\omega}_r = \frac{aC_f-bC_r}{I_Z}\beta + (\frac{a^2C_f+b^2C_r}{I_Z v_x})\omega_r - \frac{aC_f}{I_Z}\delta_f \end{cases} \quad (1)$$

Also, the vehicle kinematics model is easily obtained and shown in Formulation (2).

$$\begin{cases} \dot{y} = v(\varphi + \beta) \\ \dot{\varphi} = \omega_r \end{cases} \quad (2)$$

The vehicle parameters are demonstrated in Table 1.

Table 1. Vehicle model parameters.

Symbol	Description	Value
v_x, v_y	longitudinal/lateral speed of the vehicle centroid	/
α_f, α_r	front/rear wheel cornering angle	/

Table 1. *Cont.*

Symbol	Description	Value
a	lengths from the front axles to the vehicle mass center	1.035 m
b	lengths from the rear axles to the vehicle mass center	1.655 m
C_f	front tire cornering stiffness	−31,106.98 N/rad
C_r	rear tire cornering stiffness	−29,584.56 N/rad
m	vehicle mass	1740.7 kg
I_Z	yaw moment of inertia	3048 kg·m^2
ψ	yaw angle	/
δ_f	steering wheel angle	/

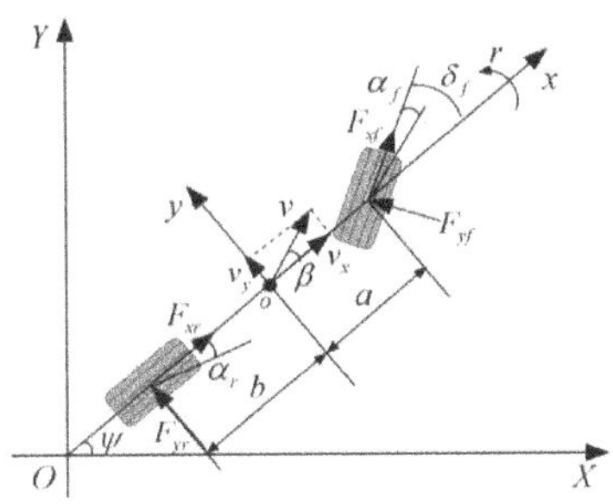

Figure 2. Two-DOF vehicle dynamic model.

In this paper, we combine the vehicle kinematics model and the two-degree-of-freedom vehicle dynamics model to establish the space–state equation, as shown in Formulation (3).

$$\begin{cases} \dot{y} = v(\varphi + \beta) \\ \dot{\varphi} = \omega_r \\ \dot{\beta} = \frac{C_f + C_r}{m v_x}\beta + (\frac{aC_f - bC_r}{m v_x^2} - 1)\omega_r - \frac{C_f}{m v_x}\delta_f \\ \dot{\omega}_r = \frac{aC_f - bC_r}{I_Z}\beta + (\frac{a^2 C_f + b^2 C_r}{I_Z v_x})\omega_r - \frac{aC_f}{I_Z}\delta_f \end{cases} \quad (3)$$

where $X = \begin{bmatrix} y & \varphi & \beta & \omega_r \end{bmatrix}^T$ gives the space–state variables of the system, which represent the lateral displacement, yaw angle, side slip angle, yaw rate, respectively; a and b are the lengths from the front and rear axles to the vehicle mass center; m is the vehicle mass; C_f and C_r are the front and rear tire cornering stiffness; I_z is the yaw moment of inertia.

$$X = \begin{bmatrix} y & \varphi & \beta & \omega_r \end{bmatrix}^T$$

In the differential equation of (3), δ_f denotes the steering wheel angle of the front wheels, and also is the control variable of the vehicle system.

$$U = \begin{bmatrix} \delta_f \end{bmatrix}$$

The relationship between steering wheel angle and front wheel steering angle is as follows:

$$\delta_f = \frac{\delta_{swa}}{i_{sw}}$$

Therein, i_{sw} is the transmission ratio of vehicle steering system.

Then, Formulation (3) is rewritten into state–space equations.

$$\dot{X} = AX + BU$$

$$A = \begin{bmatrix} 0 & v & v & 0 \\ 0 & 0 & 0 & 1 \\ 0 & 0 & \frac{C_f+C_r}{mv_x} & \frac{aC_f-bC_r}{mv_x^2} - 1 \\ 0 & 0 & \frac{aC_f-bC_r}{I_Z} & \frac{a^2C_f+b^2C_r}{I_Zv_x} \end{bmatrix}, B = \begin{bmatrix} 0 \\ 0 \\ -\frac{C_f}{mv_x} \\ -\frac{aC_f}{I_Z} \end{bmatrix}$$

By discretizing the continuous-time system (3) at a sample time of T_s and using the method of the first-order-based Taylor formulation, the discrete-time system of the space-state is obtained as follows:

$$\begin{aligned} x(k+1) &= A_c x(k) + B_c \delta_f(k) \\ y(k) &= Cx(k) \end{aligned} \tag{4}$$

Therein, the A_c, B_c, and C_c can be calculated as follows:

$$A_c = e^{A\cdot T_s} = A\cdot T_s + I,$$

$$B_c = \int_0^{T_s} e^{A\cdot\tau} d\tau \cdot B = B\cdot T_s,\ C_c = C$$

$$A_c = \begin{bmatrix} 1 & v\cdot T_s & v\cdot T_s & 0 \\ 0 & 1 & 0 & T_s \\ 0 & 0 & 1 + \frac{C_f+C_r}{mv_x}\cdot T_s & (\frac{aC_f-bC_r}{mv_x^2} - 1)\cdot T_s \\ 0 & 0 & \frac{aC_f-bC_r}{I_Z}\cdot T_s & 1 + \frac{a^2C_f+b^2C_r}{I_Zv_x}\cdot T_s \end{bmatrix} B_c = \begin{bmatrix} 0 \\ 0 \\ -\frac{C_f}{mv_x}\cdot T_s \\ -\frac{aC_f}{I_Z}\cdot T_s \end{bmatrix}$$

$$C_c = \begin{bmatrix} 1 & 0 & 0 & 0 \end{bmatrix}$$

2.3. Autonomous Vehicle Trajectory-Tracking Controller

In this section, a vehicle trajectory-tracking controller for an autonomous vehicle is developed based on the single-point preview optimal curvature model [30].

According to Formulation (1), the gain of the yaw rate with respect to the front wheel steering angle is given by Equation (5). Additionally, we aim for the vehicle to achieve steady-state steering characteristics, with the gain of the side slip angle relative to the front wheel steering angle described by Formulation (6).

$$\frac{\omega_{rd}}{\delta_f} = G_\omega \tag{5}$$

$$\frac{\beta}{\delta_f} = G_\beta \tag{6}$$

As shown in Figure 3a,b, based on the principle of "preview-follow theory" and "the preview optimal curvature model" [30], the formulation is as shown in (7)–(8). Substituting (7) and (8) into (5) to obtain the formulation of (9), then the trajectory-tracking controller for autonomous driving is designed.

$$y_{GC} = \tan(\frac{\theta}{2} + \beta)x_{GC} \tag{7}$$

$$\omega_r = \frac{\theta}{t_p} = \frac{2[\arctan(\frac{y_{GC}}{v_x t_p}) - \beta]}{t_p} \tag{8}$$

$$\delta_f^* = \frac{2\arctan(\frac{\Delta f}{v_x t_p})}{t_p G_\omega + 2G_\beta} \tag{9}$$

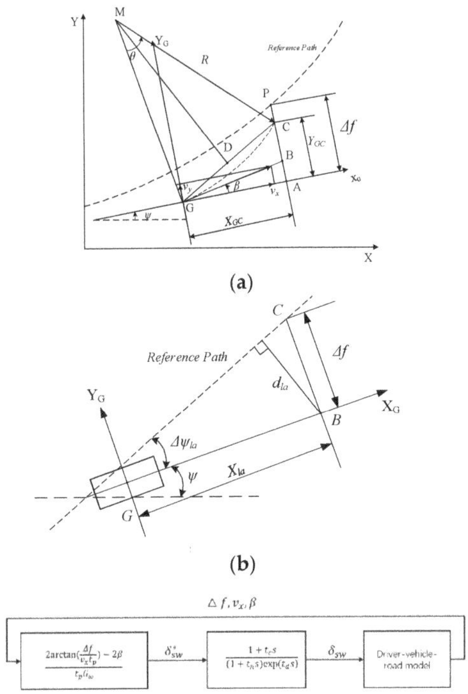

(c)

Figure 3. Diagram of preview-follow principle. (**a**) Preview-follow principle. (**b**) Error of the preview-follow model. (**c**) Diagram of the driver–vehicle–road model.

2.4. Driver Modeling

Referring to Formula (3), the driver model is designed using the principles of model predictive control (MPC). The predictive performance of MPC for future states closely resembles the preview characteristics exhibited by drivers when operating the vehicle. Thus, we employ the MPC method to model driver behavior. To accurately simulate real driver handling characteristics, we also incorporate actual constraints into the driver model [31].

To reflect the actual driver manipulation, we enhance the driver model by adding constraints on the steering wheel angle and its rate of change. This results in a new state–space equation, as outlined below:

$$\xi(k) = \begin{bmatrix} x(k) \\ u(k-1) \end{bmatrix},$$

$$\eta(k) = \begin{bmatrix} I_{Nx} & 0 \end{bmatrix} \begin{bmatrix} x(k) \\ u(k-1) \end{bmatrix} = \tilde{C}\xi(k)$$

The system output equation can be expressed as follows:

$$\begin{aligned}\xi(k+1) &= \begin{bmatrix} x(k+1) \\ u(k) \end{bmatrix} = \begin{bmatrix} A_c x(k) + B_c u(k) \\ u(k-1) + u(k) - u(k-1) \end{bmatrix} \\ &= \begin{bmatrix} A_c & B_c \\ 0_{Nu*Nx} & I_{Nu} \end{bmatrix} \begin{bmatrix} x(k) \\ u(k-1) \end{bmatrix} + \begin{bmatrix} B_c \\ I_{Nu} \end{bmatrix} \Delta u(k) \\ &= \tilde{A}\xi(k) + \tilde{B}\Delta u(k) \end{aligned} \tag{10}$$

According to the predictive modeling method, the driver model is developed based on system (10), as follows:

$$Y_d(k) = \varphi \cdot X(k) + \Theta \cdot U(k) \tag{11}$$

where

$$Y_d(k) = \begin{bmatrix} \eta(k+1) & \cdots & \cdots & \cdots & \eta(k+N_c) \end{bmatrix}^T$$

$$\varphi = \begin{bmatrix} \tilde{C}\tilde{A} & \cdots & \tilde{C}\tilde{A}^N & \cdots & \tilde{C}\tilde{A}^P \end{bmatrix}^T$$

$$\Theta = \begin{bmatrix} \tilde{C}\tilde{B} & \cdots & 0 \\ \vdots & & \vdots \\ \tilde{C}\tilde{A}^{N-1}\tilde{B} & \cdots & \tilde{C}\tilde{B} \\ \vdots & & \vdots \\ \tilde{C}\tilde{A}^{P-1}\tilde{B} & \cdots & \tilde{C}\sum_{i=0}^{P-N}\tilde{A}^i\tilde{B} \end{bmatrix},$$

$$U(k) = \begin{bmatrix} \Delta u(k) & \cdots & \cdots & \cdots & \Delta u(k+N-1) \end{bmatrix}^T$$

To solve the quadratic programming (QP) problem conveniently, we define the error variable as $E(k)$.

$$E(k) = \varphi \cdot X(k) - Y_{ref\_d}(k) \tag{12}$$

The objective function is

$$J(k) = \|\Theta \cdot \Delta U(k) + E(k)\|^2_{Q_d(k)} + \|\Delta U(k)\|^2_{R_d(k)} \tag{13}$$

In order to easily solve the numerical optimization problems, the objective function is transformed into a quadratic programming (QP) problem.

$$\begin{aligned} J &= [\Theta\Delta U(k) + E(k+1)]^T Q_d [\Theta\Delta U(k) + E(k+1)] + \Delta U(k)^T R_d \Delta U(k) + \rho\varepsilon^2 \\ &= \Delta U^T(\Theta^T Q_d \Theta + R_d)\Delta U + 2E^T Q_d \Theta \Delta U + \rho\varepsilon^2 \\ &= \frac{1}{2}\begin{bmatrix} \Delta U & \varepsilon \end{bmatrix} 2 \begin{bmatrix} \Theta^T Q_d \Theta + R_d & 0 \\ 0 & \rho \end{bmatrix} \begin{bmatrix} \Delta U & \varepsilon \end{bmatrix}^T + \begin{bmatrix} 2E^T Q_d \Theta & 0 \end{bmatrix} \begin{bmatrix} \Delta U \\ \varepsilon \end{bmatrix} \end{aligned} \tag{14}$$

Therein, ρ, ε represent weight coefficient and relaxation factor, respectively.

$$H = 2\begin{bmatrix} \Theta^T Q \Theta + R & 0 \\ 0 & \rho \end{bmatrix}$$

$$f = \begin{bmatrix} 2E^T Q \Theta & 0 \end{bmatrix}^T \tag{15}$$

To simulate driver manipulation characteristics and adhere to the actuator constraints of the vehicle chassis, we define the system constraints that include both control inputs and their increments, as specified in Formulation (16). We use MATLAB to program and solve the quadratic programming (QP) problem, employing the "*quadprog*" function to compute the optimal sequences of $\triangle U(k)$.

$$Ax \leq b \tag{16}$$

$$A = \begin{bmatrix} L & 0_{Nc*1} \\ -L & 0_{Nc*1} \end{bmatrix},\ b = \begin{bmatrix} u_{\max}(k) - u(k-1) \\ u_{\max}(k+1) - u(k-1) \\ \vdots \\ u_{\max}(k+N-1) - u(k-1) \\ -u_{\min}(k) + u(k-1) \\ -u_{\min}(k+1) + u(k-1) \\ \vdots \\ -u_{\min}(k+N-1) + u(k-1) \end{bmatrix},\ L = \begin{bmatrix} 1 & \cdots & \cdots & 0 & 0 \\ \vdots & 1 & & & 0 \\ \vdots & & \ddots & & \vdots \\ \vdots & & & \ddots & \vdots \\ 1 & \cdots & \cdots & \cdots & 1 \end{bmatrix}_{Nc\times Nc},$$

$$l_b = \begin{bmatrix} \Delta u_{\min}(k) \\ \Delta u_{\min}(k+1) \\ \vdots \\ \Delta u_{\min}(k+N-1) \\ 0 \end{bmatrix},\ u_b \leq \begin{bmatrix} \Delta u_{\max}(k) \\ \Delta u_{\max}(k+1) \\ \vdots \\ \Delta u_{\max}(k+N-1) \\ M \end{bmatrix},$$

The control increment executed currently is as follows:

$$\Delta u(k) = \begin{bmatrix} 1 & 0 & \cdots & 0 \end{bmatrix}_{1*(N_c+1)} \cdot \Delta U(k) \tag{17}$$

So the actual vehicle steering wheel angle is the sum of last time control quantity $u(k-1)$ and increment control $\Delta u(k)$, as shown in the following:

$$u(k) = \Delta u(k) + u(k-1) \tag{18}$$

3. Driver–Automated Shared Steering Controller Design Based on Model Predictive Control

The human–machine shared control for which the human-driver should be kept in the control loop, provides that the realistic front wheel steering angle is concluded by δ_d and δ_c, the ratio of them is determined by ∂, which is the driving authority of the shared control system.

$$\delta_f(k) = \partial\delta_d(k) + (1-\partial)\delta_c(k) \tag{19}$$

Therein, $\delta_d(k)$ is the front wheel angle output by the driver model, $\delta_c(k)$ is the front wheel angle output by the automated-driving system.

Substituting (19) to (3), the driver–automated shared steering control system is established; its discrete-time state–space is shown in (20):

$$\begin{aligned} x(k+1) &= A_c x(k) + B_D\partial(k) + B_c\delta_c(k) \\ y(k) &= C_{dc}x(k) \end{aligned} \tag{20}$$

Then, $B_D = B_c(\delta_d(k) - \delta_c(k))$.

$$\begin{aligned} x(k+1) &= A_c x(k) + B_c\delta_f(k) \\ &= A_c x(k) + B_c[\partial\delta_d(k) + (1-\partial)\delta_c(k)] \\ &= A_c x(k) + \partial B_c\delta_d(k) + B_c\delta_c(k) - \partial B_c\delta_c(k) \\ &= A_c x(k) + \partial B_c(\delta_d(k) - \delta_c(k)) + B_c\delta_c(k) \\ &= A_c x(k) + \partial(k)B_D + B_c\delta_c(k) \end{aligned}$$

According to the model predictive control method, the output of this system in the predictive horizon can be demonstrated as the following:

$$\begin{aligned}
y(k+1) &= C_{dc}A_c x(k) + C_{dc}B_D\partial(k) + C_{dc}B_c\delta_c(k) \\
y(k+2) &= C_{dc}A_c^2 x(k) + C_{dc}A_cB_D\partial(k) + C_{dc}B_D\partial(k+1) + \\
&C_{dc}A_cB_c\delta_c(k) + C_{dc}B_c\delta_c(k+1) \\
&\cdots \\
y(k+N) &= C_{dc}A_c^N\Delta x(k) + \\
&C_{dc}A_c^{N-1}B_D\partial(k) + C_{dc}A_c^{N-2}B_D\partial(k+1) + \cdots + C_{dc}B_D\partial(k+N-1) + \\
&C_{dc}A_c^{N-1}B_c\delta_c(k) + C_{dc}A_c^{N-2}B_c\delta_c(k+1) + \cdots + C_{dc}B_c\delta_c(k+N-1) \\
&\cdots \\
y(k+P) &= C_{dc}A_c^P\Delta x(k) + \\
&C_{dc}A_c^{P-1}B_D\partial(k) + C_{dc}A_c^{P-2}B_D\partial(k+1) + \cdots + C_{dc}A_c^{P-N}B_D\partial(k+N-1) \\
&C_{dc}A_c^{P-1}B_c\delta_f(k) + C_{dc}A_c^{P-2}B_c\delta_f(k+1) + \cdots + C_{dc}A_c^{P-N}B_c\delta_f(k+N-1)
\end{aligned}$$

In the above, N is the control horizon, P is the predictive horizon.

$$Y(k) = \begin{bmatrix} y(k+1) & \cdots & y(k+N) & \cdots & y(k+P) \end{bmatrix}^T$$

Then, the output equation of the system can be shown as follows:

$$Y(k) = S_x X(k) + S_u\Omega(k) + S_{uc}U(k) \tag{21}$$

$$\Omega(k) = \begin{bmatrix} \partial(k) \\ \partial(k+1) \\ \vdots \\ \vdots \\ \vdots \\ \partial(k+N-1) \end{bmatrix}, U(k) = \begin{bmatrix} \delta_c(k) \\ \delta_c(k+1) \\ \vdots \\ \vdots \\ \vdots \\ \delta_c(k+N-1) \end{bmatrix},$$

For this paper, we assumed that $\delta_c(k) = \delta_c(k+1) = \ldots\ldots = \delta_c(k+N-1)$ at time k.

$$S_x = \begin{bmatrix} C_{dc}A_c \\ C_{dc}A_c^2 \\ \vdots \\ C_{dc}A_c^N \\ \vdots \\ C_{dc}A_c^P \end{bmatrix}, S_u = \begin{bmatrix} C_{dc}B_D & 0 & \cdots & \cdots & 0 \\ C_{dc}A_cB_D & C_{dc}B_D & \cdots & \cdots & 0 \\ \vdots & \vdots & \vdots & \vdots & \vdots \\ C_{dc}A_c^{N-1}B_D & C_{dc}A_c^{N-2}B_D & \cdots & \cdots & C_{dc}B_D \\ \vdots & \vdots & \vdots & \vdots & \vdots \\ C_{dc}A_c^{P-1}B_D & C_{dc}A_c^{P-2}B_D & \cdots & \cdots & C_{dc}A_c^{P-N}B_D \end{bmatrix},$$

$$S_{uc} = \begin{bmatrix} C_{dc}B_c & 0 & \cdots & \cdots & 0 \\ C_{dc}A_cB_c & C_{dc}B_c & \cdots & \cdots & 0 \\ \vdots & \vdots & \vdots & \vdots & \vdots \\ C_{dc}A_c^{N-1}B_c & C_{dc}A_c^{N-2}B_c & \cdots & \cdots & C_{dc}B_c \\ \vdots & \vdots & \vdots & \vdots & \vdots \\ C_{dc}A_c^{P-1}B_c & C_{dc}A_c^{P-2}B_c & \cdots & \cdots & C_{dc}A_c^{P-N}B_c \end{bmatrix},$$

The objective function comprises four components. The first component represents the error in the vehicle's lateral displacement. The second component reflects the deviation between the actual front wheel steering angle and the driver's desired steering angle. The third component measures the discrepancy between the desired driving authority and the actual value. The fourth component captures the increment in driving authority.

$$\begin{aligned}\min_{\Delta\partial(k)} J(k) &= \Gamma_1 J_1 + \Gamma_2 J_2 + \Gamma_3 J_3 + \Gamma_4 J_4 \\ &= \left\| Z(k) - Y_{ref}(k) \right\|^2_{Q_1(k)} + \sum_{i=1}^{P} \left\| \delta_f(k+i) - \delta_d(k+i) \right\|^2_{Q_2(k)} + \\ &\sum_{i=1}^{P} \| \partial(k+i) - \partial^*(k+i) \|^2_{R_1(k)} + \sum_{i=1}^{P} \| \partial(k+i) \|^2_{R_2(k)}\end{aligned} \tag{22}$$

The optimization problem can be described as follows:

$$\begin{aligned}&\min_{\Delta\partial(k)} J(k) = \Gamma_1 J_1 + \Gamma_2 J_2 + \Gamma_3 J_3 + \Gamma_4 J_4 \\ &s.t. \begin{cases} x(k+1) = A_c x(k) + B_D \partial(k) + B_{dc} \delta_c(k) \\ \Delta\partial_{\min} \leq \Delta\partial(k+i) \leq \Delta\partial_{\max} \\ 0 \leq \partial(k+i) \leq 1 \\ y_{\min} \leq y(k+i) \leq y_{\max} \end{cases}\end{aligned} \tag{23}$$

The constraint of the system is as follows:

(a) Output constraint of the system:

$$\begin{aligned} S_{uc}\Omega(k) &\leq Y_{\max}(k+1|k) - S_x X(k) - S_{uc} U(k) \\ -S_{uc}\Omega(k) &\leq -Y_{\min}(k+1|k) + S_x X(k) + S_{uc} U(k) \end{aligned} \tag{24}$$

(b) Control variable constraint of the system:

$$\begin{bmatrix} \partial_{\min} \\ \partial_{\min} \\ \vdots \\ \partial_{\min} \end{bmatrix} \leq \begin{bmatrix} \partial(k) \\ \partial(k+1) \\ \vdots \\ \partial(k+N-1) \end{bmatrix} \leq \begin{bmatrix} \partial_{\max} \\ \partial_{\max} \\ \vdots \\ \partial_{\max} \end{bmatrix} \tag{25}$$

$$\partial_{\min} = 0,\ \partial_{\max} = 1$$

(c) Increment control variable constraint of the system:

$$\begin{bmatrix} \Delta\partial_{\min} \\ \Delta\partial_{\min} \\ \vdots \\ \Delta\partial_{\min} \end{bmatrix} \leq \begin{bmatrix} \partial(k) - \partial(k-1) \\ \partial(k+1) - \partial(k) \\ \vdots \\ \partial(k+N-1) - \partial(k+N-2) \end{bmatrix} \leq \begin{bmatrix} \Delta\partial_{\max} \\ \Delta\partial_{\max} \\ \vdots \\ \Delta\partial_{\max} \end{bmatrix} \tag{26}$$

$$\Delta\partial_{\min} = -0.1,\ \Delta\partial_{\max} = 0.1$$

All the constraints shown in the above are integrated as follows:

$$\begin{bmatrix} L \\ -L \\ E \\ -E \\ S_{uc} \\ -S_{uc} \end{bmatrix} \Omega(k) \leq \begin{bmatrix} \Delta\partial_{\max}(k) + \partial(k-1) \\ \Delta\partial_{\max}(k+1) + \partial(k) \\ \vdots \\ \Delta\partial_{\max}(k+N-1) + \partial(k+N-2) \\ -\Delta\partial_{\min}(k) - \partial(k-1) \\ -\Delta\partial_{\min}(k+1) - \partial(k) \\ \vdots \\ -\Delta\partial_{\min}(k+N-1) - \partial(k+N-2) \\ \partial_{\max}(k) \\ \partial_{\max}(k+1) \\ \vdots \\ \partial_{\max}(k+N-1) \\ -\partial_{\min}(k) \\ -\partial_{\min}(k+1) \\ \vdots \\ -\partial_{\min}(k+N-1) \\ Y_{\max}(k+1|k) - S_x X(k) - S_{uc} U(k) \\ -Y_{\min}(k+1|k) + S_x X(k) + S_{uc} U(k) \end{bmatrix} \tag{27}$$

where

$$E = \begin{bmatrix} 1 & \cdots & 0 \\ \vdots & \ddots & \vdots \\ 0 & \cdots & 1 \end{bmatrix}_{N \times N}$$

A numerical optimization method is proposed to calculate the optimal weight sequence. Following the principles of model predictive control (MPC), the first element in this weight sequence is chosen as the driving weight coefficient for the driver–automated shared steering system at the current time.

According to the constraint conditions and cost function, the system belongs to a typical nonlinear optimization system, so it is solved by "fmincon" command in MATLAB.

$$\partial(k) = \begin{bmatrix} 1 & 0 & \cdots & 0 \end{bmatrix}_{1 \times N} \cdot \Omega(k) \tag{28}$$

4. Simulation and Discussion

To verify the effectiveness of the proposed optimal driving authority strategy, simulation tests are conducted with both consistent and inconsistent driving intentions between the driver and the automated system.

4.1. Driver and Automated with Consistent Driving Intention

In the simulation tests, the vehicle speed is set to 63 km/h, and the road adhesion coefficient is 0.8. The preview time for the trajectory-tracking controller of the automated system is chosen as 0.8 s. For the single-lane-change scenario, both the driver and the automated system are configured with consistent driving intentions throughout the simulation.

In Figure 4, in a single-lane change simulation scenario, this study compared the performance of a traditional driver model with that of a driver–autonomous driving shared steering control system. The results indicate that the driver–autonomous driving shared steering control system has a significant advantage in vehicle-trajectory-tracking accuracy. Analysis of the steering wheel angle response curve revealed that when using the shared steering control system, the steering wheel angle changes more smoothly, avoiding rapid fluctuations in a short period. This response characteristic not only improves trajectory-tracking accuracy but also enhances the system's stability and safety to some extent.

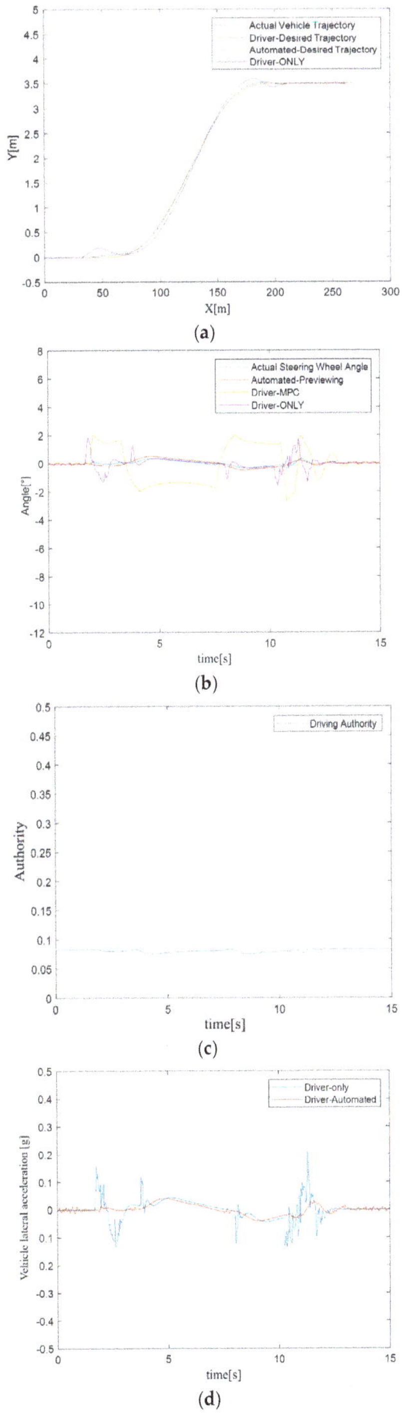

Figure 4. Results of driver and automated system with consistent driving intention. (**a**) Vehicle displacement. (**b**) Steering wheel angle. (**c**) Driving authority. (**d**) Vehicle lateral acceleration.

Although the shared steering control system improves vehicle control performance, it also slightly increases the physiological and psychological burden on the driver. This increase is moderate and manageable. During the experiment, the driver's physiological workload and psychological pressure showed a slight increase, likely because the driver remains in the control loop and needs to collaborate with the autonomous driving system. This collaboration requires the driver to make adjustments in terms of attention and decision-making speed, leading to a minor increase in psychological burden, which in turn raises the physiological burden. However, experimental data indicate that despite this increase, the driver's control authority remained below 10%, averaging around 8%. This figure suggests that within the shared steering control system, the autonomous driving system still predominantly controls the vehicle, and the increase in the driver's burden does not significantly impact the overall system performance.

Furthermore, throughout the entire single-lane change process, the driver's control authority did not exhibit noticeable fluctuations, providing favorable conditions for the driver to adapt to and trust the autonomous driving system. Although the driver's burden within the system increased, this increase remained within an acceptable range. At the same time, the efficiency of the shared steering control system further demonstrates its potential in achieving high-precision trajectory tracking and stable vehicle control. Overall, the shared steering control system, by optimizing human–machine collaboration, improves the overall system performance and lays a solid foundation for the future development of autonomous driving technology.

4.2. Driver and Automated System with Inconsistent Driving Intentions

Case 1: Change in driving intention from consistent to inconsistent, and then back to consistent (delayed lane change).

In the simulation test, the vehicle speed is set at 63 km/h, and the road adhesion coefficient is 0.8. The preview time for the automated controller is selected as 0.8 s, corresponding to a preview distance of 14 m. As depicted in Figure 5a, the driver's intention is to perform a single-lane change within 2 s (with an X-axis displacement of 50 m), represented by the "red line." In contrast, the automated system's intention is to delay the lane change until an X-axis displacement of 100 m, as shown by the "yellow line." Thus, during the simulation, the driving intention shifts from being consistent to inconsistent and then returns to being consistent.

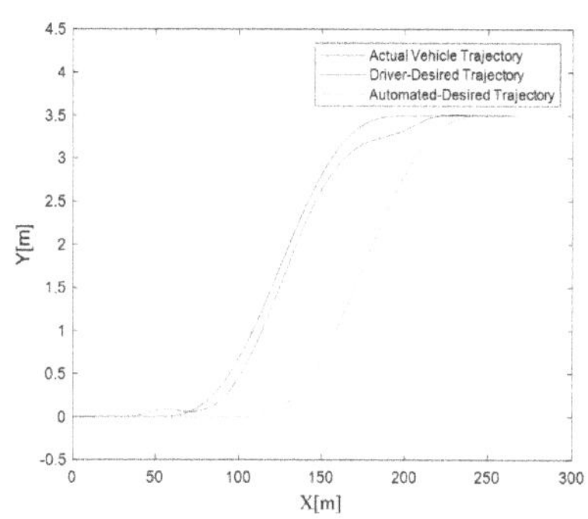

(a)

Figure 5. *Cont.*

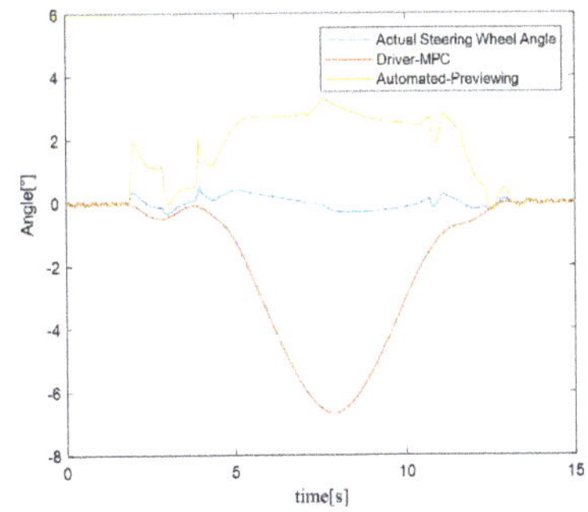

(**b**)

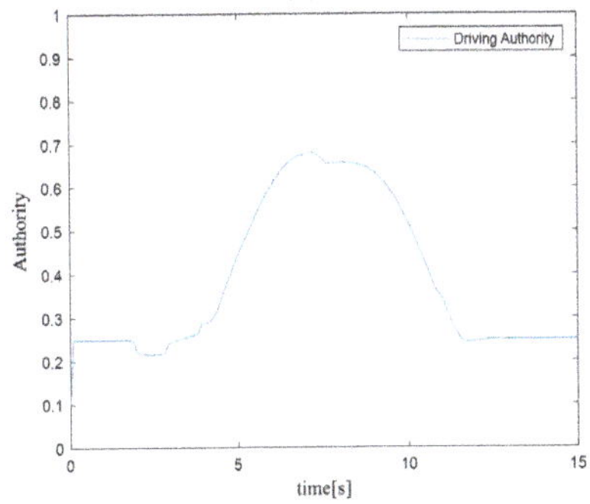

(**c**)

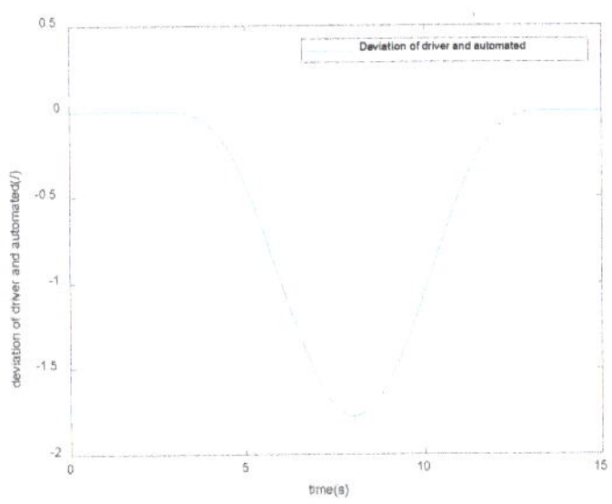

(**d**)

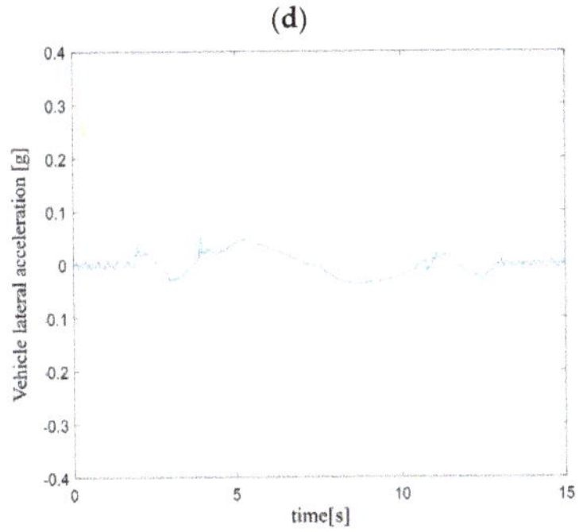

(**e**)

Figure 5. Results of driver and automated with inconsistent driving intention. (**a**) Vehicle displacement. (**b**) Steering wheel angle. (**c**) Driving authority. (**d**) Driving intention deviation between driver and automated. (**e**) Vehicle lateral acceleration.

During the simulation tests, the driving authority (as illustrated in Figure 5c) varies in response to the deviation in driving intention between the driver and the automated system, as shown in Figure 5d. The proposed driver–automated shared steering optimization framework prioritizes the driver's intentions. When the deviation in driving intention reaches its maximum value, the driver is allocated the maximum driving authority. As the deviation decreases, the driving authority correspondingly decreases. As depicted in Figure 5b, the framework effectively resolves conflicts between the driver and the automated system by adjusting the vehicle's steering wheel angle based on the allocated driving authority. Consequently, the actual steering wheel angle of the vehicle changes smoothly, and the vehicle's lateral acceleration response is also smooth, as shown in Figure 5e. This approach ensures that the driver's intentions are prioritized during conflicts, resulting in a vehicle trajectory that closely aligns with the driver's desired path.

During the phases from 0 to 4 s and 12 to 16 s in the simulation, the driver–automated shared steering optimization framework maintains the driving authority at approximately 0.25, with the automated system handling most of the driving workload. In these phases, both the driver and the automated system have consistent driving intentions, focusing on driving in a straight line.

In the interval from 4 to 12 s, the driving authority gradually increases and then decreases. During this period, the optimal framework transitions driving authority from the automated system to the driver and back to the automated system. Initially, around 4 s, the driver intends to change lanes, but the automated system continues to drive straight, resulting in a gradual increase in driving authority. By approximately 8 s, the driver gains the majority of the driving control authority due to the maximum deviation. After 8 s, the automated system changes lanes, reducing the deviation in driving intention between the driver and the automated system. Consequently, the driving authority, controlled by the optimal framework, decreases and is gradually transferred from the driver back to the automated system.

Case 2: Driving intention change from consistent to inconsistent.

During the simulation test, the vehicle speed is set to 63 km/h, and the road adhesion coefficient is 0.8. The preview time for the automated controller is selected as 0.8 s, corresponding to a preview distance of 14 m. As depicted in Figure 6a, the driver's intention is to perform a single-lane change, represented by the "red line", while the automated system's intention is to continue driving straight, shown as the "yellow line". This indicates that the driving intention between the driver and the automated system remains consistently inconsistent from the moment the driver expects to initiate the lane change.

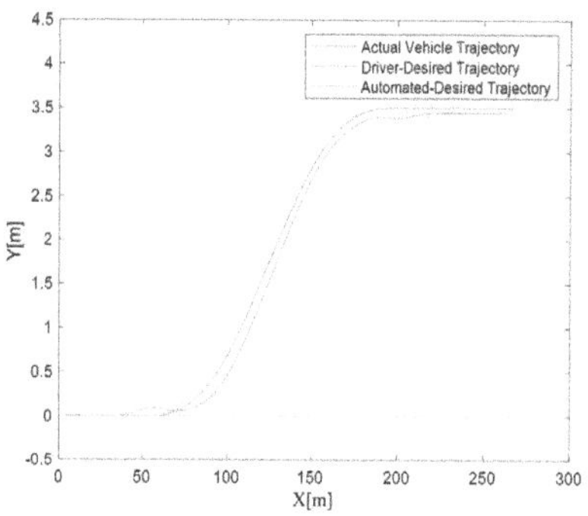

(a)

Figure 6. *Cont.*

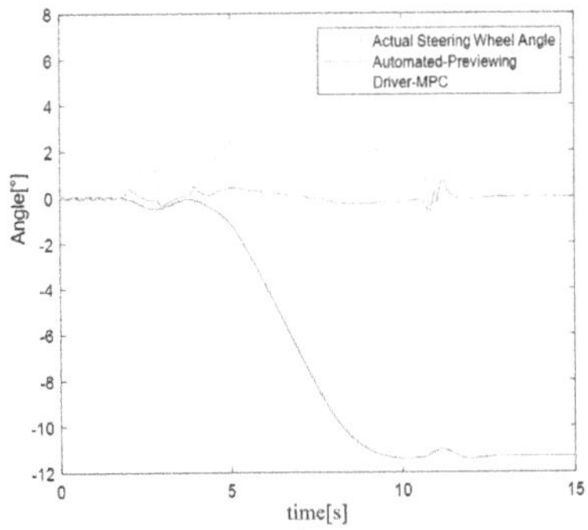

(b)

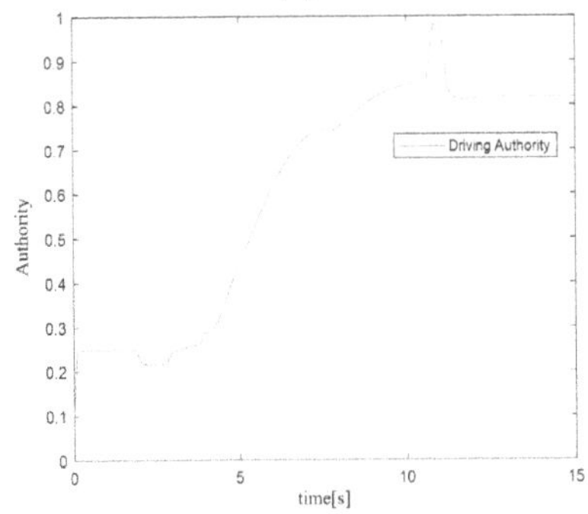

(c)

Figure 6. Results of driver and automated system with inconsistent driving intention. (**a**) Vehicle displacement. (**b**) Steering wheel angle. (**c**) Driving authority.

From Figure 6a,b, it can be observed that, in conflict situations between the driver and the automated system, the intelligent vehicle completes the single-lane change according to the driver's intention. During this lane change task, the automated driving system maintains a straight-line driving intention, while the driver aims to execute the lane change, as shown in Figure 6a. In the simulation, driving authority is gradually transferred from the automated system to the driver to align with the driver's intention, as illustrated in Figure 6c. This demonstrates the effectiveness of the driving authority optimization framework. As the vehicle travels a longitudinal distance of 50 m, the deviation between the driver's intention and the vehicle's intention begins to increase. By the time the longitudinal displacement reaches approximately 200 m, the deviation between the driver's intention and the vehicle's intention reaches its maximum value and remains constant until the end of the simulation. Subsequently, the driver maintains the majority of the driving authority, allowing the intelligent vehicle to complete the lane change task. Following a brief period of fluctuation, the driving authority stabilizes around 0.8, indicating that the driver retains most of the driving authority.

Additionally, Figure 6b shows that the error between the front wheel steering angle commanded by the driver and the actual front wheel steering angle of the vehicle is smaller than the corresponding error in the automated system. This demonstrates that the driving authority optimization framework effectively adjusts the driving authority in accordance with the driver's intentions. Furthermore, the framework retains a portion of the vehicle's driving authority to provide necessary assistance to the driver and to prevent instability or fluctuations in the vehicle's response due to changes in driving authority between the driver and the automated system.

5. Driver-in-the-Loop Experiment

This section of the experiment aims to validate the effectiveness of the proposed optimization framework for driving permissions in the autonomous driving system. By integrating the designed controller into a driver-in-the-loop simulation platform, comprehensive testing and validation are conducted to assess the performance of the driving permission optimization framework in real driving scenarios. The experimental results demonstrate the application of the " Double driving with double control" mode in the L2 autonomous driving system, highlighting the operational mechanism of shared vehicle control between the driver and the automated system. In this section, to validate the effectiveness of the proposed driver–automated system driving authority optimization framework, a human–machine co-driving simulation platform is constructed using MATLAB/Simulink 2021b, CarSim 2019.1, PreScan 8.5 joint simulation software, and the SensoWheel force feedback system, as shown in Figure 7. The SensoWheel force feedback system includes hardware components such as a steering wheel, torque motor, steering wheel torque sensor, and pedals, capable of providing realistic road feel feedback. The torque motor has a maximum torque of 16 N·m, a torque resolution of 0.03 N·m, and a response time of 1 ms. In this experiment, the driver operates the steering wheel to generate the control angle for the system, while the steering wheel sensor measures the input torque from the driver.

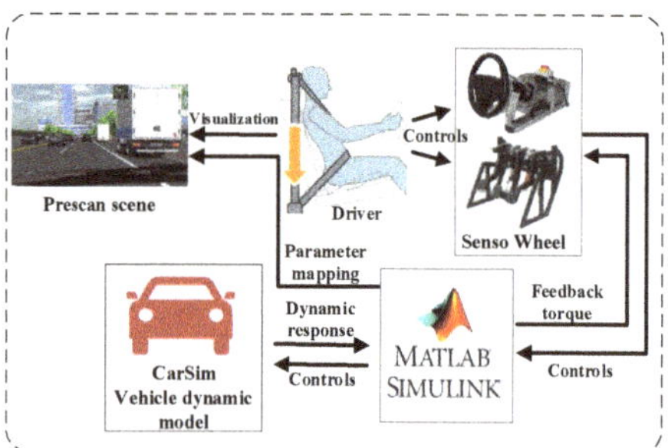

Figure 7. Human–machine co-piloting simulation platform.

CarSim is a simulation software designed specifically for automotive dynamics. In the experiment, CarSim's vehicle dynamics model is integrated into the Simulink model to provide realistic automotive dynamics responses for both the driver and the automated system. PreScan software is used to create virtual test scenarios that interact with the CarSim-provided vehicle model, offering drivers a virtual scene view, as illustrated in Figure 8.

Figure 8. Driver-in-the-loop platform; ① Prescan visualization ② SensoWheel force feedback system ③ Human–machine co-driving algorithm.

Two types of testing scenarios are designed to validate the effectiveness of the driver–automated system driving authority optimization framework: consistent driving intent and conflicting driving intent. These scenarios are used to assess the framework's performance in terms of trajectory-tracking accuracy and the allocation of driving authority.

5.1. Driver and Automated with Inconsistent Driving Intention Driver and Automated System Operating under Inconsistent Driving Intentions

Case 1. The driving intentions of the driver and the automated system are aligned.

During the experiment, the vehicle speed was set to 63 km/h, the road surface adhesion coefficient was 0.8, and the automated system's look-ahead distance was 10 m. Both the driver and the automated system were aligned in their target trajectory, with the goal of performing a lane change. The experimental results are illustrated in Figure 9.

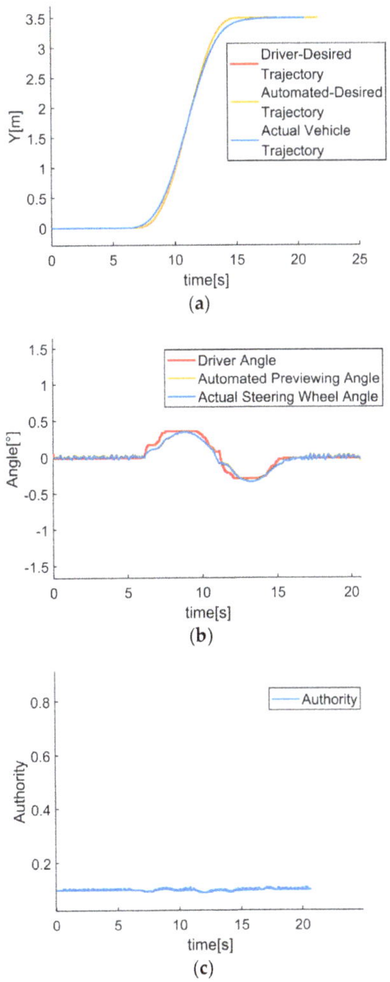

Figure 9. *Cont.*

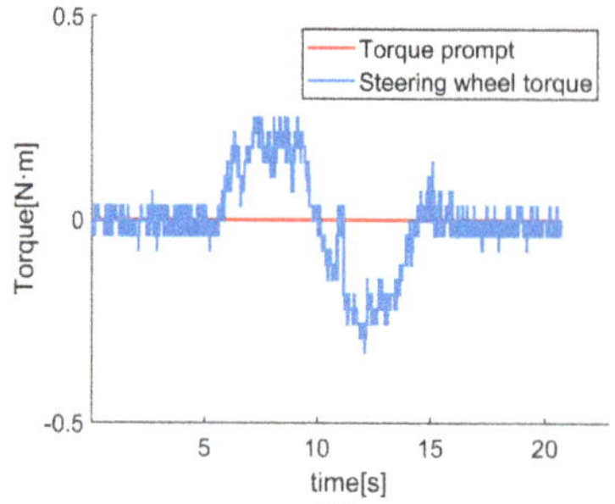

(d)

Figure 9. Results of driver and automated system with consistent driving intention. (**a**) Vehicle displacement. (**b**) Steering wheel angle. (**c**) Driving authority. (**d**) Driver torque.

In the lane-changing scenario, as shown in Figure 9a,b, the vehicle closely follows the desired trajectory with minimal fluctuations in the actual front wheel angle, and the response curve remains smooth. This demonstrates that the proposed driver–automated system authority optimization framework enhances the vehicle's trajectory-tracking precision. As depicted in Figure 9c, the driver's authority weight averages around 10% with minimal fluctuations throughout the lane change, facilitating the driver's adaptation to the proposed framework. The automated system is equipped with a torque reminder function to alert users during conflicts between human and automated intentions. During the lane-changing process, when the objectives of the driver and the automated system are aligned, the torque reminder is zero, as illustrated in Figure 9d.

5.2. *The Driving Intentions of the Driver and the Automated System Are in Conflict*

Case 1: Driving intention changes from consistent to inconsistent.

During the experiment, the vehicle speed is set to 63 km/h, the road surface adhesion coefficient is 0.8, and the automated system's look-ahead distance is 10 m. The driving intentions of the driver and the automated system transition from being consistent to inconsistent. Initially, the driver's target trajectory is to drive straight and then change lanes at 7 s, as indicated by the "red line" in Figure 10a. In contrast, the automated system's target trajectory is to maintain a straight course throughout, as shown by the "yellow line" in Figure 10a. From the moment the driver initiates the lane change, the driving intentions of the driver and the automated system become inconsistent, as demonstrated by the experimental results in Figure 10.

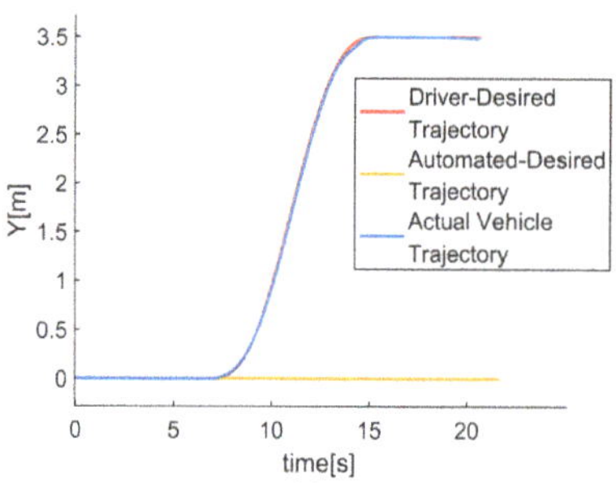

(a)

Figure 10. *Cont.*

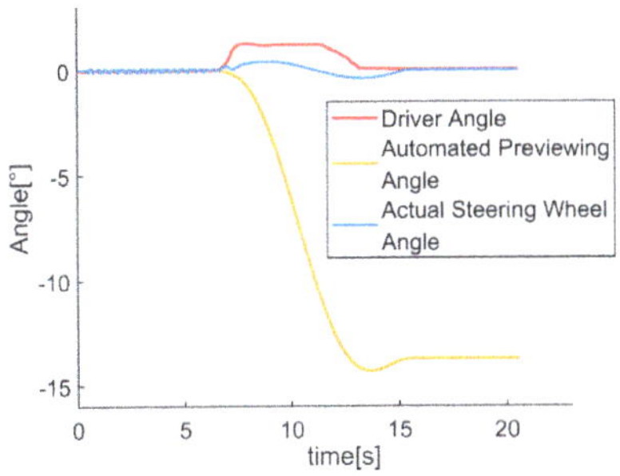

(b)

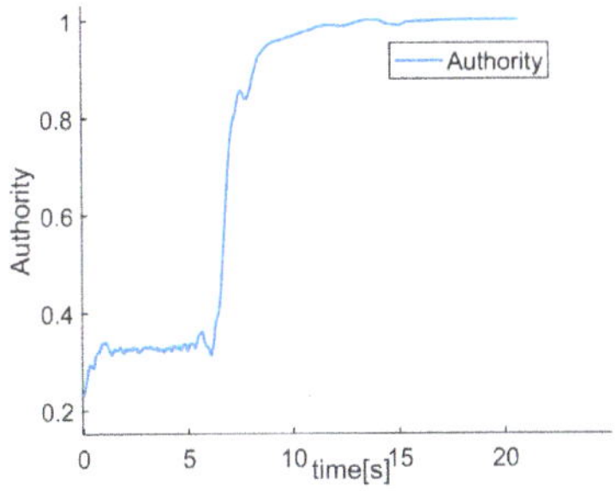

(c)

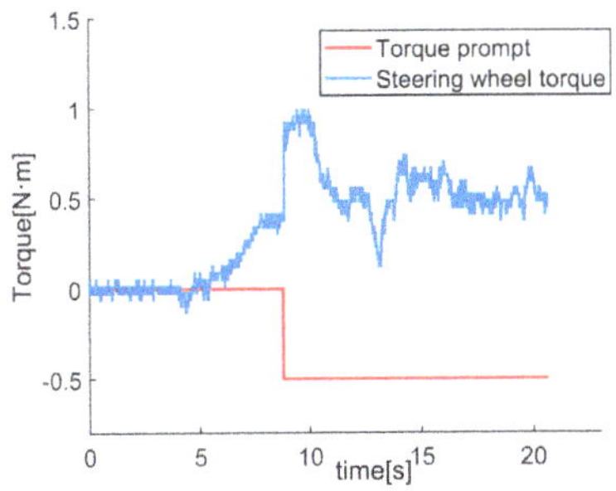

(d)

Figure 10. Results of driver and automated system with inconsistent driving intention. (**a**) Vehicle displacement. (**b**) Steering wheel angle. (**c**) Driving authority. (**d**) Driver torque.

As shown in Figure 10a,c, during the initial phase of straight driving, the intentions of the driver and the automated system are aligned, with the driving authority predominantly controlled by the automated system. At 7 s, when the driver initiates a lane change, the deviation between the driver's intentions and those of the automated system gradually increases, resulting in a swift transfer of control to the driver. This demonstrates that the proposed driver–automated system authority optimization framework prioritizes the driver's intentions during conflicts between human and machine. Figure 10 illustrates that the deviation between the driver's target steering angle and the actual vehicle steering angle is smaller compared to that of the automated system. This indicates that the framework effectively adjusts the balance between the driver's target steering angle and the automated system's target steering angle, ensuring that the driver's intentions are met and enhance the vehicle's trajectory-tracking accuracy.

Figure 10d shows that during the lane-changing process, when the driver's and the automated system's intentions are inconsistent, the automated system provides a torque of 0.5 N·m as a reminder to the driver to help prevent accidents and ensure vehicle safety.

Case 2: Driving intention changes from consistent to inconsistent, and then to consistent (delaying lane change).

During the experiment, the vehicle speed is set to 63 km/h, the road surface adhesion coefficient is 0.8, and the automated system's look-ahead distance is 10 m. The driver's target trajectory is to drive straight initially and then change lanes at 7 s, as indicated by the "red line" in Figure 11a. In contrast, the automated system's target trajectory is to drive straight initially but to delay the lane change until 12 s, as shown by the "yellow line" in Figure 11a. The experimental results are illustrated in Figure 11.

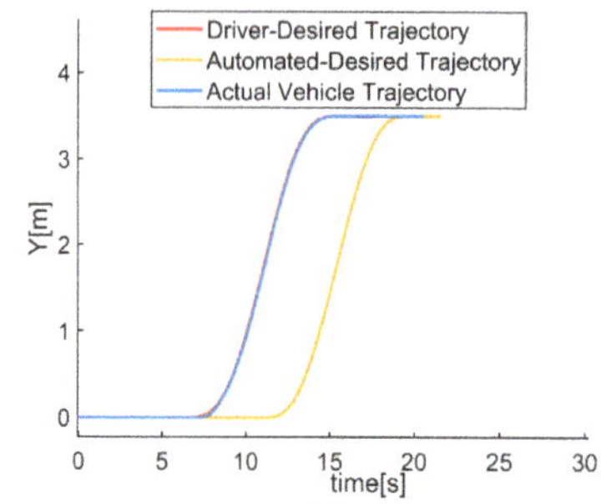

(a)

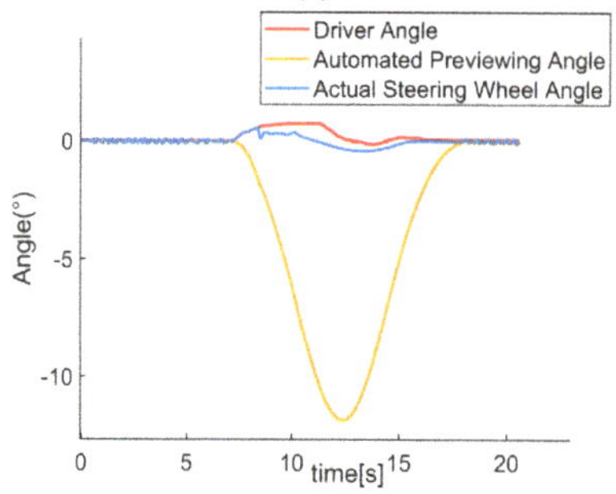

(b)

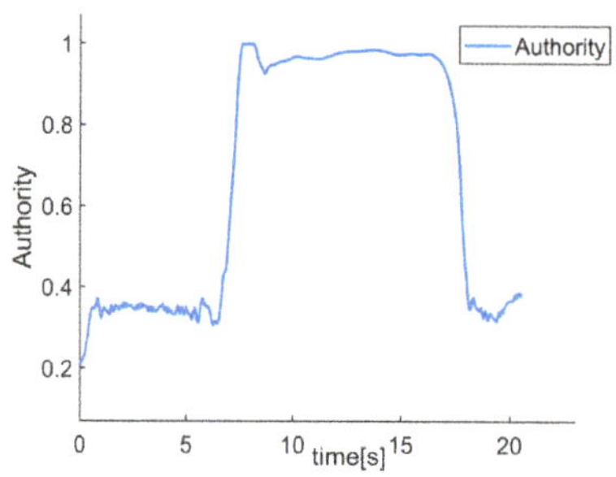

(c)

Figure 11. *Cont.*

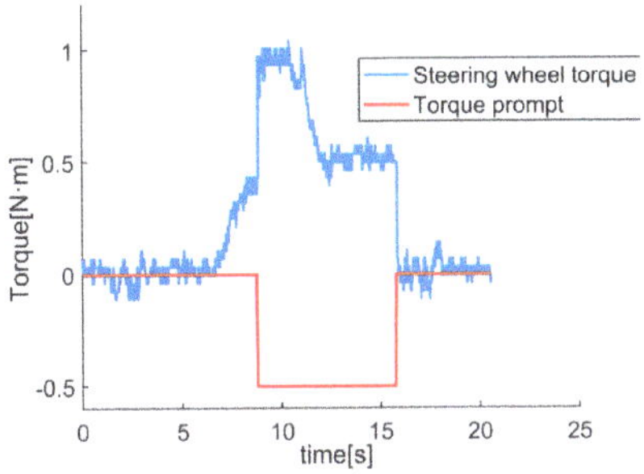

(d)

Figure 11. Results of driver and automated system with inconsistent driving intention. (**a**) Vehicle displacement. (**b**) Steering wheel angle. (**c**) Driving authority. (**d**) Driver torque.

As depicted in Figure 11a,c, at 7 s, when the driver initiates a lane change while the automated system continues to drive straight, the driver's control weight gradually increases, and the vehicle aligns with the driver's intentions. When the deviation between the driver's and the automated system's driving intentions reaches its maximum, the driver gains the highest control weight over the vehicle. This demonstrates that the proposed driver–automated system authority optimization framework prioritizes the driver's intentions during conflicts between human and machine. As the automated system begins its lane change, the deviation between the driving intentions of the driver and the automated system decreases, leading to a reduction in the driver's control weight. This results in the vehicle's actual front wheel angle aligning more closely with the automated system's target steering angle, thereby making the actual vehicle trajectory more aligned with the automated system's desired trajectory. Furthermore, as shown in Figure 11b, the actual front wheel angle is allocated reasonably, based on the adjustment of driving authority, with a smooth response curve. This indicates that the authority optimization framework effectively balances the ratio between the driver's target steering angle and the automated system's target steering angle, ensuring accurate vehicle tracking.

As illustrated in Figure 11d, when the driver begins to change lanes, the inconsistency between human and machine intentions prompts the automated system to provide a torque reminder of 0.5 N·m. When both the driver's and the automated system's intentions are to drive straight, the objectives align, resulting in a zero torque reminder.

6. Conclusions

The driver–automated driving authority optimization framework for a shared steering control pattern is uniquely presented in this manuscript. This framework adjusts the driving authority between the autonomous vehicle system and the human driver. The design motivation behind this optimization framework is to automatically allocate driving authority based on the deviation between the driver's and the automated system's intentions. This approach aims to ensure the safety of intelligent vehicle driving and mitigate conflicts between the driver and the automated system. Simulation results across different scenarios highlight the rationale and benefits of the proposed driver–automated driving authority optimization strategy. The findings indicate that conflicts can be resolved by adjusting the driving authority when the driver and the automated system have inconsistent driving intentions. During a conflict, the driving authority weight ranges from 0.31 to 0.96 within 3 s. Additionally, the vehicle trajectory-tracking accuracy and driving safety improve when both the driver and the automated system share similar driving intentions. The maximum lateral error is 0.1245 m. The effectiveness of the proposed framework in terms of trajectory-tracking accuracy and driving authority allocation was further validated through driver-in-the-loop experiments.

Future work should focus on developing improved strategies to further reduce driver workload and burden when the driver and automated system have consistent driving intentions. Additionally, considerations should be given to drivers with varying skill levels. The driving authority should be varied within a reasonable range to ensure more rational assistance performance.

Author Contributions: Conceptualization, S.Y.; methodology, S.Y.; software, Q.W.; validation, X.X.; formal analysis, Q.W.; data curation, X.X.; writing—original draft preparation, Q.W.; writing—review and editing, Q.W., X.X., D.Z., X.L.; visualization, X.L.; supervision, X.X.; project administration, Q.W.; funding acquisition, X.X. All authors have read and agreed to the published version of the manuscript.

Funding: Open Fund of the State Key Laboratory of Automotive Safety and Energy of Tsinghua University (KFY2211); Hebei Natural Science Foundation (F2021203107).

Data Availability Statement: Data is unavailable due to privacy or ethical restrictions.

Conflicts of Interest: The authors declare no conflict of interest.

Abbreviations

Acronym	Full name
MPC	Model predictive control
AD	Autonomous driving
ABS	Anti-lock braking system
ADAS	Advanced driving assistance system
LKA	Lane-keeping assist
ARS	Active rear steering
LTV	Linear time-varying
CAVs	Automated vehicles
LSTM	Long short-term memory
QP	Quadratic programming

References

1. He, H.; Shi, M.; Li, J.; Cao, J.; Han, M. Design and experientialtest of a model predictive path following control with adaptive preview for autonomous buses. *Mech. Syst. Signal Process.* **2021**, *157*, 107701. [CrossRef]
2. Ma, Y.; Wang, Z.; Yang, H.; Yang, L. Artificial intelligence applications in the development of autonomous vehicles: A survey. *IEEE/CAA J. Autom. Sin.* **2020**, *7*, 315–329. [CrossRef]
3. Kebria, P.M.; Khosravi, A.; Salaken, S.M.; Nahavandi, S. Deep imitation learning for autonomous vehicles based on convolutional neural networks. *IEEE/CAA J. Autom. Sin.* **2019**, *7*, 82–95. [CrossRef]
4. Duan, X.; Jiang, H.; Tian, D.; Zou, T.; Zhou, J.; Cao, Y. V2I based environment perception for autonomous vehicles at intersections. *China Commun.* **2021**, *18*, 1–12. [CrossRef]
5. Paret, D.; Rebaine, H. *Autonomous and Connected Vehicles: Network Architectures from Legacy Networks to Automotive Ethernet*; John Wiley & Sons: Hoboken, NJ, USA, 2022.
6. Casner, S.M.; Hutchins, E.L.; Norman, D. The challenges of partially automated driving. *Commun. ACM* **2016**, *59*, 70–77. [CrossRef]
7. Wang, W.; Na, X.; Cao, D.; Gong, J.; Xi, J.; Xing, Y.; Wang, F.Y. Decision-making in driver-automation shared control: A review and perspectives. *IEEE/CAA J. Autom. Sin.* **2020**, *7*, 1289–1307. [CrossRef]
8. Aoki, S.; Lin, C.-W.; Rajkumar, R. Human-robot cooperation for autonomous vehicles and human drivers: Challenges and solutions. *IEEE Commun. Mag.* **2021**, *59*, 35–41. [CrossRef]
9. Feng, D.; Haase-Schütz, C.; Rosenbaum, L.; Hertlein, H.; Glaeser, C.; Timm, F.; Wiesbeck, W.; Dietmayer, K. Deep multi-modal object detection and semantic segmentation for autonomous driving: Datasets, methods, and challenges. *IEEE Trans. Intell. Transp. Syst.* **2020**, *22*, 1341–1360. [CrossRef]
10. *SAE J3016_202104*; Taxonomy and Definitions for Terms Related to Driving Automation Systems for On-Road Motor Vehicles. SAE International: Warrendale, PA, USA, 2021.
11. Ma, B.; Liu, Y.; Na, X.; Liu, Y.; Yang, Y. A shared steering controller design based on steer-by-wire system considering human-machine goal consistency. *J. Frankl. Inst.* **2019**, *356*, 4397–4419. [CrossRef]
12. Jayan, K.; Muruganantham, B. Advanced Driver Assistance System technologies and its challenges toward the development of autonomous vehicle. In *Intelligent Computing and Applications: Proceedings of ICICA 2019*; Springer: Singapore, 2021; pp. 55–72.

13. Han, J.; Zhao, J.; Zhu, B.; Song, D. Adaptive steering torque coupling framework considering conflict resolution for human-machine shared driving. *IEEE Trans. Intell. Transp. Syst.* **2021**, *23*, 10983–10995. [CrossRef]
14. Lu, Y.; Liang, J.; Yin, G.; Xu, L.; Wu, J.; Feng, J.; Wang, F. A shared control design for steering assistance system consid-ering driver behaviors. *IEEE Trans. Intell. Veh.* **2022**, *8*, 900–911. [CrossRef]
15. Mulder, M.; Abbink, D.A.; Boer, E.R. Sharing control with haptics: Seamless driver support from manual to automatic control. *Hum. Factors* **2012**, *54*, 786–798. [CrossRef] [PubMed]
16. Liu, J.; Dai, Q.; Guo, H.; Guo, J.; Chen, H. Human-oriented online driving authority optimization for driver-automation shared steering control. *IEEE Trans. Intell. Veh.* **2022**, *7*, 863–872. [CrossRef]
17. Huang, C.; Hang, P.; Hu, Z.; Lv, C. Collision-probability-aware human-machine cooperative planning for safe automated driving. *IEEE Trans. Veh. Technol.* **2021**, *70*, 9752–9763. [CrossRef]
18. Li, S.; Deng, L.; Hu, J.; Kang, S.; Qiu, J.; Li, Q. A comprehensive approach to evaluate human–machine conflicts in shared steering systems. *Accid. Anal. Prev.* **2024**, *207*, 107758. [CrossRef]
19. Biondi, F.; Alvarez, I.; Jeong, K.A. Human–vehicle cooperation in automated driving: A multidisciplinary review and appraisal. *Int. J. Hum. Comput. Interact.* **2019**, *35*, 932–946. [CrossRef]
20. Sentouh, C.; Nguyen, A.T.; Benloucif, M.A.; Popieul, J. Driver-automation cooperation oriented approach for shared control of lane keeping assist systems. *IEEE Trans. Control Syst. Technol.* **2018**, *27*, 1962–1978. [CrossRef]
21. Pinheiro, D.J.; Faber, J.; Micera, S.; Shokur, S. Human-machine interface for two-dimensional steering control with the auricular muscles. *Front. Neurorobotics* **2023**, *17*, 1154427. [CrossRef]
22. Hang, P.; Chen, X.; Wang, W. Cooperative control framework for human driver and active rear steering system to advance active safety. *IEEE Trans. Intell. Veh.* **2020**, *6*, 460–469. [CrossRef]
23. Chen, J.; Sun, D.; Li, Y.; Zhao, M.; Liu, W.; Jin, S. Human–machine cooperative scheme for car-following control of the connected and automated vehicles. *Phys. A Stat. Mech. Its Appl.* **2021**, *573*, 125949. [CrossRef]
24. Qu, T.; Wang, Q.; Cong, Y.; Chen, H. Modeling driver's longitudinal and lateral control behavior: A moving horizon control modeling scheme. In Proceedings of the 2018 Chinese Control and Decision Conference (CCDC), Shenyang, China, 9–11 June 2018; IEEE: New York, NY, USA, 2018; pp. 3260–3264.
25. Qu, T.; Chen, H.; Cao, D.; Guo, H.; Gao, B. Switching-based stochastic model predictive control approach for modeling driver steering skill. *IEEE Trans. Intell. Transp. Syst.* **2014**, *16*, 365–375. [CrossRef]
26. Zuo, Z.; Yang, X.; Li, Z.; Wang, Y.; Han, Q. MPC-based cooperative control strategy of path planning and trajectory tracking for intelligent vehicles. *IEEE Trans. Intell. Veh.* **2020**, *6*, 513–522. [CrossRef]
27. Li, R.; Li, Y.; Li, S.E.; Zhang, C.; Burdet, E.; Cheng, B. Indirect shared control for cooperative driving between driver and automation in steer-by-wire vehicles. *IEEE Trans. Intell. Transp. Syst.* **2020**, *22*, 7826–7836. [CrossRef]
28. Zimmermann, M.; Schopf, D.; Lütteken, N.; Liu, Z.; Storost, Z.; Baumann, M.; Happee, R.; Bengler, K. Carrot and stick: A game-theoretic approach to motivate cooperative driving through social interaction. *Transp. Res. Part C Emerg. Technol.* **2018**, *88*, 159–175. [CrossRef]
29. Li, W.; Li, Q.; Li, S.E.; Li, R.; Ren, Y.; Wang, W. Indirect shared control through non-zero sum differential game for cooperative automated driving. *IEEE Trans. Intell. Transp. Syst.* **2022**, *23*, 15980–15992. [CrossRef]
30. Ding, H.T.; Guo, K.H.; Li, F.; Zhan, J.W. Arbitrary path and speed following driver model based on vehicle acceleration feedback. *J. Mech. Eng.* **2010**, *46*, 116–120. [CrossRef]
31. Guo, H.; Cao, D.; Chen, H.; Sun, Z.; Hu, Y. Model predictive path following control for autonomous cars considering a measurable disturbance: Implementation, testing, and verification. *Mech. Syst. Signal Process.* **2019**, *118*, 41–60. [CrossRef]

processes

MDPI

Article

Synergizing Phenomenological and AI-Based Models with Industrial Data to Develop Soft Sensors for a Sour Water Treatment Unit

Danielle Gradin Queiroz [1], Francisco Davi Belo Rodrigues [1,2], Júlia do Nascimento Pereira Nogueira [1], Príamo Albuquerque Melo [3] and Maurício B. de Souza, Jr. [1,3,*]

1 Escola de Química, EPQB, Universidade Federal do Rio de Janeiro, P.O. Box 68542, Rio de Janeiro 21941-909, RJ, Brazil; daniellegradin@eq.ufrj.br (D.G.Q.); davibelo@eq.ufrj.br (F.D.B.R.); julia_18nogueira@hotmail.com (J.d.N.P.N.)
2 Petróleo Brasileiro S.A. Refinaria Duque de Caxias, Rodovia Washington Luiz, km 113.7, Campos Elíseos, Duque de Caxias 25225-010, RJ, Brazil
3 Programa de Engenharia Química, PEQ/COPPE, Universidade Federal do Rio de Janeiro, P.O. Box 68502, Rio de Janeiro 21941-972, RJ, Brazil; melo@peq.coppe.ufrj.br
* Correspondence: mbsj@eq.ufrj.br

Abstract: Sour waters are one of the main aqueous byproducts generated during petroleum refining and require processing in sour water treatment units (SWTUs) to remove contaminants such as H_2S and NH_3 in compliance with environmental legislations. Therefore, monitoring the composition of SWTU effluxents, including acid gas, ammoniacal gas, and treated water, is essential. This study aims to present an AI (artificial intelligence) hybrid-based methodology to develop soft sensors capable of real-time prediction of H_2S and NH_3 mass fractions in the effluents of SWTUs and validate them using real data from industrial units. Initially, a new database based on the dynamic simulation of a two-stripping-column SWTU phenomenological model, developed in Aspen Plus Dynamics® V10, was generated, aiming at non-faulty runs, unlike our previous work. Ensemble methods (decision trees), such as gradient boosting and random forest, and support vector machines were compared for soft sensor creation using these simulated data. The best outcome was the development of six soft sensors based on random forest with R^2 greater than 0.87, MAE less than 0.12, MSE less than 0.17, and RMSE less than 0.41. Variable importance analysis revealed that the temperature of the second stage of Column 1 significantly influences the thermodynamic equilibrium of H_2S and NH_3 separation from sour waters, being critical for five of the six soft sensors. After this initial stage using data from the phenomenological model, data from an industrial-scale SWTU were used to develop real soft sensors. The results proved the effectiveness of the conjugated use of a physical model and industrial data approach in the development of soft sensors for two-column SWTUs.

Keywords: sour water; soft sensor; environmental protection; machine learning; database; Aspen Plus Dynamics®; random forest; hybrid modeling

1. Introduction

One of the most important challenges of our times is to reduce the climate impacts of industrial processes. The reduction in harmful gas emissions and the prevention of soil and water contamination play an important role in that purpose.

In the context of petroleum refining process, steam and water streams are used in various units such as hydroprocessing, delayed coking, and fluidized catalytic cracking, just to name a few systems. One of the effluents from these units is referred to as sour water, as it is contaminated with various weak electrolytes such as phenol, ammonia (NH_3), hydrogen sulfide (H_2S), and possible traces of carbon dioxide (CO_2) [1,2].

After the refining stage, sour water is directed to the sour water treatment unit (SWTU) to reduce contaminant levels, focusing primarily on the removal of NH_3 and H_2S. The stream undergoes a stripping process, being subjected to a heating and rectification system where the necessary heat is provided to reduce the partial pressures of gases and separate the contaminants [3].

This process is usually carried out with two columns when ammonia concentration in sour water is high, as illustrated in Figure 1. In this configuration, the first column produces a top stream rich in H_2S, known as acid gas, which is directed to the sulfur recovery unit (SRU). The second column is responsible for separating the bottom output from the first one into two streams: a top stream rich in NH_3, known as ammoniacal gas, which is sent to the ammonia incinerator, and a bottom stream referred to as treated water.

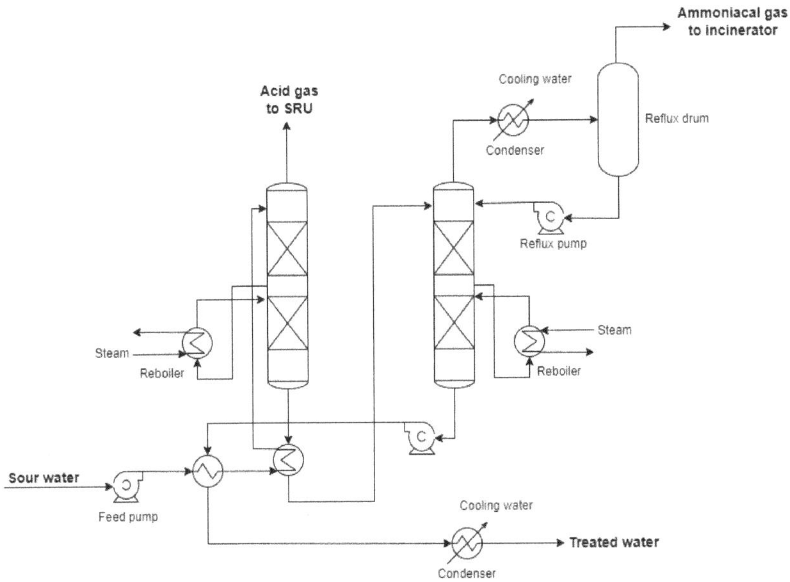

Figure 1. Simplified scheme of an sour water treatment unit with two columns.

According to Brazilian environmental legislation, at least 90% of the incoming H_2S load in sour water must be removed and sent to the sulfur recovery unit (SRU). This is a critical point in the process because if H_2S residues are sent to the second column, they will be eliminated along with ammoniacal gas (NH_3), producing SOX, which is environmentally undesirable as it contributes to air pollution and is a precursor to acid rain [4–6]. Nevertheless, a high efficiency in H_2S removal can have negative consequences on the operation of the sulfur recovery unit (SRU). An increase in H_2S removal can lead to an increase in NH_3 concentration in the acid gas, causing operational issues in the unit. This problem may result in efficiency loss due to line blockages and the formation of NOx, a compound highly detrimental to the environment and public health, as it is also a precursor to acid rain and photochemical smog [7,8].

Therefore, maintaining a high H_2S recovery in the first tower while simultaneously reducing the NH_3 content in the acid gas represents conflicting goals and characterizes an operation with a narrow tolerance range for units with this configuration. Furthermore, small disturbances that may cause variations in the STWU feed streams composition can result in faults that ultimately lead to the emission of environmentally harmful compounds above regulatory limits.

Currently, a major operational challenge is to monitor the effluent composition of these units online. Soft sensors play an important role in solving these difficulties. Soft sensors are mathematical models or algorithms that allow for estimating the value of a variable of interest based on available process information, without the need for direct measurements of that variable [9]. They can be employed for real-time monitoring of process variables, enabling trend and anomaly identification, and triggering alerts and notifications when potential problems are detected. Additionally, they offer substantial cost savings compared to traditional sensors, as they eliminate the need for physical installation and maintenance and are not affected by obsolescence due to corrosion in industrial environments [10]. Soft sensors can also assume the role of physical sensors in situations of unexpected failure. This ensures that operators have continuous access to the necessary information to monitor and control processes, reducing the risk of unplanned disruptions due to lack of essential information [2]. Finally, for process control applications, they eliminate the measurement delay issues associated to complex laboratory analyses [11]. Thus, the development of soft sensors for an SWTU is an advantageous solution for investment and precision issues compared to online analyzers and laboratory analyses.

There are three main approaches used to estimate variables based on other measurements. These are (i) model-driven (or white-box), in which a model based on fundamental principles is employed; (ii) data-driven (or black-box), which relies solely on historical process data, allowing for models to be created using machine learning techniques; and (iii) gray-box, which is a combination of the two previous methods [12,13].

The main objective of the present study was the development of soft sensors for predicting the composition of all main effluents from a SWTU with a two striping tower configuration using a gray-box or hybrid approach.

In the initial phase of the investigation, a new database was developed to support this research, utilizing a modified version of the dynamic model of an SWTU with two stripping columns previously developed in Aspen Plus Dynamics® V10 by [1]. Artificial intelligence approaches, including ensemble methods (decision trees) such as gradient boosting [14] and random forest [15,16], along with support vector machines [17,18], were initially compared for soft sensor creation using the simulated data generated by the phenomenological model. Subsequently, the best AI paradigm from the initial comparison was selected to develop six models for predicting H_2S and NH_3 concentrations in each output stream of the unit. Hyperparameter optimization was applied in the development of each AI model. The importance of the input variables for the prediction of each soft sensor was determined and analyzed.

In the second part of this work, the methodology was validated by applying it to a real industrial-scale unit. The goal was to utilize the available information about the process provided by the phenomenological model to enhance and calibrate the data-driven AI-based model. This approach allowed for the combination of theoretical understanding with the flexibility of data-driven modeling. Compared to other techniques that implicitly integrate machine learning techniques with physical equations [19], the present proposal aims to be more intuitive and natural, as the intention was to develop a tool to assist the operator in a human-centered Industry 5.0 context [20].

Although developments in soft sensors for other wastewater treatments are more common [21], the number of applications specifically addressing SWTUs is scarce. In [22], soft sensor models were developed to estimate the removal efficiency of H_2S in an SWTU. The collected data spanned a two-year operational period of a sour water stripper processing non-phenolic acid water. Only steady-state data were employed, so that a total of 144 different operating conditions were considered. A simplified phenomenological approach developed in AspenOne® and a linear statistical model did not show reliability in the results obtained. On the other hand, a multilayered perceptron (MLP) neural network proved suitable for the desired application. In [23], a semi-supervised learning approach based on deep neural networks was employed to develop two soft sensors for estimating the concentrations of H_2S and NH_3 in the wastewater of a SWTU with only one column. This study addresses the difficulty of labeling data, as, in certain industrial scenarios, input variables are constantly measured, while output variables are quantified only once or twice a day. The results of the deep neural network-based sensors used were compared with MLP network models. The sensors developed with deep neural networks exhibited better performance and proved to be an effective strategy for industrial applications where a shortage of labeled data can significantly impact the performance of data-driven soft sensors.

The environmental importance of treating sour water in refineries, combined with the operational challenges of SWTUs and safety concerns, establishes the need for real-time predictions of H_2S and NH_3 mass fractions in the effluent. The soft sensors proposed here will aid in the operation, monitoring, optimization, and control of the process.

The major contribution of this work is the development of six soft sensors for a two-column SWTU. To achieve this, intermediary contributions were made, including the generation of dynamic operational data representative of the units, a comprehensive analysis of the produced data, and the inclusion of industrial validation. Methodologically, the synergistic use of a phenomenological model alongside data-based models for the industrial case is an important contribution from a practical standpoint.

The present work is divided into six sections: Section 1 encompasses the motivation, objectives, and a review of previous works. Section 2 presents the materials and methods used, including the phenomenological model in Section 2.1, the development of the database in Section 2.2, and the modeling of the soft sensors based on AI tools in Sections 2.3–2.6. Sections 3 and 4 present and discuss the results, respectively. Finally, Section 5 addresses the conclusions.

2. Materials and Methods

2.1. The Phenomenological Model

In [1], the author conducted the dynamic simulation of an SWTU in Aspen Plus Dynamics® V10 using the gas process association sour water equilibrium (GPSWAT) model, considered the most suitable for sour water applications. In this simulation, there are thirty-four material streams and seven different compositions, three heat streams, and twenty-three pieces of equipment: twelve valves, three heat exchangers, three splitters, two mixers, one pump, and two columns. The block diagram in Figure 2 summarizes all the streams and key equipment in the simulation, where H1, H2, and H3 are the heat exchangers, and C1 and C2 are the stripping columns.

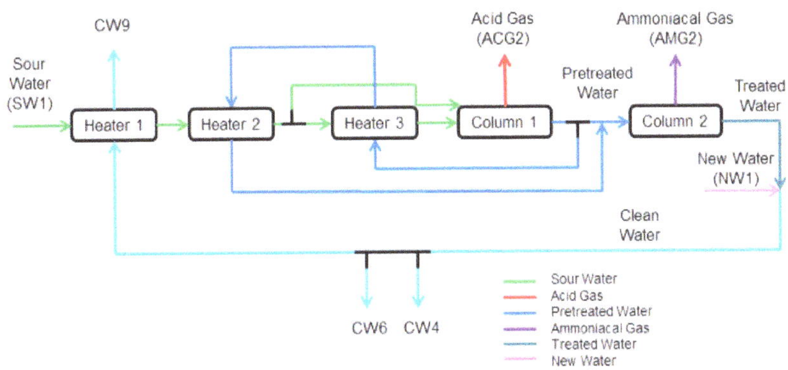

Figure 2. Block diagram of process simulation.

The seven different compositions described in the process are sour water, which is the incoming stream to be treated: acid gas, which is the overhead stream from the first column rich in H_2S; ammoniacal gas, which is the overhead stream from the second column rich in NH_3; pre-treated water, which is the bottom stream from the first column, hence rich in NH_3; treated water, which is the bottom stream from the second column, containing small amounts of dissolved H_2S and NH_3 contaminants; new water, which is an aqueous stream without contaminants; and finally, clean water, which is the mixture of new water with treated water. Table 1 correlates the abbreviations of these streams with their compositions.

Table 1. Legend of the streams in the dynamic simulation.

Abbreviation	Stream Composition
SW	Sour water
PW	Pre-treated water
TW	Treated water
NW	New water
CW	Clean water
ACG	Acid gas
AMG	Ammoniacal gas

The two input streams of the process are SW1 and NW1, while the five output streams are ACG2, AMG2, CW4, CW6, and CW9. SW1 is the sour water to be treated, and NW1 is the new uncontaminated water source that dilutes the treated water stream after passing through the two columns. Table 2 displays the physical properties of the input streams.

Table 2. Physical properties of the input streams of the system.

Property	Unity	SW1	NW1
Temperature	°C	38.8	122.4
Pressure	bar	10.33	2.29
Mass flow	t/h	SW1F	38.5% de SW1F
Mass fraction of H_2O	-	0.9928	1.0000
Mass fraction of NH_3	-	0.0032	0.0000
Mass fraction of H_2S	-	0.0040	0.0000

-: dimensionless.

These streams had their mass flow described as a function of SW1, defined as SW1F. These values were anonymized due to the confidentiality of information that originated the simulations developed by [1].

Finally, ACG2 and AMG2 are the outputs of acid gas and ammoniacal gas, respectively. CW4, CW6, and CW9 are the outputs of clean water with the same composition.

The simulation begins with the SW1 stream, which is preheated in heat exchangers H1 and H2 through energy integration with streams CW7 and PW5, respectively, becoming SW4. SW4 is divided into SW5 and SW7. SW5 becomes SW6 after passing through valve V2 and enters the top of column C1. SW7 is preheated once again in heat exchanger H3 through energy integration with PW4 and becomes SW9 after passing through valve V3, subsequently entering the lower feed of column C1. Therefore, streams SW6 and SW9 feed C1.

Column C1 is heated by the thermal load Q1 from the reboiler, separating the overhead stream ACG1 (acid gas) from the bottom stream PW1 (pre-treated water). ACG1, after valve V4, becomes ACG2 and is removed from the system, while PW1 is split into two streams: PW2 and PW4. PW4 is cooled in heat exchanger H3 through energy integration with SW7 and becomes PW5. PW5 is further cooled in H2 through energy integration with SW3 and becomes PW6. PW2 becomes PW3 after passing through valve V5 and is mixed with PW6, forming PW7. It is noteworthy that PW7 has the same mass flow as PW1, as the streams that split reunite without losses. After valve V6, PW7 becomes PW8 and enters the feed of column C2.

Column C2 has a condenser and a reboiler, with thermal loads Q2 and Q3, respectively. The overhead stream from Column 2 is AMG1 (ammoniacal gas), and the bottom stream is TW1 (treated water). AMG1, after valve V7, becomes AMG2 and is removed from the system. TW1, after valve V8, becomes TW2. TW2 is mixed with NW2, becoming clean water (CW1). After passing through pump P1, CW1 is then called CW2. CW2 is divided into three streams: CW3, CW5, and CW7. CW3, after valve V10, becomes CW4 and is removed from the system. CW5, after passing through valve V11, becomes CW6 and is removed from the system. Meanwhile, CW7 is cooled in heat exchanger H1 through energy integration with SW2, becoming CW8. CW8, after valve V12, becomes CW9 and is removed from the system. A more complete view of the process is in Figure 3, showing the static simulation developed by [1].

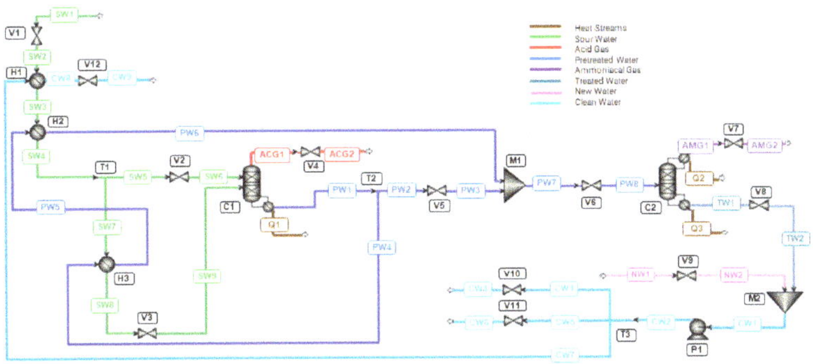

Figure 3. Static process simulation.

There are two 'RadFrac' type columns and three 'HeatX' type heat exchangers in the simulation. Column C1 consists of 5 stages, and Column C2 consists of 6 stages. The feed occurs at stages 1 (SW6) and 2 (SW9) in C1, and at stage 2 (PW8) in C2. The two mixers were configured to receive both liquid and vapor phases. Valves V4 and V7 were configured only to receive vapor, while the other valves accept only the liquid phase. All valves have an associated pressure drop.

The dynamic model by [1] includes 14 controllers added to ensure the operability and stability of the process, as well as aid in the convergence of the dynamic simulation. The controllers in the simulation are the following: two for pressure at the tops of the columns, four for level at the bottom and top of the columns, three for temperature, including one before the first column and one in its second stage, and another at the top of the second column; four mass flow controllers, two before the first column and two after the second column; and an integrated control of mass flow and thermal load in the second column. Table 3 describes the controlled and manipulated variables for each of these controllers. The tuning of the controllers was performed based on general rules described in the literature and through industry expert advice [1].

The first step in creating the new database for the development of soft sensors involved the inclusion of six controllers not previously used in the dynamic simulation of the SWTU with two stripping columns by [1]. Among these controllers, two were incorporated into the acid gas stream (ACG1) at the top of Column 1 (ACID_G_H_2S and ACID_G_NH_3), another two were added to the ammoniacal gas stream (AMG1) at the top of Column 2 (AMON_G_H_2S and AMON_G_NH_3), and two more (WATER_H_2S and WATER_NH_3) were introduced into the treated water stream (TW1) withdrawn from the bottom of Column 2. The controllers present in the original simulation can be seen in Table 3. Figure 4 includes all the controllers.

To enable the collection of dynamic data for the fractions of H_2S and NH_3 in the effluents described, the PID (proportional, integral, and derivative) controllers included were adjusted with a gain of 0, an integral time of 1 min, and a derivative time of 0. This configured them to operate solely as 'indicators', without effectively performing control functions during the simulation, only storing the dynamic data.

Table 3. Controllers present in the original dynamic simulation.

Controller	Controlled Variable	Manipulated Variable
SW1-FC	SW1 mass flow rate	V3 opening %
SW5-FC	SW5 mass flow rate	V2 opening %
CW5-FC	CW5 mass flow rate	V11 opening %
CW8-FC	CW8 mass flow rate	V12 opening %
SW8-TC	SW8 temperature	V5 opening %
C1S2-TC	C1 2nd Stage temperature	C1 reboiler load
AMG1-TC	AMG1 temperature	C2 condenser load
C1S-LC	C1 bottom level	V6 opening %
C2D-LC	C2 Reflux drum level	Mass flow rate of C2 reflux drum
C2S-LC	C2 bottom level	V9 opening %
C2S-LC2	C2 bottom level safeguard	V10 opening %
ACG1-PC	AGC1 pressure	V4 opening %
AMG1-PC	AMG1 pressure	V7 opening %
C2HD-IC	Ratio between the C2 reboiler Load and PW7 mass flow rate	C2 reboiler load

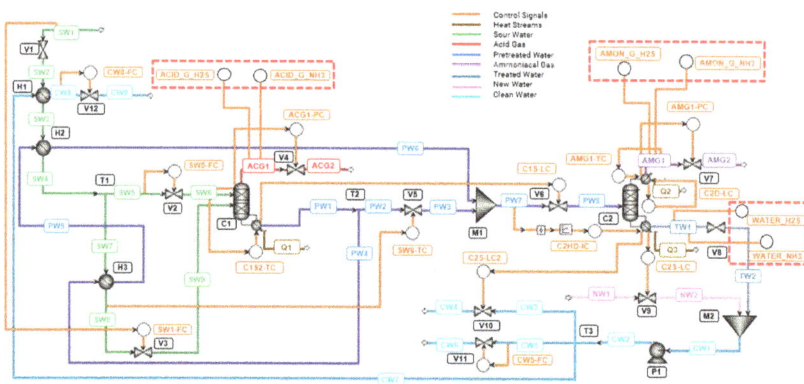

Figure 4. Dynamic simulation with the inclusion of the new six controllers: ACID_G_H_2S, ACID_G_NH_3, AMON_G_H_2S, AMON_G_NH_3, WATER_H_2S, and WATER_NH_3.

2.2. Development of the Database and Phenomenological Study of the Process

The simulations, calculations, and codes were developed and executed on a computer with an Intel(R) Core™ i5 7200U with a 2.50 GHz–2.70 GHz CPU, 8.00 GB installed memory (RAM), and Windows 10 64-bit operating system. The dynamic simulation of the SWTU was performed using Aspen Plus Dynamics® V10. All data processing and artificial intelligence algorithms were implemented in Python 3.11.4 within the integrated development environment Visual Studio Code 1.79.2. The libraries NumPy (1.25.2), Matplotlib (2.0.2), joblib (1.3.2), scikit-learn (0.19.0), pandas (2.1.0), Seaborn (0.12.2), and Optuna (3.3.0) were extensively used.

The dynamic simulation, after the changes discussed in Section 2 and illustrated in Figure 4, was tested in various scenarios of normal operation with disturbances in five variables of interest, namely, SW1 mass flow rate, SW5 mass flow rate, SW1 temperature, and NH_3 and H_2S mass fractions in SW1. These variables were chosen because they are input process variables that could undergo variations in a real process. SW1 is the sour water entering the system for treatment, and SW5 represents 10% of SW1, which, after passing through the heat exchangers, enters the top of column C1. Normal operation is characterized when events such as complete valve opening or closing, overflows, or lack of level in the columns do not occur.

2.3. Comparison of Artificial Intelligence Algorithms

To create the six soft sensors, AI tools such as random forests, gradient boosting, and SVM with linear and RBF kernels models were evaluated using the same training and test datasets. The methods of gradient boosting (GradientBoostingRegressor function), random forest (RandomForestRegressor), and SVM (SVR function; linear and RBF kernels) from the scikit-learn library (version 0.19.0) were employed. These kernels were chosen because they demonstrated the best results and the shortest computational times in a previous study [24]. These tools were chosen because they are simple, shallow machine learning methods that are not computationally demanding and can be easily adopted in industrial environments.

The Optuna library was used for hyperparameter optimization of the RF algorithm, aiming to obtain the most effective and accurate model for each of the six sensors. The optimized hyperparameters included the number of estimators (rf_n_estimators), representing the number of trees in the forest, the maximum depth of the trees (rf_max_depth), and the minimum number of samples required to form a leaf (terminal node) of the tree (rf_min_samples_leaf). The lower and upper limits were defined for each of the RF hyperparameters and are shown in Table 4. This methodology enhances the efficiency of the optimization by eliminating the need to consider extreme or impractical values and helps to prevent overfitting or underfitting issues in the model and resulting in computational time savings. Similarly, the number of iterations (or number of trials) for the hyperparameter optimization process was set to 40.

Table 4. Lower and upper limits defined for each hyperparameter.

Hyperparameters	Limits
rf_n_estimators	1 to 50
rf_max_depth	5 to 10
rf_min_samples_leaf	1 to 10

Analogously to the RF, the same hyperparameters mentioned in Table 4 were optimized for the GB algorithm using the same limits and number of iterations, while for the SVM models, there was no optimization of the hyperparameters (C and gamma), as optimizing them did not yield significant gains as shown by [24].

Data processing takes place after storing the dynamic data from all 20 controllers after each simulation (or simply 'run'). Concatenating this information within normal operation resulted in the development of the database, which was later divided into training and testing data to be used in the development of AI-based soft sensors. To guarantee the representativeness and efficient training of the models, it was ensured that each database had at least one example of each simulation in normal operation. Additionally, all data related to a single run were consolidated in the same database to prevent data leakage, especially considering that the simulations contain temporal data. Furthermore, data shuffling, separation into inputs and outputs, as well as normalization were performed.

After dividing the database, the training set consisted of 45,483 samples (59%), while the test set contained 31,525 samples (41%). A sampling time of 36 s was considered. The listing with the description of each simulation that composed each set can be found in Tables 5 and 6.

To clarify the division of runs between training and testing databases, each database contains one example of each simulation. Although they share the same description, each simulation is unique, featuring various versions in terms of intensity, amplitude, and timing of the disturbances. The normalized database can be found in [25].

Table 5. Listing of the runs present in the training database.

Run	Training Database Description
1	Global run with disturbances between −1% and +1% on the set points of the 5 standard variables—version A
2	Global run with disturbances between −30% to 40% of the set point of the mass flow rate of SW5—version A
3	Global run with more numerous disturbances between −30% to 40% of the set point of the mass flow rate of SW5—version A
4	Global run with disturbances between −25% to 30% in the temperature of SW1—version A
5	Global run with disturbances between −25% to 30% in the H_2S mass fraction in SW1—version A
6	Global run with disturbances between −25% to 30% in the NH_3 mass fraction in SW1—version A
7	Global run with positive disturbances in a gradually increasing ramp in the temperature of the 2nd stage of C1—version A
8	Global run with random positive and negative disturbances in the temperature of the 2nd stage of C1—version A
9	Partial run of the mass flow rate of SW1—version A
10	Partial run of the mass flow rate of SW1—version B
11	Partial run of the mass flow rate of SW5—version A
12	Partial run of the mass flow rate of SW5—version B
13	Partial run of the temperature of SW1—version A
14	Partial run of the temperature of SW1—version B
15	Partial run of the NH_3 mass fraction in SW1—version A
16	Partial run of the NH_3 mass fraction in SW1—version B
17	Partial run of the H_2S mass fraction in SW1—version A
18	Partial run of the H_2S mass fraction in SW1—version B

Table 6. Listing of the runs present in the test database.

Run	Test Database Description
1	Global run with disturbances between −1% and +1% on the set points of the 5 standard variables—version B
2	Global run with disturbances between −30% to 40% of the set point of the mass flow rate of SW5—version B
3	Global run with disturbances between −25% to 30% in the temperature of SW1—version B
4	Global run with disturbances between −25% to 30% in the H_2S mass fraction in SW1—version B
5	Global run with disturbances between −25% to 30% in the NH_3 mass fraction in SW1—version B
6	Global run with positive disturbances in a gradually increasing ramp in the temperature of the 2nd stage of C1—version B
7	Global run with random positive and negative disturbances in the temperature of the 2nd stage of C1—version B
8	Partial run of the mass flow rate of SW1—version C
9	Partial run of the mass flow rate of SW5—version C
10	Partial run of the temperature of SW1—version C
11	Partial run of the NH_3 mass fraction in SW1—version C
12	Partial run of the H_2S mass fraction in SW1—version C

2.4. Soft Sensors Using Random Forests

A RF model was developed for each of the six virtual sensors: ACID_G_H_2S, ACID_G_NH_3, AMON_G_H_2S, AMON_G_NH_3, WATER_H_2S, and WATER_NH_3. Each algorithm was trained, validation was conducted with 20% of the training data, and the final evaluation of the best model for each sensor was performed using the test data. Hyperparameter optimization was conducted for all sensors using the lower and upper limits specified in Table 4.

2.5. Global Analysis of the Importance of Variables and Phenomenological Study of the Process

The randomForest package provides valuable information for phenomenologically evaluating a process: the measurement of the importance of predictor variables (variable importance—VI). The significance of a variable arises from its complex interaction with other variables. The algorithm assesses the importance of a variable by observing the extent to which the prediction error increases when the data for that variable is altered, while keeping all other variables unchanged. The necessary calculations are conducted tree by tree as the random forest is constructed [26]. This functionality enables the reduction of the dataset of interest by eliminating less important variables, ensuring the simplification and optimization of the model, along with a reduction in computational processing time [27,28].

Therefore, an analysis of the most relevant variables was conducted for each of the sensors. Variables with importance below 2% were excluded, and the predictive capacity of the retrained sensors was evaluated using the same presented metrics (R2, MAE, and RMSE). This reduction of the input dataset, achieved by removing less important variables, enables the simplification and optimization of the model, thus increasing the analysis speed.

2.6. Implementation to a Real SWTU

In this section, the knowledge gained with the approach based on the phenomenological model is employed to develop random forest models using data from a real SWTU of a Brazilian oil refinery. The chosen unit for this development has the same flowsheet configuration of the simulated plant, making it possible to use the same inputs as in the simulated modelling.

The WATER_NH_3 sensor was selected for modeling. For this sensor, the plant utilizes an online analyzer that produces results approximately every 20 min. Unfortunately, analysis for acid and ammoniacal gases were not conducted on the site and only a limited number of laboratory results were available for H_2S on treated water. This was expected due to the hazards involved in sampling acid gas and that H_2S is not a primary concern on treated water given that H_2S is easier to remove than NH_3 in the second column, making more sense to control only the NH_3 content on treated water.

Even though the online analyzer generates results every 20 min, it is worth to mention that any control application to be implemented on the plant would need information at least on a minute basis, justifying the importance of modeling this kind of sensor even in a plant with online analyzer.

Given that operating conditions of the industrial plant did not precisely match the intervals used in the simulation runs, it was not prudent to directly use the models obtained with simulation data. Therefore, the sensor was retrained with real plant data. Variables SW5-FC, SW1-FC, SW8-TC, ACG1-PC, C1S2-TC, C1S-LC, AMG1-TC, AMG1-PC, C2D-LC, and C2S-LC were considered as input variables. Variables CW8_FC, C2HD-IC, and CW5-FC were not included, as these measurements were not available in the real plant.

The first step of data acquisition involved retrieving results from the analyzer, starting from its installation date, covering a period of 22 months, yielding 40,338 results. Subsequently, rather than acquiring input data, at the same timestamps of the analyzer results, they were collected applying a 20 min time average prior to those timestamps. This approach is due to the need to minimize not only the effects of dynamic and dead times but also disturbances characteristics of any industrial plant. It is believed that using the input values on exactly the timestamps of the analyzer results would not correctly represent the information that generated the outputs.

Before modelling, an interquartile range method was applied to remove outliers. The scale value used was 1.7, which means that this procedure discarded any value greater than 3 standard deviations from the respective mean. The final useful data were composed of 21,433 rows. The data were normalized using the method StandardScaler from the scikit-learn Python module. The normalized data were split into 10% for test and 90% for training from which 20% was reserved for validation.

Model training was executed with the same methodology used in training with simulated data, which consisted of using Python's Optuna library to explore the hyperparameters of random forest regressor, i.e., rf_min_samples_leaf (1 to 10), rf_max_depth (5 to 10), and rf_n_estimators (1 to 50). The best model hyperparameters were, respectively, 1, 10, and 48.

3. Results

3.1. Database and Phenomenological Analysis of the Process

A preliminary study of integration methods [29] revealed that the standard implicit Euler integrator was the most suitable when compared to other available options, such as Runge–Kutta 4 and gear integration methods. The observed integration errors could be attributed to the inherent complexity of the differential equations in the system.

The initial strategy for building the database was to develop a global run consisting of four random disturbances ranging between +1% and −1% of the set point value for each of the five variables subjected to disturbances. Simultaneously, partial runs were developed that featured only the disturbances of a specific variable at the same time they occurred in the global run. That is, the partial run of the mass flow rate of SW1, for example, refers to the global run in which all disturbances not related to the mass flow rate of SW1 were removed from the programmed task.

The results from the global run, which includes random disturbances ranging from −1% to +1%, along with their respective partial runs, are shown in Figure 5. Similar studies were also conducted for the other effluent streams, considering H_2S and NH_3 [29]. After interpreting the simulations, it is concluded that the mass flow rates of SW1 and SW5, and the temperature of SW1, are extremely important variables for the thermodynamic separation process in the two stripping columns. This is because they directly interfere with the fractions of H_2S and NH_3 in the acid gas (effluent from Column 1), ammoniacal gas, and treated water (effluents from Column 2). This is a consistent result, as these are characteristics linked to the input stream of the process.

However, despite the applied deviations, it was observed that the disturbances caused only a slight variation in the fractions of H_2S and NH_3 in the effluents, especially in the H_2S fraction of the acid gas, which presented only a 0.12% variation. This result was not ideal, as it would imply low variability in the data subsequently used in the soft sensors training.

Therefore, new global runs were developed with substantially more intense disturbances (ranging from −25% to +30% of the set point value) for each of the five variables studied individually, while maintaining disturbances between +1% and −1% for the other input variables. The objective was to achieve, at least, a 1% variation in the H_2S fraction of the acid gas and to keep the simulation within normal operation. Furthermore, the aim was also to determine which of the analyzed variables were most relevant to the variation in H_2S fraction.

The global runs with increased magnitude disturbances in the SW1 mass flow rate, SW5 mass flow rate, SW1 temperature, and NH_3 and H_2S mass fractions in SW1 still showed a small variation in acid gas H_2S fraction, as displayed in Table 7. On the other hand, the global run of the mass flow rate of SW5 proved to be more effective among all. Therefore, a new global run was created for the mass flow rate of SW5 with disturbances of higher magnitude than the previous ones (ranging from -30% to +40% of the set point value for the variable), resulting in a 1% variation in acid gas H_2S fraction, while keeping normal operation. The general behavior observed reflects the effectiveness of the regulatory control implemented, which adhered to industrial standards and minimized large variations in the controlled variables.

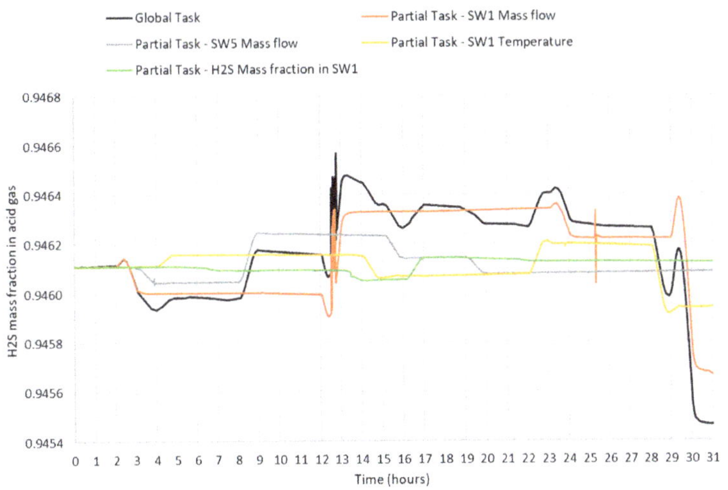

Figure 5. Important variables for determining the fraction of H_2S in the acid gas.

Table 7. Variations in acid gas H_2S mass fraction for each global run.

Global Runs	Variation in Acid Gas H_2S Mass Fraction
SW1 mass flow rate	0.15
SW5 mass flow rate	0.66
SW1 temperature	0.29
H_2S mass fraction in SW1	0.24
NH_3 mass fraction in SW1	0.12

Aiming at further expanding the database diversity and contributing to the future development of more robust sensors, an exploratory analysis was conducted within the dynamic simulation to identify the variable with the greatest potential to modify the H_2S fraction in the acid gas. The attempt was to 'trick' the controller's set point to understand how the other variables in the system would be affected. For this purpose, a splitting element was added to the simulation, where Input 1 was the mass fraction of H_2S in the acid gas stream, and the output was the remote set point (SP) of the ACID_G_H_2S controller. While the simulation was running, the value of Input 2 was altered to set a new SP outside the normal operation of the controller so that the responses of the other variables involved in the process could be evaluated.

This process involved trying to emulate multiple scenarios and disturbances in the dynamic simulation and observing the long-term effect they had in the composition of the outlet process streams. This was important in the database development step of the soft sensors' creation, aiming to ensure a more realistic and diverse database, encompassing several different normal operational conditions, without anomalies.

After the test, it was identified that the variable most affected by the change was the temperature of the second stage of Column 1, a variable controlled by the C1S2-TC controller. Based on this, new runs were generated with a focus solely on evaluating this variable. One of them consisted of a gradual ramp-up of the temperature of the second stage of Column 1, by applying only positive disturbances (between 2% to 15%) to the set point of the C1S2-TC controller. This resulted in a variation in the H_2S fraction in the acid gas of 2.3%, which is more than double of the value obtained thus far.

3.2. Resuts of the Comparison of AI Algorithms

Figure 6 shows the comparison of the coefficient of determination (R^2) among the three evaluated methods (RF, GB, and SVM) for each of the six sensors, while, in Figure 7, the comparison of the root mean square error (RMSE) is presented.

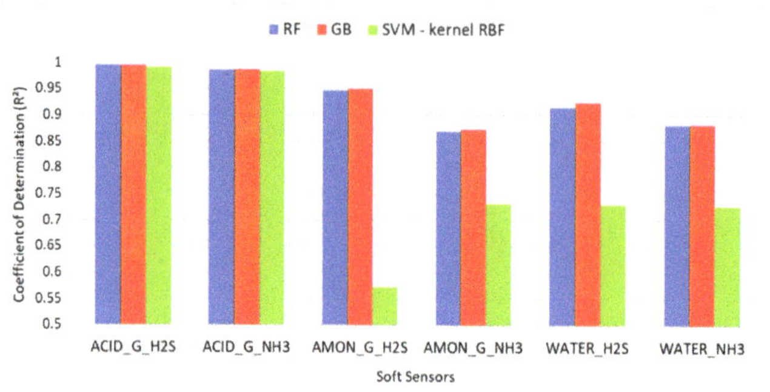

Figure 6. Comparison of the coefficient of determination (R^2) among RF, GB, and SVM with RBF kernel models.

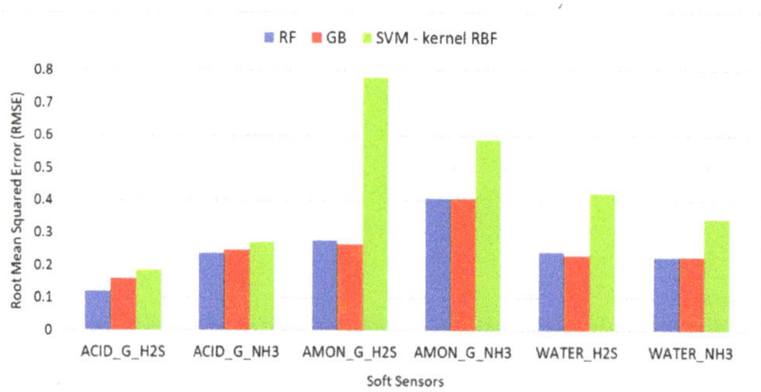

Figure 7. Comparison of the root mean square error (RMSE) among RF, GB, and SVM with RBF kernel models.

An initially relevant conclusion was that the SVM method using the linear kernel did not fit well to the data from the developed database, as it is evident from the prediction shown in Figure 8. Additionally, the computational time for processing the ACID_G_H_2S sensor algorithm was significantly higher, exceeding 20 min. The coefficient of determination (R^2) obtained was 0.9388, which is lower than the results achieved by other methods as shown in Figure 7. Therefore, the linear kernel and its results were discarded and, in Figures 7 and 8, only the data related to the SVM method with RBF kernel were presented.

From Figures 6 and 7, it became evident that the R^2 and RMSE metrics, for all SVM sensors, showed poorer performance compared to the other two methods. This is attributed here to the fact that GB and RF have several decision trees that improved the performance, and ensemble methods tend to present better results. Additionally, it is important to emphasize that this method does not provide information about variable importance, a significant aspect for understanding the process. On the other hand, the metrics for GB are very similar to those found for RF in all sensors. In addition to the metrics, the prediction and residuals graphs, and the overall results of variable importance are very similar for both methods. These results were not presented in this section to avoid hindering the readability, but they can be found in [29].

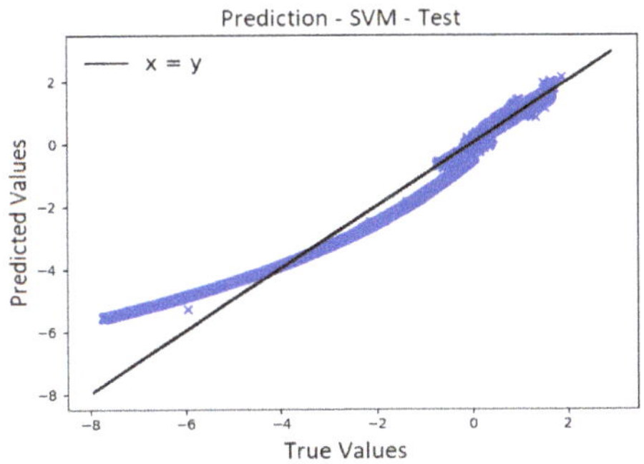

Figure 8. Prediction of the ACID_G_H_2S sensor using SVM—linear kernel.

3.3. *Results of the Soft Sensors Using Random Forests*

The hyperparameter optimization results are detailed in Table 8. The most effective model for all sensors required the maximum tree depth (rf_max_depth), which is 10. A decision was made not to increase this number to avoid overly complex models, as a satisfactory predictive performance was already obtained with this value. As for the other hyperparameters, i.e., number of trees (rf_n_estimators) and number of samples to split a node (rf_min_samples_leaf), intermediate values were typically obtained by Optuna, except for two cases: rf_min_samples_leaf in AMON_G_NH_3 and WATER_H_2S.

Table 8. Optimized hyperparameters for the sensors using RF.

Sensor	Hyperparameters	Optimal Value
ACID_G_H_2S	rf_min_samples_leaf	6
	rf_n_estimators	33
ACID_G_NH_3	rf_min_samples_leaf	3
	rf_n_estimators	47
AMON_G_H_2S	rf_min_samples_leaf	3
	rf_n_estimators	30
AMON_G_NH_3	rf_min_samples_leaf	1
	rf_n_estimators	30
WATER_H_2S	rf_min_samples_leaf	1
	rf_n_estimators	34
WATER_NH_3	rf_min_samples_leaf	2
	rf_n_estimators	47

For the evaluation of all sensors, predictions can be assessed in Figure 9, where the true values from the training and test databases were compared with the values predicted by the sensor. It is evident that most sensors exhibit excellent data prediction capability, due to the high R^2 values. Furthermore, the low error values for MAE (mean absolute error) and RMSE (root mean squared error) suggest that the models can make accurate predictions, and the individual predictions are generally very close to the actual values.

The sensors to predict H_2S and NH_3 in acid gas showed better performance than the others. The reliable prediction range of acid gas NH_3 fraction is notably broader when compared to the range of the H_2S fraction, which varies between −2.28% and 0.53% of normal operation. This behavior is supported by the results observed in Table 5, where the acid gas H_2S fraction varied by approximately −2.17%, while the NH_3 fraction varied around 143.81% during the temperature increase in the second stage of Column 1 from 120.3 to 141.9 °C.

This highlights that, in the evaluated normal operating situations, the acid gas NH_3 mass fraction is notably more sensitive to disturbances, especially concerning the temperature of the second stage of the first stripping column. On the other hand, the acid gas H_2S mass fraction is more stable, showing smaller variations in the presence of disturbances. This analysis emphasizes the relative importance of these two variables in the contaminants separation process and provides valuable insights for process control and optimization.

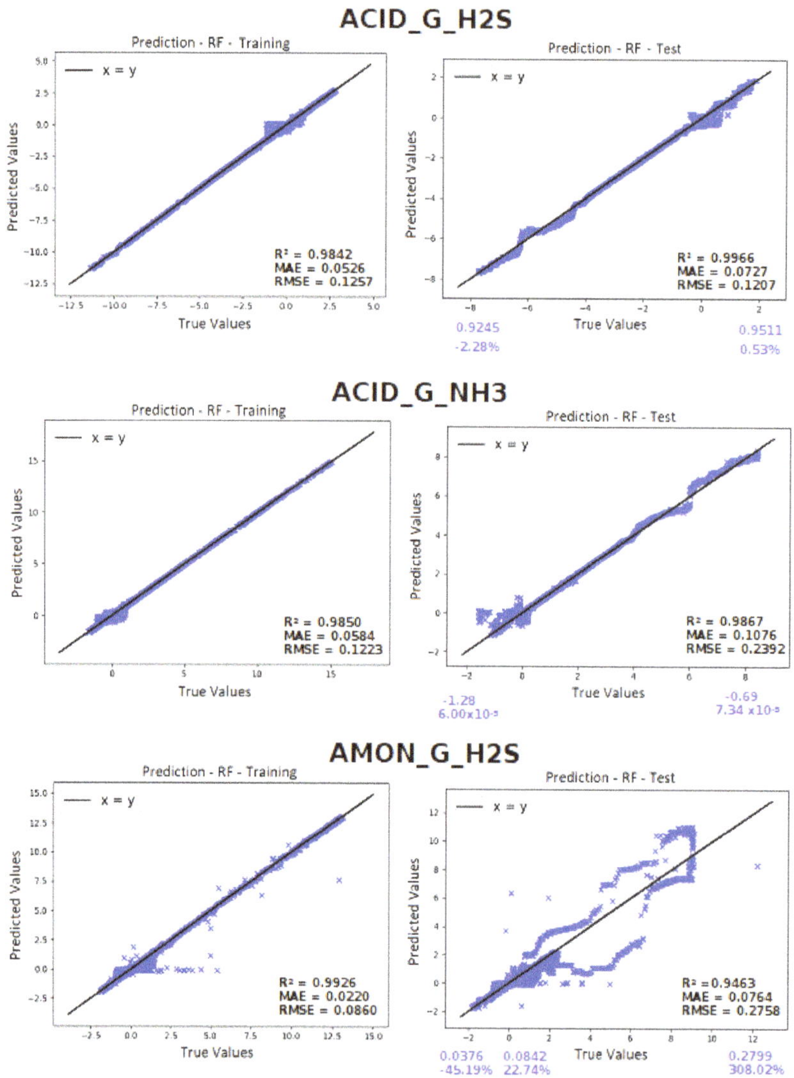

Figure 9. *Cont.*

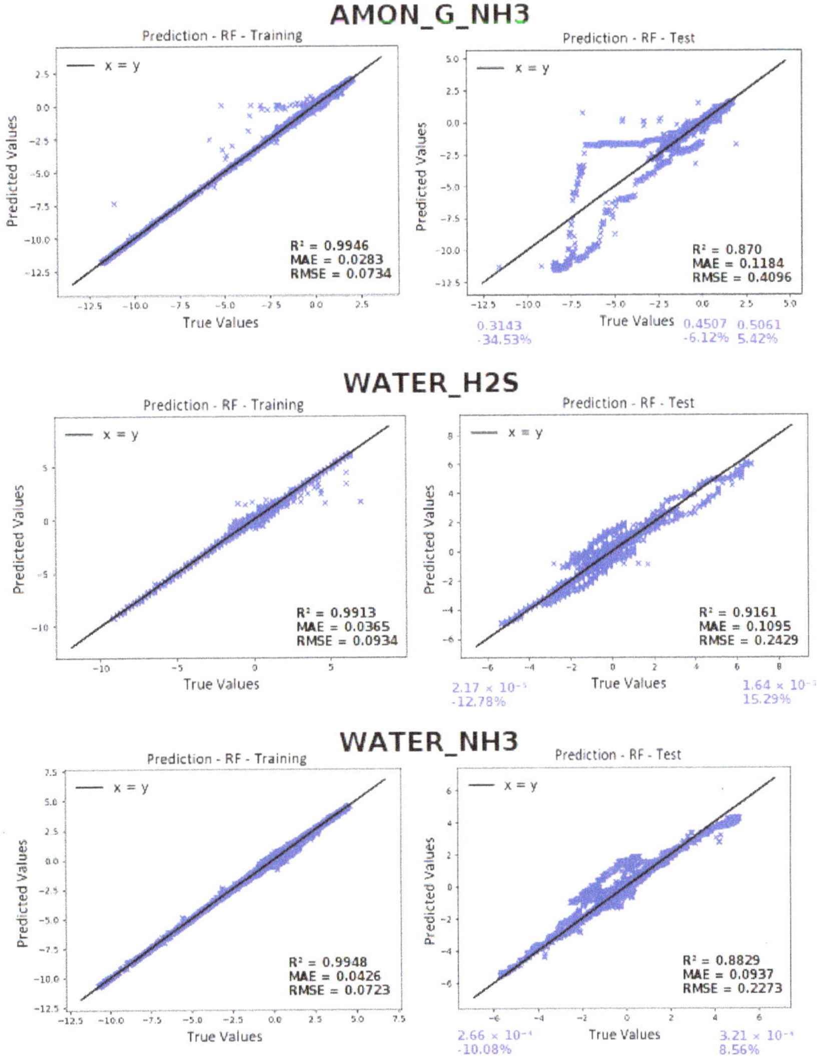

Figure 9. Prediction sensors using RF with training and test data. Models metrics are presented on the respective plots. The denormalized validated interval is in blue in the axis, along with the relative interval considering the normal operation value as parameter.

Initially, when analyzing the metrics in conjunction with the prediction of the AMON_G_H_2S sensor, it is observed that, unlike what happened with the acid gas sensors, the results for the training dataset were quite different from those found for the test dataset. Despite the R^2 value between the two sets showing only a slight absolute decrease of 4.89%, the results of MAE, MSE, and RMSE for the test set exhibited a significant absolute increase when compared to the training one. A sort of data hysteresis was observed in the test results, which was not as evident in the training ones.

This hysteresis indicated that the model predicted different values of H_2S fraction in ammoniacal gas for the same observed value present in the test set. On the other hand, the reverse was also observed, where the same value was predicted for different and distant observed values. This hysteresis is likely to have occurred due to the transient dependence on the previous state, influencing the current state.

In order to better evaluate this behavior, each of the 12 simulations from the test bank listed in Table 7 was tested individually. The best model, already optimized for the AMON_G_H_2S sensor, was used to assess the contribution of each simulation to the overall model. Examining each simulation in isolation revealed that the dataset from Run 7 played a role in the hysteresis of the results, leading to less precise predictions within the range of values from 0.86 to 12.17, as depicted in Figure 10. Notably, this specific simulation accounts for only 11% of the test dataset. Consequently, the remaining 89% of the data appears to have contributed to the model attaining a high R^2 value of 0.9463. Meanwhile, the dispersion caused by the 11% of data from Run 7 was apparent primarily in error metrics like MAE, MSE, and RMSE.

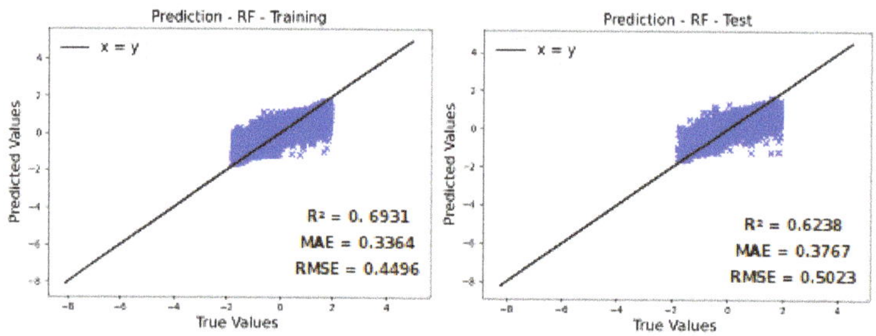

Figure 10. Prediction of the WATER_NH_3 sensor with data of real SWTU using RF: training (**left**) and test data (**right**).

Similar to the AMON_G_H_2S sensor, the results for the AMON_G_NH_3 sensor exhibited notable differences between the training and test sets. As before, it became evident that the dataset from Run 7 contributed to the hysteresis of the results, leading to less accurate predictions in the range of values from −11.56 to −1.89. As previously highlighted, this specific simulation represents only 11% of the comprehensive test set. Nevertheless, these data significantly impacted not only the R^2 value but also error metrics, including MAE, MSE, and RMSE, obtained for the AMON_G_NH_3 sensor, resulting in the lowest R^2 among all sensors following the evaluation with the test set. In order to deal with these transient effects, we recommend the use of memory-based approaches, such as recurrent neural networks (RNNs) or long short-term memory (LSTM) networks [30].

The predictive performance of this sensor demonstrated only reasonable quality in comparison to the top-performing acid gas sensors developed. The R^2 value for the training set marginally surpassed that of the acid gas sensors, while for the test set, it exhibited a slight decrement. Conversely, the error results were found to be comparable to those observed for the ACID_G_NH_3 sensor.

The analysis of the WATER_NH_3 sensor revealed a slightly lower coefficient of determination (R^2) for the test set compared to its counterpart, the WATER_H_2S sensor. However, it is noteworthy that error metrics such as MAE, MSE, and RMSE yielded, in most cases, more favorable results for the WATER_NH_3 sensor in both the training and test datasets. These findings suggest that, although the WATER_NH_3 sensor may have a relatively lower capacity to explain the variation in input data, its predictions are more accurate in terms of proximity to actual values.

3.4. *Results of the Global Analysis of the Importance of Variables*

For four out of the six developed sensors, the reduction generally led to metrics improvements, except for the AMON_G_H_2S and WATER_NH_3 sensors. In the case of the AMON_G_H_2S sensor, the MAE values increased by around 13.22%, which is still considered acceptable for practical use. This allows for the removal of variables without compromising the model's performance for this sensor. However, for the WATER_NH_3 sensor, error metrics increased substantially, with the MAE, for example, rising by approximately 90.72%. Therefore, unlike the other sensors, WATER_NH_3 requires the use of the original set of input variables to maintain the confidence of the results.

The complete analysis is summarized in Table 9, which shows the selected variables and the reduction in RMSE for the test data compared to the prediction before variable reduction.

Table 9. Global analysis of the importance of variables in all sensors after reduction.

Variables/Sensors	ACID_G_H_2S	ACID_G_NH_3	AMON_G_H_2S	AMON_G_NH_3	WATER_H_2S	WATER_NH_3
CW8_FC						
SW5-FC	X	X				
SW1-FC						
SW8-TC	X		X			
ACG1-PC						
C1S2-TC	X	X	X	X	X	
C1S-LC				X		
AMG1-TC				X	X	X
AMG1-PC			X			X
C2D-LC						
C2S-LC						
C2HD-IC						
CW5-FC						
MAE % change	−22.69	−13.48	+13.22	−10.72	−9.41	+90.72

The results found in Table 9 reinforce that the process of removing H_2S and NH_3 contaminants from sour waters is largely dependent on temperature. All temperature measurements in the simulation (highlighted in Table 9)—in the first column feed stream (SW8), the second stage of Column 1 (C1S2), and ammoniacal gas (AMG)—emerged as important variables among the six virtual sensors, indicating the fundamental role of temperature in controlling the thermodynamic separation process of H_2S and NH_3.

3.5. Results of the Implementation to the Real SWTU

In Figure 10 the model results are presented as well as their metrics. Additionally, Figure 11 presents the residual values for both the training and testing datasets. An R^2 of 0.69 for training and 0.62 for testing indicates that the model explains a substantial portion of the variance in the data and generalizes reasonably well to new, unseen data. However, the presence of industrial data introduces additional variability. The frequency histogram shows deviations from normality, particularly in the high tail, as confirmed by the Shapiro–Wilk test (test statistic of 0.99 with a p-value near zero). These deviations are likely due to extreme values. Despite these challenges, the model's performance suggests it is useful for practical applications within this industrial context.

An analysis of the most relevant variables was conducted for the sensor. Figure 12 presents the variables' importance resulting from the random forest modelling considering all the ten available measurements as inputs.

Figure 12 evidences the reduced importance of ACG1-PC and AMG1-PC (<2%). This was not primarily expected as AMG1-PC has great importance for the WATER_NH_3 sensor modeled with simulated data. Nonetheless, it was noted that the acquired pressure data for the real SWTU had the lowest coefficient of variation compared to other variables (i.e., ACG1-PC, 3.38%, and AMG1-PC, 2.00%, while C1S2-TC, 7.50%, and AMG1-TC, 6.26%). As random forests are an ensemble of decision trees, they are expected to exhibit robust prediction behavior. However, the impact of correlations, dynamic effects, and other factors on importance measures is a question of great interest. In this specific case, the reduced importance of ACG1-PC and AMG1-PC may be attributed to their dynamic behavior—specifically, the minimal variance they exhibited due to tight automatic control. In general, other methods that account for dynamic behavior [30], multicollinearity [31], and other factors should also be investigated.

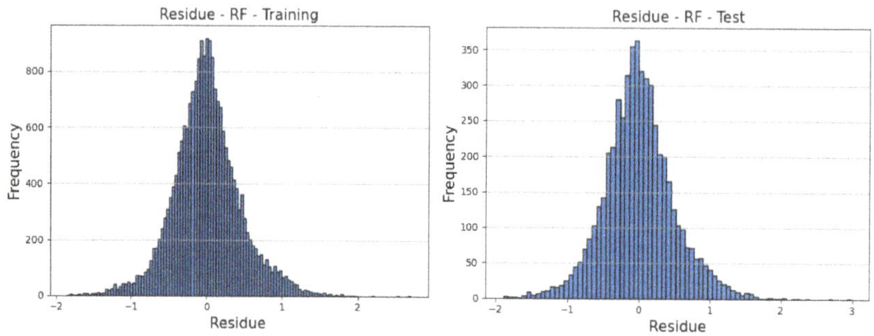

Figure 11. Training (**left**) and test (**right**) prediction residue.

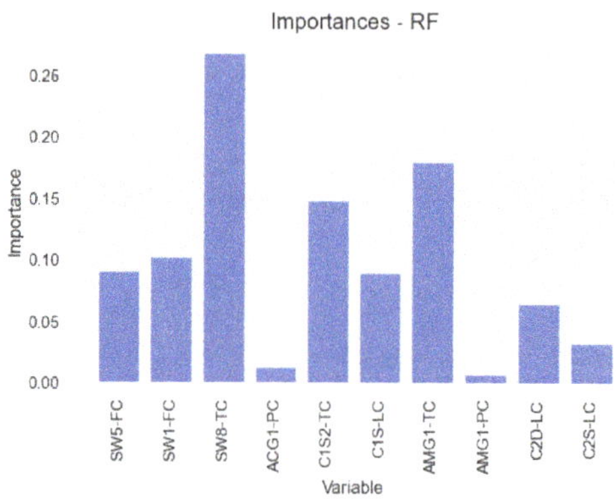

Figure 12. Variable importance for the sensor based on a random forest model obtained with data from a real SWTU.

Hence, a reduced model was produced excluding these variables from the input data. Figure 13 shows the predictions for this simplified model, also including the metrics. It was possible to achieve slightly better results removing the inputs with marginal importance. Finally, Figure 14 shows the final input variables' importance, where the three most important variables are temperatures. Hence, the quantitative analysis confirmed the relevance of the temperature variables in the industrial SWTU operation.

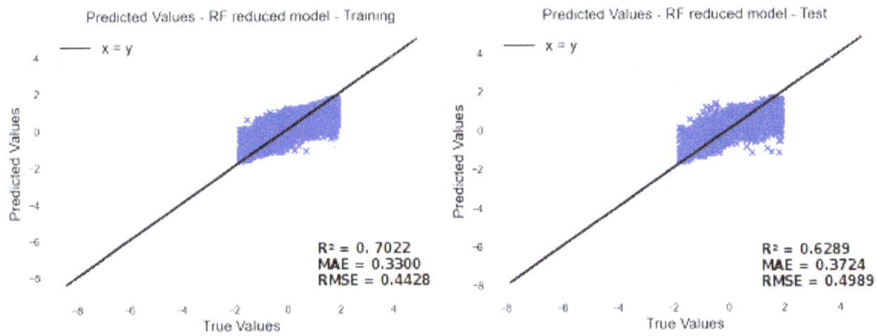

Figure 13. Prediction of the WATER_NH_3 sensor with reduced dataset of real SWTU using RF with training (**left**) and test data (**right**).

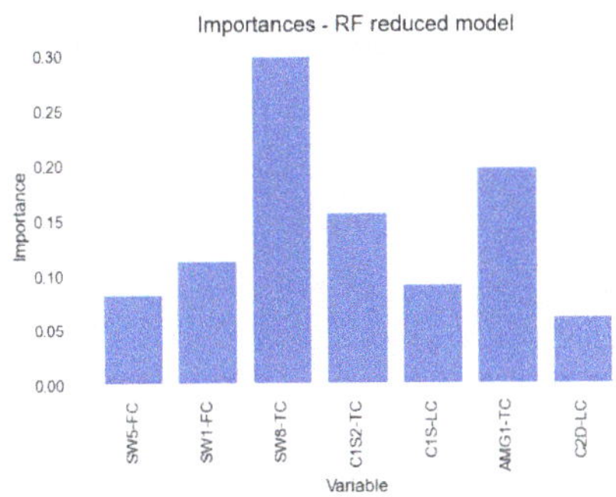

Figure 14. Variable importance for the sensor based on random forest model obtained with reduced data from real SWTU.

4. Discussion

4.1. Analysis of the Process

Due to its detected relevance, the influence of the temperature of the second stage of Column 1 on the separation process of contaminants NH_3 and H_2S was evaluated during the simulation at specific points. The total mass flow rate of the acid gas stream (ACG1) was verified to calculate the recovery of H_2S in the acid gas. The information obtained from this simulation can be found in Table 10.

Table 10. Results of the exploratory analysis on the temperature variable of the second stage of Column 1.

Time (h)	Acid Gas H_2S Mass Fraction	Acid Gas NH_3 Mass Fraction	Temperature in the 2nd Stage of C1 (°C)	Acid Gas H_2S Recovery (%)	Variation between the Recovery Values (%)
2.00	0.9476	7.67×10^{-5}	120.30	81.25	-
7.30	0.9461	8.19×10^{-5}	123.32	89.54	9.25
15.00	0.9445	8.79×10^{-5}	126.65	91.10	1.71
30.00	0.9395	1.10×10^{-4}	132.94	92.68	1.70
45.00	0.9360	1.28×10^{-4}	136.21	93.18	0.54
60.00	0.9277	1.82×10^{-4}	141.54	93.83	0.70
70.00	0.9270	1.87×10^{-4}	141.90	93.85	0.02

Through the dynamic data and information from Table 10, it was possible to prepare Figure 15. The results obtained highlighted that the H_2S removal efficiency increases with the rise in the temperature of the second stage of Column 1, particularly until achieving approximately 90% recovery, a value required by Brazilian environmental legislation. Beyond this point, it is noted that the continuous increase in column temperature results in small increments in H_2S recovery variation. On the other hand, a more pronounced vaporization of NH_3 begins, leading to an acid gas with a higher ammonia content. Elevated levels of NH_3 in the acid gas sent to the SRU can result in deposits of ammonium salts in the cold spots of the unit and an increase in NOx emissions, which is environmentally undesirable.

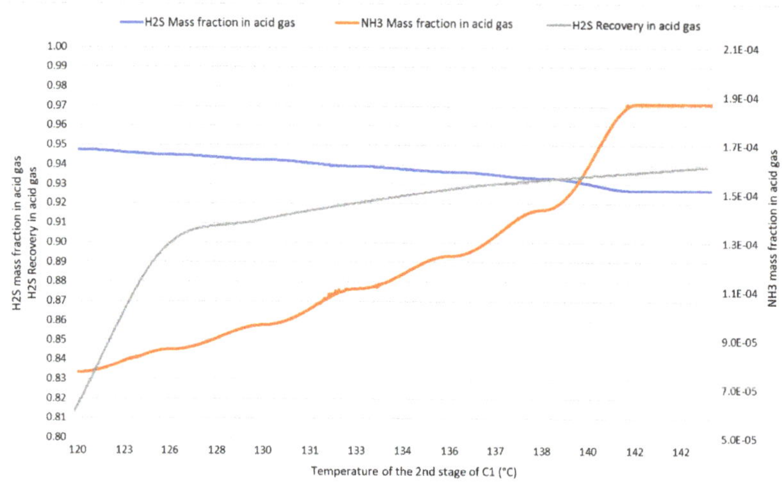

Figure 15. Relationship between H_2S recovery in the acid gas and temperature of the 2nd stage of Column 1.

Thus, the need for effective control over the first column thermal load is emphasized to prevent H_2S recovery efficiency overspecification and unnecessary energy expenditure in the reboiler. This analysis highlights the potential optimization gain of a process with soft sensors to analyze compositions of effluents from an SWTU.

4.2. Discussion on the Comparison of AI Algorithms

Based on the results comparison, it became clear that GB is an equivalent method to RF, and SVM proves to be less accurate in modeling the specific database. Therefore, the RF models were selected, and all subsequent analyses of virtual sensors were conducted based on this algorithm. This method also proved to be the most effective choice for the dataset of [1], achieving an accuracy of 95.2% after variable reduction in the fault detection and diagnosis (FDD) study of the same SWTU.

4.3. Discussion on the Importance of Variables

The temperature of the second stage in Column 1 (C1S2) was identified as the most significant in five out of the six virtual sensors. This result emphasizes that this is a critical process variable, which should be constantly monitored, as it affects both the recovery of H_2S in the acid gas from the first column and, consequently, the recovery of NH_3 in the ammoniacal gas from the second column. Additionally, it corroborates with the results found in Section 3.1, where the findings indicated the need for effective control over the thermal load of the first column to prevent overspecification of the H_2S recovery efficiency. This, besides causing unnecessary energy expenditure in the reboiler, can lead to operational issues in the SRU due to an increase in acid gas NH_3 content.

The controller C1S2-TC acts on the thermal load of C1's reboiler, which is the manipulated variable, to regulate the temperature of the C1's second separation stage. Consequently, it is consistent that the SW5 mass flow (i.e., the stream entering C1's first stage) and SW8 temperature (i.e., the temperature of stream entering C1's second stage) are considered important. Disturbances in the mass flow or temperature of streams feeding the first column will directly affect the thermal load of C1's reboiler, as it will work to maintain the thermodynamic equilibrium within the column.

Evaluating the effluents from Column 2 (i.e., ammoniacal gas and treated water), a significant influence of temperature is also observed, as indicated by the temperatures of ammoniacal gas (AMG) and C1's second stage (C1S2). Additionally, the pressure of ammoniacal gas (AMG) emerges as an important variable in both AMON_G_H_2S and WATER_NH_3. This is due to the need to maintain a lower pressure in Column 2 than in Column 1, to facilitate the selective removal of NH_3 in the ammoniacal gas, resulting in lower levels of NH_3 in the treated water. Deviations from this pressure balance could potentially increase both the H_2S fraction in the ammoniacal gas and the NH_3 fraction in the treated water, which are undesirable for the process.

The bottom level of Column 1 (C1S) emerges as an essential variable for the AMON_G_NH_3 sensor. The associated controller, C1S-LC, directly modulates the opening percentage of valve V6, regulating the flow of pre-treated water entering Column 2. The mass within the column significantly impacts the thermal load needed to maintain thermodynamic separation equilibrium. This factor is essential to achieve the maximum possible concentration of NH_3 in the ammoniacal gas, representing the primary objective of Column 2 operation.

Finally, it is important to highlight that, among the 13 variables analyzed, only 6 were considered important for five out of the six virtual sensors developed. This observation suggests that modeling for these sensors in different SWTUs can be simplified by using only the six essential variables as main inputs, aiming to optimize resources and seek improvements in the achieved results. However, this does not apply to the WATER_NH_3 sensor.

5. Conclusions

The first contribution of this study involves the modifying a phenomenological model of a two-column SWTU based on [1]. New simulated data were then generated for sensor development, expanding the Aspen Plus Dynamics® V10 model with six additional controllers for acid gas, ammoniacal gas, and treated water effluents. This resulted in a new database comprising 30 simulations, over 77 thousand samples, set to be publicly shared as a benchmark for virtual sensor development studies.

The data generation involved applying disturbances to five input variables: mass flow of SW1 and SW5, temperature of SW1, and mass fractions of H_2S and NH_3 in SW1. A key finding using the phenomenological investigation indicates that the operational variable most influencing the H_2S fraction in the acid gas is the temperature of the second stage of Column 1. It was also observed that increasing this temperature improves the recovery of H_2S in the acid gas. However, additional increments in the temperature of Column 1, after the system achieves 90% recovery, as required by Brazilian environmental legislation, result in minimal gains in H_2S removal efficiency and pronounced volatilization of NH_3 in the acid gas, which may lead to operational issues in the SRU. This highlights the importance of precise control of thermal load in the first column to optimize the process and the possibility of using virtual sensors to analyze effluents from an SWTU and may be characterized as the second contribution of this work.

A third contribution of this manuscript is the comparative analysis of the RF, GB, and SVM algorithms, which revealed that the first two are equivalent, exhibiting very similar metrics, while the SVM method with RBF kernel showed the worst performance. Thus, RF was chosen as the standard AI methodology for the development of virtual sensors.

This leads to the core advancement of this research, which is the development of six soft sensors (ACID_G_H_2S, ACID_G_NH_3, AMON_G_H_2S, AMON_G_NH_3, WATER_H_2S, and WATER_NH_3). All sensors exhibited a coefficient of determination (R^2) greater than

0.87 and RMSE less than 0.41 before reducing the input variables. The study of variable importance showed that the sensors continued to present good metrics after the exclusion of the input variables with a contribution of less than 2%, proving useful for improving computational processing time without compromising model performance. The only exception was the WATER_NH_3 sensor, which revealed the need for the original set of input variables to maintain result reliability. The overall analysis of the variable importance highlighted the significant influence of temperature on the process of removing H_2S and NH_3 contaminants from sour waters. The results indicated that all temperature variables were relevant for controlling the thermodynamic separation process of these compounds, with the temperature of the second stage of Column 1 being particularly important.

The fifth contribution of this work is particularly challenging as it validated the approach through its application to a real-world process. In this case, the application referred to an industrial unit where there was available data for the NH_3 fraction in treated water. The developed algorithms were applied to the actual unit data, yielding satisfactory results from the practical point of view. This is noteworthy, considering that the data originated from process historians, thus being subject to treatments (such as compression and exception handling) that impact their intrinsic variability, making modeling more challenging.

The soft sensors developed here serve for multiple purposes, including monitoring, control, and 'what-if' analysis. The approach adopted considered the synergistic use of a phenomenological and AI-based models. The simulation study provided valuable physical insights, guiding the selection of the most appropriate input variables for the empirical simplified models. As a result, the most significant advancement achieved here is the facilitation of soft sensors development for any SWTU with the presented configuration.

Author Contributions: Conceptualization, P.A.M. and M.B.d.S.J.; Data curation, D.G.Q., F.D.B.R. and J.d.N.P.N.; Formal analysis, J.d.N.P.N., P.A.M. and M.B.d.S.J.; Investigation, D.G.Q., F.D.B.R. and J.d.N.P.N.; Methodology, J.d.N.P.N., P.A.M. and M.B.d.S.J.; Resources, J.d.N.P.N. and M.B.d.S.J.; Software, D.G.Q., F.D.B.R. and J.d.N.P.N.; Supervision, P.A.M. and M.B.d.S.J.; Validation, F.D.B.R.; Writing—original draft, D.G.Q. and F.D.B.R.; Writing—review and editing, J.d.N.P.N., P.A.M. and M.B.d.S.J. All authors have read and agreed to the published version of the manuscript.

Funding: Professor Maurício B. de Souza, Jr. is grateful to financial support from CNPq (Grant No. 311153/2021-6) and Fundação Carlos Chagas Filho de Amparo à Pesquisa do Estado do Rio de Janeiro (FAPERJ) (Grant No. E-26/201.148/2022).

Data Availability Statement: The following supporting information can be downloaded at: https://github.com/danigradin/UTAA_Soft_Sensors, Database for UTAA Soft Sensors, Reference [25].

Conflicts of Interest: Author Francisco Davi Belo Rodrigues was employed by the company Petróleo Brasileiro S.A. Refinaria Duque de Caxias. The remaining authors declare that the research was conducted in the absence of any commercial or financial relationships that could be construed as a potential conflict of interest.

References

1. Nogueira, J.N.P.; Melo, P.A.; Souza, M.B., Jr. Faulty scenarios in sour water treatment units: Simulation and AI-based diagnosis. *Process Saf. Environ. Prot.* **2022**, *165*, 716–727. [CrossRef]
2. Ho, C.-D.; Chen, Y.-H.; Chang, C.M.; Chang, H. Evaluation of Process Control Schemes for Sour Water Strippers in Petroleum Refining. *Processes* **2021**, *9*, 363. [CrossRef]
3. Soares, A.d.F.; Penteado, E.D.; Diniz, A.A.R.; Komesu, A. Influence of operational parameters in sour water stripping process in effluents treatment. *J. Water Process Eng.* **2021**, *41*, 102012. [CrossRef]
4. Brazil, Ministry of the Environment, N.E.C., Resolution Number 382 of December 26, 2006. Official Gazette, No. 01 of January 2, Section 1. 2007. Available online: https://conama.mma.gov.br/images/conteudo/CONAMA-ingles.pdf (accessed on 2 August 2024).
5. Quinlan, M.; Hati, A. Processing NH3 Acid Gas in Sulphur Recovery Unit. In Proceedings of the 2010 Laurance Reid Gas Conditioning Conference, Norman, OK, USA, 21–24 February 2010; pp. 45–55.

6. CONAMA (National Environmental Council). RESOLUTION No. 436, OF DECEMBER 22. 2011. Available online: https://www.udop.com.br/download/legislacao/meio/institucional_site_juridico/Conama_Resolucao%20436.pdf (accessed on 20 February 2023). (In Portuguese).
7. EAP (United States Environment Protection Agency). What is Acid Rain? 2021. Available online: https://www.epa.gov/acidrain/what-acid-rain (accessed on 20 February 2023).
8. Colbeck, I.; Mackenzie, A.R. *Air Pollution by Photochemical Oxidants*; Elsevier: Amsterdam, The Netherlands; New York, NY, USA, 1994.
9. Kadlec, P.; Gabrys, B.; Strandt, S. Data-Driven Soft Sensors in the Process Industry. *Comput. Chem. Eng.* **2009**, *33*, 795–814. [CrossRef]
10. Barros, D.J.S. Investigation of the Effect of Process Variables on H2S Removal Efficiency in a Two-Stage Sour Water Treatment Unit. Master's Thesis, Programa de Pós-Graduação em Engenharia Química—PPGEQ UFPR, Curitiba, Brazil, 2016. (In Portuguese)
11. Joucowski, J.; Ndiaye, P.M.; Corazza, M.L.; Lenzi, M.K. Inferring Light-cycle-oil Stream Properties Using Soft Sensors. *Chem. Biochem. Eng. Q.* **2013**, *27*, 289–296.
12. Fortuna, L.; Graziani, S.; Rizzo, A.; Xibilia, M.G.; OTHERS. *Soft Sensors for Monitoring and Control of Industrial Processes*; Springer: Berlin/Heidelberg, Germany, 2007; Volume 22.
13. Lin, B.; Recke, B.; Knudsen, J.K.; Jørgensen, S.B. A systematic approach for soft sensor development. *Comput. Chem. Eng.* **2007**, *31*, 419–425. [CrossRef]
14. Zhou, Z.H. *Ensemble Methods: Foundations and Algorithms*; CRC Press, Taylor & Francis Group, LLC: Washington DC, USA, 2012.
15. Breiman, L.; Friedman, J.; Olshen, R.A.; Stone, C.J. *Classification and Regression Trees*; Taylor & Francis: Washington DC, USA, 1984.
16. Breiman, L. Random Forests. *Mach. Learn.* **2001**, *45*, 5–32. [CrossRef]
17. Travalis, T.B.; Ince, H. Support vector machine for regression and applications to financial forecasting. In Proceedings of the IEEE-INNS-ENNS International Joint Conference on Neural Networks. IJCNN 2000. Neural Computing: New Challenges and Perspectives for the New Millennium, Como, Italy, 27 July 2000; Volume 6, pp. 348–353.
18. Ivanciuc, O. Applications of Support Vector Machines in Chemistry. In *Reviews in Computational Chemistry*; Chapter 6; John Wiley & Sons, Ltd.: Hoboken, NJ, USA, 2007; pp. 291–400.
19. Queiroz, L.H.; Santos, F.P.; Oliveira, J.P.; Souza, M.B. Physics-Informed deep learning to predict flow fields in cyclone separators. *Digit. Chem. Eng.* **2021**, *1*, 100002. [CrossRef]
20. Offermans, T.; Szymanska, E.; Souza, F.A.A.; Jansen, J.J. Process expert knowledge is essential in creating value from data-driven industrial soft sensors. *Comput. Chem. Eng.* **2024**, *83*, 108602. [CrossRef]
21. Bagheri, M.; Mirbagheri, S.A.; Ehteshami, M.; Bagheri, Z. Modeling of a sequencing batch reactor treating municipal wastewater using multi-layer perceptron and radial basis function artificial neural networks. *Process Saf. Environ. Prot.* **2015**, *93*, 111–123. [CrossRef]
22. Barros, D.J.S.; Barros, E.S.; Zanoelo, E.F. Soft-sensor models to estimate the efficiency of H2S removal from an oil refinery stream of nonphenolic sour water. *Chem. Eng. Commun.* **2018**, *205*, 1050–1059. [CrossRef]
23. Graziani, S.; Xibilia, M.G. A deep learning based soft sensor for a sour water stripping plant. In Proceedings of the IEEE International Instrumentation and Measurement Technology Conference (I2MTC), Turin, Italy, 22–25 May 2017.
24. Nogueira, J.N.P. Fault Detection and Diagnosis of a Two-Column Sour Water Treatment Unit Based on Artificial Intelligence Algorithms. Master's Thesis, EPQB—Programa de Pós-Graduação em Engenharia de Processos Químicos e Bioquímicos da EQ/UFRJ, Rio de Janeiro, Brazil, 2021.
25. Queiroz, D.G. A Database for UTAA Soft Sensors. 2023. Available online: https://github.com/danigradin/UTAA_Soft_Sensors (accessed on 1 May 2024).
26. Li, A.; Wiener, M. Classification and Regression by Random forest. *R News* **2002**, *2*, 18–22.
27. Belgiu, M.; Dragut, L. Random Forest in remote sensing: A review of applications and future directions. *ISPRS J. Photogramm. Remote Sens.* **2016**, *114*, 24–31. [CrossRef]
28. Speiser, J.L.; Miller, M.E.; Tooze, J.; Ip, E. A comparison of random forest variable selection methods for classification prediction modeling. *Expert Syst. Appl.* **2019**, *134*, 93–101. [CrossRef] [PubMed]
29. Queiroz, D.G. Development of Virtual Sensors for Prediction the Composition of Effluents from a Sour Water Treatment Unit. Bachelor's Thesis, Federal University of Rio de Janeiro, Rio de Janeiro, Brazil, 2023. (In Portuguese)
30. Lima, F.A.R.D.; Moraes, M.G.F.; Secchi, A.R.; Souza, M.B., Jr. Development of a recurrent neural networks-based NMPC for controlling the concentration of a crystallization process. *Digit. Chem. Eng.* **2022**, *5*, 100052. [CrossRef]
31. Chong, I.; Jun, C. Performance of some variable selection methods when multicollinearity is present. *Chemom. Intell. Lab. Syst.* **2005**, *78*, 103–112. [CrossRef]

processes

MDPI

Article

A PLC-Embedded Implementation of a Modified Takagi–Sugeno–Kang-Based MPC to Control a Pressure Swing Adsorption Process

Teófilo Paiva Guimarães Mendes [1,2], Ana Mafalda Ribeiro [3,4], Leizer Schnitman [2] and Idelfonso B. R. Nogueira [5,*]

1 Área de Sistemas Elétricos e Computacionais, Centro de Ciências Exatas e Tecnológicas, Universidade Federal do Recôncavo Baiano, R. Rui Barbosa, 710–Centro, Cruz das Almas 44380-000, BA, Brazil; teopgmendes@ufrb.edu.br
2 Programa de Pós-Graduação em Mecatrônica, Escola Politécnica (Polytechnic School), Universidade Federal da Bahia, R. Prof. Aristides Novis, 2–Federação, Salvador 40210-630, BA, Brazil; leizer@ufba.br
3 LSRE-LCM—Laboratory of Separation and Reaction Engineering-Laboratory of Catalysis and Materials, Faculty of Engineering, University of Porto, Rua Dr. Roberto Frias, 4200-465 Porto, Portugal; apeixoto@fe.up.pt
4 ALiCE—Associate Laboratory in Chemical Engineering, Faculty of Engineering, University of Porto, Rua Dr. Roberto Frias, 4200-465 Porto, Portugal
5 Department of Chemical Engineering, Norwegian University of Science and Technology, 7491 Trondheim, Norway
* Correspondence: idelfonso.b.d.r.nogueira@ntnu.no; Tel.: +47-73-59-23-99

Abstract: The paper presents a case study that applies a model predictive control (MPC) approach in a Micro850 programmable logic controller (PLC) to a laboratory pressure swing adsorption (PSA) process used for separating gas mixtures of CO_2 and CH_4. PLC is an industrial hardware characterized by its robustness to hazardous environments and limited computational capacities, which poses computational challenges for MPC implementation. This paper's main contribution is the application of the modified Takagi–Sugeno–Kang-based MPC (MTSK-MPC) algorithm to this PSA unit, which provides features to investigate and implement feasible MPC designs in PLCs. The investigation consists of a sensitivity analysis of how some design parameters influence the PLC memory and the MPC implementation and a comparative evaluation of the computational processing from different MPC algorithms and simulations. The comparison comprises software-in-the-loop simulations with three algorithms in the PC: an implicit MPC, an explicit MPC, and the MTSK-MPC. Additionally, it includes a hardware-in-the-loop simulation with the implemented MTSK-MPC in Micro850. The results show that the MPC algorithms achieve close performance, tracking setpoint changes and rejecting output disturbances, with the MTSK-MPC presenting the lower processing time among the MPCs in the PC. The study concludes that the implementation of MTSK-MPC in the Micro850 is feasible.

Keywords: pressure swing adsorption process; model predictive control; programmable logic controller; Takagi–Sugeno–Kang model

1. Introduction

Pressure Swing Adsorption (PSA) is a method used for complex gas-phase separations since the late 1950s. The PSA approach has significantly expanded in scale, versatility, and complexity. Currently, PSA-type processes are employed in various settings, ranging from compact laboratory units used in medical applications, which capture atmospheric air and separate nitrogen (N_2) and oxygen (O_2) using one or two adsorbent beds [1], to large-scale industrial refinery units that purify hydrogen (H_2) by using 16 adsorbent beds [2].

The PSA-type processes are based on the properties of solids to preferentially fix some molecules of a gas mixture on their surface. The installations of these processes utilize fixed-bed adsorbers, using one or more separation columns, which change process steps periodically. Each cycle contains at least two steps: high-pressure adsorption and low-pressure desorption. Due to the cyclic nature of their operation, PSA processes do not reach a fixed steady-state condition. In contrast to other dynamical systems, these processes go to a cyclic steady state (CSS) when each step time remains constant, given the feed load, temperature, and pressure profiles of each step. That is, CSS represents a stable operational regime in which all differential and algebraic variables have the same values at each period during the cycles. Besides the peculiar cyclical behavior, PSA processes are multivariable and nonlinear within each step, with strong interactions between variables. In addition, the optimal operation point must simultaneously meet conflicting objectives, such as productivity/energy consumption and product purity/recovery, as well as operational and safety constraints. Therefore, although this process is a phenomenon well understood and mathematically modeled, it still presents control problems to investigate [3–8].

Several authors have proposed approaches to improve the operability of PSA-type processes through control methods. These approaches range from classical control, mainly utilizing proportional-integral-derivative (PID) controllers, to modern control theory, such as linear quadratic regulator (LQR) and model predictive controllers (MPC).

The pioneering investigation to control the PSA process explored use of a PID controller to handle single-input and single-output (SISO) cases and PID control with feedforward actions to address multiple-input and single-output (MISO) cases with two inputs and one output [1]. Among the studies with PID controllers and PSA processes, it is important to mention those that combined other methods, such as the utilization of $H\infty$ criteria to address the PSA multi-variable nature [9] and the application of Hammerstein models [9–12] and fuzzy PID tuning [13] to deal with the PSA nonlinearities. Nevertheless, this control approach has considerable drawbacks, as it requires significant study time and is change-sensitive to modeling parameters and operating points.

In the control context, the MPC-based approaches are emerging as promising solutions for the challenging characteristics of PSA processes [8]. Comparative works in SISO cases have shown that MPC demonstrates superior tracking performance to linear PID controllers relying on feedback action [3,4,13,14]. In the same way, the results from comparative study in multiple-input and multiple-output (MIMO) cases between PID and LQR controls and PID and MPC controls have shown that the modern control approaches are more effective to handle the multivariable nature of the PSA process [11,15,16]. As with the classical control approach, authors have combined MPC with other methods to deal with PSA's nonlinearities using multiple models and robust MPC [6,8,15] and using MPC with deep neural networks [7,16]. Some works in the literature are notable regarding dimensions of MPC applied to MIMO cases of PSA systems, with up to four manipulated and two controlled variables [1,17–19]. Other reports stand out by presenting MPC control laws with guaranteed stability and compliance with economic criteria [6–8]. It is also worth mentioning the use of MPC in other processes with challenging characteristics, such as those found in PSA processes. For instance, MPC has been applied in SMB processes [20,21] which also operate in a CSS, and in polymerization processes which display highly nonlinear behavior [22].

In the implementation context of automation and control systems, the early MPC applications involved a hierarchical control structure in high-scale process industries. This structure consisted of a low layer with regulatory control, implemented on industrial-embedded hardware, and a high layer with modern control strategies, operating on non-industrial hardware with high computational capacity. A current research trend seeks to implement MPC in embedded hardware, integrating the control into a single layer. In specific environments, such as industries and hospitals, the automation hardware must prioritize robustness to installation characteristics over computational resources regarding hardware requirements. In such cases, programmable logic controllers (PLCs) are the most suitable hardware for these applications [23–25].

The classical control approach presents some implementation advantages compared to modern control strategies. The PID controller is a standard programming resource for PLCs, as it requires low memory for its parameters and is also computationally inexpensive. In addition, it is simple to consider PIDs in the PSA process control design and optimization stages [5,26].

On the other hand, only one distinguished work has successfully implemented an MPC approach in embedded hardware for small-scale PSA units [4]. However, the hardware used was a Raspberry Pi-3, which does not meet industrial installation requirements and is more computationally powerful than common PLCs. All the other works cited on PSA control using the MPC approach were investigations based on virtual simulated environments. The usual experimental setup comprises simulating the PSA process and the controller in the same software or simulating them in separate programs, communicating in a software-in-the-loop (SIL) integration. In these cases, the MPC controllers run on personal computers or servers with greater computing capacity than most industrial hardware. Therefore, the control literature applied to PSA still presents a gap in MPC applications implemented in PLCs, which is challenging due to this hardware's limited memory and processing capacity.

It is well-known that MPC is computationally demanding, requiring a careful selection of algorithms regarding computational efficiency. This choice concerns evaluating implicit MPC (iMPC) and explicit MPC (eMPC) approaches [27,28]. There are two main criteria to consider for this decision, among others. The first criterion is that general purpose PLCs lack the advanced processing and programming capabilities required to implement implicit-type MPC algorithms. The second criterion is that while explicit MPC algorithms do not require such resources, they tend to exponentially increase memory requirements with longer prediction horizons and additional constraints [23,24]. While the advantages and disadvantages of choosing either PID or MPC to control the PSA process are evident, this is not the case regarding the selection between iMPC and eMPC approaches. In the previously mentioned MPC reports, the authors' treatment of these alternatives is cursory, relying predominantly on qualitative arguments for justification. Consequently, there is another gap in the control literature applied to PSA concerning comparative analyses between implicit and explicit MPC approaches.

This paper presents a case study that addresses these gaps in the current knowledge of PSA control. The case study focuses on a laboratory PSA process, the Micro850 PLC, and the Modified Takagi–Sugeno–Kang-based MPC (MTSK-MPC) [29]. This PSA is a compact unit that separates gas mixtures of carbon dioxide (CO_2) and methane (CH_4). This PLC model is a general purpose industrial hardware used in modular process skid packages and is a representative example of an automation system of PSA units. The MTSK-MPC is a recently introduced MPC algorithm, well-suited for PLC implementation as it requires just the programming languages and function blocks from the IEC-61131-3 automation standard [30] and includes memory-reducing features through fuzzy logic rules. Therefore, this paper's main contribution is the application of the MTSK-MPC algorithm to a PSA process, which provides the means to investigate and implement computationally feasible MPC designs in PLCs.

In this sense, the investigation consists of two steps. One step comprises a sensitivity analysis of how the control and operation constraints parameters influence the PLC memory of the PSA unit under study. The aim is to identify which MTSK-MPC designs are embeddable in the PLC. The other step evaluates the computational processing of different MPC algorithms through a comparative investigation of SIL simulations. The objective is to ensure the timing requirement of the PSA cycle and identify the MPC algorithm with faster processing. The comparison includes three algorithms: an implicit MPC with interior point optimization, an explicit MPC with multiparametric optimization, and the MTSK-MPC.

The implementation is completed by programming an embedded MTSK-MPC function block into the Micro850 and performing a closed-loop validation test. The experimental setup of this test comprises a hardware-in-the-loop (HIL) simulation, where the gPROMS software, version 7, simulates the process on a PC, the PLC executes the MTSK-MPC algorithm, and these applications communicate via MATLAB and the Modbus protocol. This test compares the MTSK-MPC of the HIL simulation with the MPC of the SIL simulations. In addition, it assesses the capacity of the MTSK-MPC to accurately track setpoint changes and reject some input disturbances, as well as the PLC scan execution time.

The subsequent sections of this paper are structured as follows: Section 2 describes the PSA process considered in the experiments. Section 3 presents the MPC formulation applied to control the PSA, executed by three schemes: implicit MPC, explicit MPC, and the modified Takagi–Sugeno–Kang model-based MPC. Section 4 shows the results and discussion concerning the computational evaluation and experiments applied to the small-scale PSA study case. Finally, Section 5 presents the conclusions.

2. Description of the PSA Process

The PSA under investigation comprises a unit installed in the Laboratory of Separation and Reaction Processes (LSRE) at the Faculty of Engineering of the University of Porto (FEUP). This unit can run in single-column or two-column configurations, alternating steps of each column automatically. Figure 1 shows the operating screen with a simplified process diagram of the process used in the simulations.

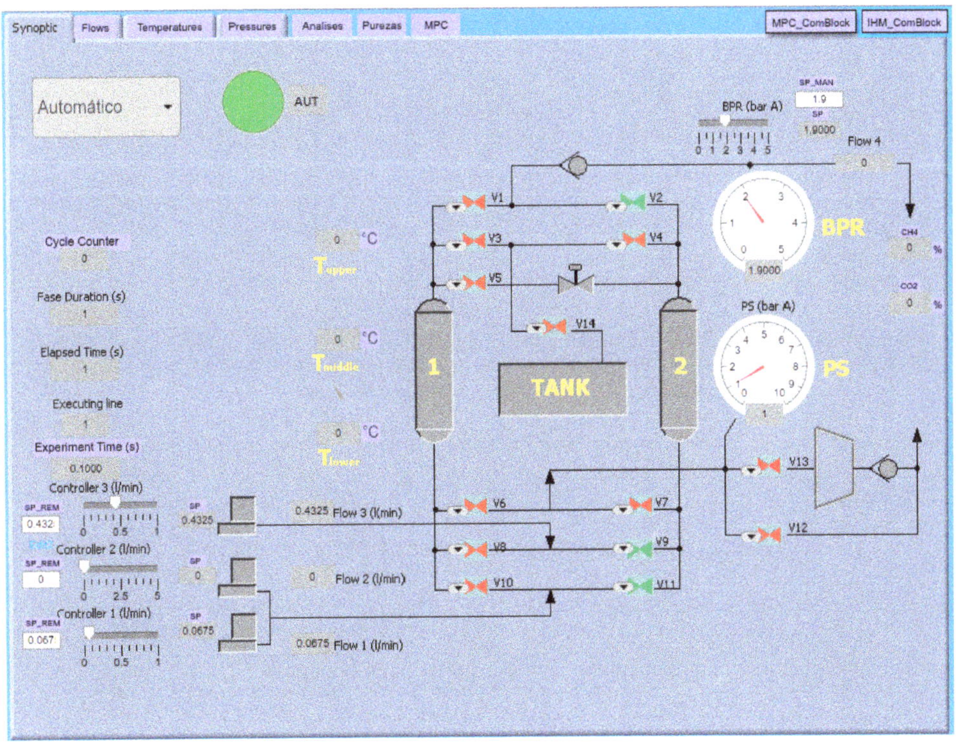

Figure 1. Operation screen with simplified process diagram of the PSA simulated (adapted from the LSRE laboratory plant).

The PSA process evaluated in this work consists of separating a mixture of CO_2 and CH_4. Mixtures with this composition are found in different process scales and column configurations, for example, in biogas production plants [31], syngas purification units [32], and Brazilian pre-salt oil exploration wells [33]. The CO_2 gas inhibits the direct use of mixtures with this composition in combustion engines. Additionally, the CO_2 contributes to the greenhouse effect when released into the atmosphere, requiring proper treatment. Therefore, the outputs of this separation process are a stream with CH_4, which could be used as fuel, and a stream with CO_2, which is captured and considered an impurity.

A phenomenological model of the PSA plant under study was described in the literature by [34]. The separation studies were experimentally validated by [32]. In the work referred to, the authors carried out breakthrough experiments and estimated the kinetic and equilibrium parameters. Furthermore, they presented concentration profiles over time showing the CO_2/CH_4 interaction with the BPL adsorbent. The simulations in the present work consider a column packed bed with commercial activated carbon (BPL), 10 cm long, 2.1 cm in diameter, and with 13.8 g of adsorbent. Table 1 shows the fitting parameters of the dual-site Langmuir model for the CH_4 and CO_2 adsorption equilibrium on the BPL-activated carbon. These references provide details on the nominal process conditions, pressure adsorption curves, and other relevant process data used in this paper. The reader can gain a more comprehensive understanding of the process by consulting the cited works.

Table 1. Fitting parameters of the dual-site Langmuir model for CH_4 and CO_2 adsorption equilibrium on the BPL activated carbon.

Species	$q_{m,1}$ (mol × kg^{-1})	$q_{m,2}$ (mol × kg^{-1})	$b_{0,1} \times 10^4$ (bar^{-1})	$b_{0,1} \times 10^4$ (bar^{-1})	$-\Delta H_1$ (J × mol^{-1})	$-\Delta H_2$ (J × mol^{-1})
CH_4	4.72	0.40	1.76	2.98	16,310	20,595
CO_2	7.12	0.32	0.30	0.80	22,199	28,433

The PSA unit of the LSRE includes a set of solenoid valves that operate in a coordinated manner, following a preconfigured program in the automation system. These valves open and close according to basic process control to determine the flow direction through the columns depending on the distinct step of the PSA process cycle [10,17,18,35]. In the PSA process studied, the column goes through four cyclic steps: 1—pressurization with feed (adsorption), 2—rinse, 3—depressurization (desorption), and 4—purging.

The cycle initiates with the adsorbent column pressurization to super-atmospheric conditions using a feed charge containing mixed gases. In this first step, the load continues to be supplied through the column at constant pressure, causing a concentrated gas stream of the desired product to leave the column and the other gas adsorption by the bed. Following this, the rinse step introduces a purging gas into the adsorption bed to displace and remove any remaining impurities. In the sequence, the depressurization step consists of supplying gas to the column in the reversed direction, reducing the bed pressure, and starting to cause the gas adsorbed in the column to be desorbed and expelled into the atmosphere or sent to waste storage. In the last step, during the purging, a reverse stream of part of the produced gas quickly lowers the column pressure and cleans the remaining gases previously adsorbed in the bed.

Table 2 shows the scheduled discrete positions per solenoid valve, and Table 3 shows the programmed continuous values in each step of the PSA cycle. The total inlet flow in volume per minute changes for each PSA step according to Table 3, maintaining a fixed molar fraction of 0.865 CH_4 and 0.135 CO_2 when it is non-zero. Note that the single-column configuration does not need all the available instruments in the LSRE laboratory plant, and some valves remain unchanged in such cases.

Table 2. Scheduling of valve discrete position in the PSA cycle steps.

Step	V1	V2	V3	V4	V5	V6	V7	V8	V9	V10	V11	V12	V13
1	0	1	0	0	0	0	0	0	1	0	1	0	0
2	0	1	0	0	0	0	0	0	0	0	1	0	0
3	0	0	0	0	0	0	1	0	0	0	0	1	0
4	0	0	1	1	0	0	1	1	0	0	0	1	0

Table 3. Scheduling of continuous values in the PSA cycle steps.

Step	Inlet Flow (L/min)	Back Pressure (Pascal)
1	0.5	1.9×10^5
2	0.1	1.9×10^5
3	-	0.3×10^5
4	0.2	0.3×10^5

As the purpose of this PSA is to produce a certain amount of separated gas with a determined concentration, the controlled variables selected for the MPC were the purity of CH_4 gas produced and the CO_2 recovered. The total cycle time and the individual duration of its steps considerably affect the concentration of the separated products. These times determine the changes and concentration limits of the products at the end of the adsorber beds. Therefore, the step times considered as the manipulated variables of the MPC were the feeding time and the rinse time [3,4,9,10,13,17–19,34].

3. The Compared Model-Based Predictive Control Approaches

The primary purpose of this work is to embed an MPC algorithm in the PLC of a PSA unit. This hardware has notable limitations in memory and programming capabilities in contrast to standard PCs. Therefore, this case study focuses on the following linear MPC approaches.

Notations: A vector $\boldsymbol{x}$ composed by n_x real numbers is denoted by $\boldsymbol{x} \in \mathbb{R}^{n_x}$. The component j of the vector $\boldsymbol{x}$ is denoted by x_j and the vector value at time instant k by $\boldsymbol{x}(k)$. The vector $\boldsymbol{X} = [\boldsymbol{x}(1), \boldsymbol{x}(2), \ldots, \boldsymbol{x}(N)]$ is a column vector formed by the concatenation of vectors $\boldsymbol{x}(1)$ to $\boldsymbol{x}(N)$. Given a matrix $\boldsymbol{M} \in \mathbb{R}^{n \times m}$, $\boldsymbol{M}^T$ denotes its transpose. For a vector $\boldsymbol{x}$ and a quadratic matrix $\boldsymbol{Q}$, the weighted Euclidean norm is $\|\boldsymbol{x}\|_{\boldsymbol{Q}}^2 \triangleq \boldsymbol{x}^T \boldsymbol{Q} \boldsymbol{x}$. Given a table of **IF-THEN** rules, the vector $\boldsymbol{r}$ and the matrix $\boldsymbol{M}$ of the rule antecedent i are, respectively, denoted by $\boldsymbol{r}^i$ and $\boldsymbol{M}^i$, and the consequence affine function is $\boldsymbol{f}^i(\boldsymbol{x})$. Given two vectors $\boldsymbol{v}$ and $\boldsymbol{r}$, $\boldsymbol{v} \leq (\geq) \boldsymbol{r}$ denotes component-wise inequalities. Given a dynamical system, the subscript ss denotes its steady state x_{ss}, and the superscripts st, un, and in denote, respectively, the stable, $\boldsymbol{x}^{st}$, unstable, $\boldsymbol{x}^{un}$, and integrating, $\boldsymbol{x}^{in}$, states of this system. The system's original states, without the model on the incremental form of inputs, are denoted by the vector $\boldsymbol{x}'$. The system's estimated states for time k+1 at time k are denoted by $\hat{\boldsymbol{x}}(k+1/k)$. The superscript "$*$" indicates an optimal value at the current instant time and the superscript "$\sim$" indicates an optimal value inherited from the last instant. The notation n_u means the number of inputs $\boldsymbol{u}$, n_x^{un} is the number of unstable states, and n_x^{in} is the number of integrating states. An identity matrix of dimension n is denoted by $\boldsymbol{I}_n$.

3.1. Implicit MPC Formulation

This paper follows an infinite prediction horizon MPC approach in the control problem formulation [29,35]. In this regard, consider a discrete linear time-invariant model operating around an equilibrium point, which is controllable and represented by a state-space model, as follows:

$$\boldsymbol{x}'(k+1) = \boldsymbol{A}'\boldsymbol{x}'(k) + \boldsymbol{B}'\boldsymbol{u}(k) \tag{1}$$

$$\boldsymbol{y}(k) = \boldsymbol{C}'\boldsymbol{x}'(k) \tag{2}$$

where the vectors $\boldsymbol{x}'(k) \in \mathbb{R}^{n_x}$ represent the states, $\boldsymbol{y}(k) \in \mathbb{R}^{n_y}$ the outputs, and $\boldsymbol{u}(k) \in \mathbb{R}^{n_u}$ the inputs, and $\boldsymbol{A}'$, $\boldsymbol{B}'$, and $\boldsymbol{C}'$ are matrices of appropriate dimensions.

These equations can be rewritten in the incremental form of inputs, as follows:

$$\boldsymbol{x}(k+1) = \boldsymbol{A}\boldsymbol{x}(k) + \boldsymbol{B}\Delta\boldsymbol{u}(k) \tag{3}$$

$$\boldsymbol{y}(k) = \boldsymbol{C}\boldsymbol{x}(k) \tag{4}$$

in which the extended states, $x(k)$, contain the vectors $\left[x'(k)^{T}, u(k-1)^{T}\right]^{T}$.

Furthermore, consider the objective function that centers the system states predictions, $x(k+j)$, from current time $j=0$, to infinite prediction $j=+\infty$, and the system inputs increments, $\Delta u(k+j)$, from current time $j=0$, to the control horizon minus one, $j=N-1$, into the artificial steady states, $x_{ss}(k)$. Moreover, this objective function has an element that penalizes the difference between the reference, y_r, from the output at the artificial steady state, $y_{ss}(k)$, as follows:

$$V(k)=\sum_{j=0}^{N-1}\|x(k+j)-x_{ss}(k)\|_{Q}^{2}+\|\Delta u(k+j)\|_{R}^{2}+\|x(k+N)-x_{ss}(k)\|_{\tilde{Q}}^{2}+\|y_{r}-y_{ss}(k)\|_{S_{y}}^{2} \tag{5}$$

where N is the control horizon of the inputs, $Q\in\mathbb{R}^{n_x\times n_x}$ is a matrix used to weigh the state's predictions distance from the artificial states, $R\in\mathbb{R}^{n_u\times n_u}$ is a matrix used to weigh the inputs increments, $S_y\in\mathbb{R}^{n_y\times n_y}$ is a matrix used to achieve offset-free control, and $\tilde{Q}$ is the matrix of the state terminal cost weight, given by:

$$\tilde{Q}=W^{st^{T}}\times\overline{Q}\times W^{st} \tag{6}$$

with $\overline{Q}$ calculated using the Lyapunov equation:

$$\overline{Q}=\sum_{i=0}^{+\infty}\left(J^{st^{j}}\right)^{T}\times V^{st^{T}}\times Q\times V^{st}\times\left(J^{st^{j}}\right)=V^{st^{T}}\times Q\times V^{st}+J^{st^{T}}\times\overline{Q}\times J^{st} \tag{7}$$

where $W\in\mathbb{R}^{n_x\times n_x}$ is the generalized eigenvectors matrix, $V=W^{-1}$, $J\in\mathbb{R}^{n_x\times n_x}$ is a matrix of Jordan blocks eigenvalues, and the superscript indexes st indicate eigenvalues and rows of W and V associated with the stable modes.

The system must operate within constraints such that the input states and increments, the predicted states and outputs, and the artificial states are within the following limits:

$$\mathbb{X}_b=\left\{x\in\mathbb{R}^{n_x}\middle|\begin{bmatrix}x'_{min}\\u_{min}\end{bmatrix}\le\begin{bmatrix}x'\\u\end{bmatrix}\le\begin{bmatrix}x'_{max}\\u_{max}\end{bmatrix}\right\} \tag{8}$$

$$\Delta\mathbb{U}=\{\Delta u\in\mathbb{R}^{n_u}|\Delta u_{min}\le\Delta u\le\Delta u_{max}\} \tag{9}$$

$$x(k+j)\in\mathbb{X}_b,\ j=0,\ldots,N+k_2 \tag{10}$$

$$\Delta u(k+j)\in\Delta\mathbb{U},\ j=0,\ldots,N-1 \tag{11}$$

$$x_{ss}(k)\in\mathbb{X}_{ss} \tag{12}$$

where the bounds of the extended states, x, in (8) comprise the stack of bounds of the original and input states due to the incremental form of the model used in (3).

The artificial states are defined as the system equilibrium points inside $\mathbb{X}_b$ as:

$$\mathbb{X}_{ss}=\{x_{ss}(k)\in\mathbb{X}_b|(I_{n_x}-A)\times x_{ss}(k)=0\} \tag{13}$$

Finally, consider the following relaxed terminal constraint on non-stable modes of the system to limit the infinite sum of the objective function in (5):

$$\begin{bmatrix}W^{un}\\W^{in}\end{bmatrix}\times(x(k+N)-x_{ss}(k))=0 \tag{14}$$

where the indexes *un* and *in* indicate, respectively, the rows of generalized eigenvectors matrix **W** associated with the unstable and integrating modes.

Therefore, the applied MPC formulation can be resumed as the convex quadratic optimization problem as follows:

Problem P0:

$$\min_{\Delta U(k), x_{ss}(k)} V(k) = \sum_{j=0}^{N-1} \|x(k+j) - x_{ss}(k)\|_Q^2 + \|\Delta u(k+j)\|_R^2 + \|x(k+N) - x_{ss}(k)\|_{\tilde{Q}}^2 + \|y_r - y_{ss}(k)\|_{S_y}^2 \quad (15)$$
$$s.t. : (3), (4), (8), (9), (10), (11), (13) \text{ and} (14)$$

where $x(k)$ is the state observed at k current instant and is the initial condition of **Problem P0**. The decision variables are the artificial steady states, $x_{ss}(k)$, and the vector of future input increments, $\Delta U(k)$, composed of the calculated input increments at time step k up to control horizon N, stacked as follows:

$$\Delta U(k) = \left[\Delta u(k)^T, \ldots, u(k+N-1)^T\right]^T \quad (16)$$

in which only the first element, $\Delta u(k)$, is effectively applied to the process due to the receding horizon control strategy.

The implicit MPC considered in this article concerns the solution of **Problem P0** at each sampling instant by a quadratic optimization algorithm, such as, for example, the interior point method. In this sense, **Problem P0** can be rewritten in the conventional matrix notation as follows:

$$\min_{\Delta U(k), x_{ss}(k)} V(k) = \begin{bmatrix} \Delta U(k) \\ x_{ss}(k) \end{bmatrix}^T \times H \times \begin{bmatrix} \Delta U(k) \\ x_{ss}(k) \end{bmatrix} + \begin{bmatrix} \Delta U(k) \\ x_{ss}(k) \end{bmatrix}^T \times F \times \begin{bmatrix} x(k) \\ y_r(k) \end{bmatrix} + \begin{bmatrix} x(k) \\ y_r(k) \end{bmatrix}^T \times Y \times \begin{bmatrix} x(k) \\ y_r(k) \end{bmatrix}$$
$$s.t. : L \times \begin{bmatrix} \Delta U(k) \\ x_{ss}(k) \end{bmatrix} \leq W + E \times \begin{bmatrix} x(k) \\ y_r(k) \end{bmatrix} \quad (17)$$

where matrices **H**, **F**, **Y**, **L**, **W**, and **E** have appropriate dimensions and can be obtained from system modeling and controller tuning parameters in (15).

3.2. Explicit MPC Formulation

The explicit MPC formulation is the result of applying a multiparametric quadratic optimization to solve **Problem P0** [28], which obtains a lookup table of **IF-THEN** rules with R^i rules, $i = 1, \ldots p$, as follows:

$$\begin{bmatrix} \Delta U^*(k) \\ x_{ss}^*(k) \end{bmatrix} = \begin{cases} R^1 : \begin{cases} If & x(k) \in \mathbb{K}^1 \\ Then & \begin{bmatrix} \Delta U^{*1}(k) \\ x_{ss}^{*1}(k) \end{bmatrix} = G^1 \times \begin{bmatrix} x(k) \\ y_r(k) \end{bmatrix} + g^1 \end{cases} \\ \vdots \\ R^p : \begin{cases} If & x(k) \in \mathbb{K}^p \\ Then & \begin{bmatrix} \Delta U^{*p}(k) \\ x_{ss}^{*p}(k) \end{bmatrix} = G^p \times \begin{bmatrix} x(k) \\ y_r(k) \end{bmatrix} + g^p \end{cases} \end{cases} \quad (18)$$

in which G^i and g^i are parameters of the local function from the rule i, and each local function i holds only in a critical region $\mathbb{K}^i$, defined as a polyhedral set as:

$$M^i \times \begin{bmatrix} x(k) \\ y_r(k) \end{bmatrix} \geq r^i \quad (19)$$

where both the matrix M^i and the vector r^i have r^i_{max} total rows, with each row corresponding to a boundary hyperplane of critical region $\mathbb{K}^i$.

Parameters G^i, g^i, M^i, r^i and r^i_{max} depend on H, F, Y, L, W, and E matrices from (17). For a comprehensive calculation procedure of such calculus, the reader is referred to Bemporad et al. (2002) [28].

3.3. Modified Takagi–Sugeno–Kang Explicit MPC Formulation

In the MTSK-MPC [5], the set of all boundary hyperplanes from the critical regions divides the input space of this model into a polyhedral partition of fuzzy sets. In addition, a boundary hyperplane from the critical region $\mathbb{K}^i$ in (19) defines the scheduling variable $\alpha^i_j(x(k), y_r(k))$ as follows:

$$\alpha^i_j(x(k)) = \left[m^i_j, -r^i_j\right] \times \left[x(k)^T, y_r(k)^T, 1\right]^T \tag{20}$$

where m^i_j and r^i_j are, respectively, elements from the row j of the matrix M^i and vector r^i of rule i, and $x(k)$ and $y_r(k)$ are, respectively, the state and reference from the explicit MPC solution in (18).

As a result, one can state each critical region as the intersection of polyhedral fuzzy sets, determined from the respective boundary hyperplanes, as follows:

$$\mathbb{K}^i = \bigcap\nolimits_{j=1}^{r^i_{max}} \mathbb{A}^i_j \tag{21}$$

with the polyhedral fuzzy sets $\mathbb{A}^i_j$ defined by:

$$\mathbb{A}^i_j = \left\{\alpha^i_j, \mu^i_j\left(\alpha^i_j\right) \in \mathbb{R} \middle| \alpha^i_j \geq 0, \mu^i_j\left(\alpha^i_j\right) \in [0,1]\right\} \tag{22}$$

and where α^i_j is a short notation of $\alpha^i_j(x(k), y_r(k))$ used to simplify the representation.

Using these definitions and the product as a fuzzy intersection operator [36], one can derive the critical region membership function as:

$$\beta^i\left(a^i_j\right) = \prod\nolimits_{j=1}^{r^i_{max}} \mu^i_j\left(a^i_j\right) \tag{23}$$

where $\mu^i_j\left(a^i_j\right)$ are the boundary hyperplanes fuzzy membership functions, which are Heaviside-type functions.

The MTSK-MPC computes the explicit MPC solution using a fuzzy inference system with six layers, differing from the classical Takagi–Sugeno–Kang model [37,38], which contains only five layers. Figure 2 illustrates both fuzzy inference systems, highlighting the first layer of the MTSK-MPC, where the distinction between them lies.

The first layer of this fuzzy inference system consists of batch processing all scheduling variables according to:

$$^1O(x(k), y_r(k)) = \alpha_{par} \times \begin{bmatrix} x(k) \\ y_r(k) \\ 1 \end{bmatrix} \tag{24}$$

where ${}^1O \in \mathbb{R}^{\sum_{i=1}^{p} r^i_{max}}$ is the output vector of the first layer and α_{par} consists of stacking the parameters of the boundary hyperplanes from all critical regions and rules into a matrix as follows:

$$\alpha_{par} = \begin{bmatrix} m^1_1 & r^1_1 \\ \vdots & \vdots \\ m^1_{r^1_{max}} & r^1_{r^1_{max}} \\ m^2_1 & r^2_1 \\ \vdots & \vdots \\ m^2_{r^2_{max}} & r^2_{r^2_{max}} \\ \vdots & \vdots \\ m^p_{r^p_{max}} & r^p_{r^p_{max}} \end{bmatrix} \tag{25}$$

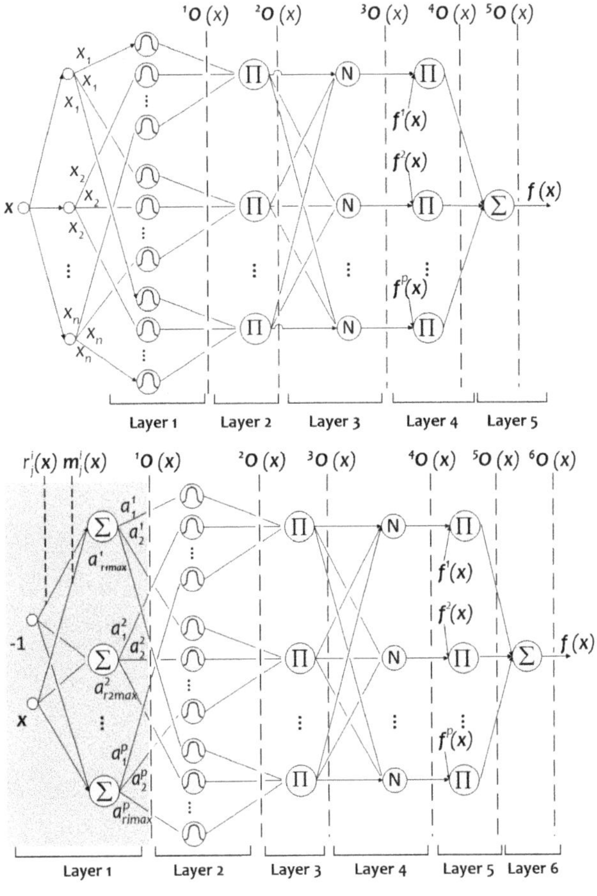

Figure 2. Fuzzy inference system of the: (**Top**) Classical TSK model and (**Bottom**) MTSK-MPC model [29].

The second layer batch evaluates the membership function of the boundary hyperplanes by the Heaviside function as:

$$ {}^{2}O(x,y_r) = 0.5 \times \left(1 + \frac{{}^{1}O(x,y_r)}{|{}^{1}O(x,y_r)|}\right) \tag{26} $$

where ${}^{2}O \in \mathbb{R}^{\sum_{i=1}^{p} r^{i}_{max}}$ is the output vector of the second layer; the time instant notation is dropped to simplify the representation.

The third layer calculates the membership functions of the critical regions using Equation (23), which corresponds to the weights of the **IF-THEN** rule, as follows:

$$ {}^{3}O(x,y_r) = \begin{bmatrix} {}^{3}O_1(x,y_r) \\ {}^{3}O_2(x,y_r) \\ \vdots \\ {}^{3}O_p(x,y_r) \end{bmatrix} = \begin{bmatrix} \prod_{n=\beta^1_{start}}^{\beta^1_{stop}} {}^{2}O_n(x,y_r) \\ \prod_{n=\beta^2_{start}}^{\beta^2_{stop}} {}^{2}O_n(x,y_r) \\ \vdots \\ \prod_{n=\beta^p_{start}}^{\beta^p_{stop}} {}^{2}O_n(x,y_r) \end{bmatrix} \tag{27} $$

where ${}^{3}O \in \mathbb{R}^{p}$ is the output vector of the third layer, ${}^{3}O_i$ is the element of the i row of this vector, ${}^{2}O_n$ is the element n of the second layer's output vector, and β^{i}_{start} and β^{i}_{stop} are indexes that correspond to the first and the last elements of the ${}^{2}O$ vector that compose the critical region i. These indexes come from the third layer's matrix of parameters, β_{par}, formed as:

$$ \beta_{par} = \begin{bmatrix} \beta^1_{start} & \beta^1_{stop} \\ \beta^2_{start} & \beta^2_{stop} \\ \vdots & \vdots \\ \beta^p_{start} & \beta^p_{stop} \end{bmatrix} = \begin{bmatrix} 1 & r^1_{max} \\ r^1_{max}+1 & r^1_{max}+r^2_{max} \\ \vdots & \vdots \\ \left(\sum_{i=1}^{p-1} r^i_{max}\right)+1 & \sum_{i=1}^{p} r^i_{max} \end{bmatrix} \tag{28} $$

In sequence, the fourth layer calculates the normalized rule weights as follows:

$$ {}^{4}O(x,y_r) = \frac{1}{\sum_{i=1}^{p} {}^{3}O_i(x,y_r)} \times {}^{3}O(x,y_r) = \begin{cases} 0 & outside\ of \mathbb{K}^i \\ 0 < and < 1 & boundary\ of \mathbb{K}^i \\ 1 & inside\ of \mathbb{K}^i \end{cases} \tag{29} $$

where ${}^{4}O \in \mathbb{R}^{p}$ is the output vector of the fourth layer, and ${}^{3}O^{i}$ is the element i of the third layer's output vector.

The fifth layer computes the local functions from all critical regions as the following vector:

$$ {}^{5}O(x,y_r) = f_{par} \times \begin{bmatrix} x(k) \\ y_r(k) \\ 1 \end{bmatrix} \tag{30} $$

where ${}^{5}O \in \mathbb{R}^{p}$ is the output vector of the fifth layer, and f_{par} is the fifth layer's matrix of parameters, which consists of stacking the parameters of the local functions into a matrix as follows:

$$ f_{par} = \begin{bmatrix} G^1 & g^1 \\ \vdots & \vdots \\ G^p & g^p \end{bmatrix} \tag{31} $$

and G^i and g^i are the matrix and vector of parameters from the local affine functions.

Finally, the sixth layer calculates the MTSK-MPC's final output as the weighted sum of the local functions' contributions, according to:

$$^{6}O = \begin{bmatrix} \Delta U^{*}(k) \\ x^{*}_{ss}(k) \end{bmatrix} = \sum_{i=1}^{p} {}^{4}O^{i}(x, y_r) \times {}^{5}O^{i}(x, y_r) \tag{32}$$

where $^{6}O \in \mathbb{R}^{p}$ is the output vector of the sixth layer, $^{4}O^{i}$ is the element i of the fourth layer's output vector, corresponding to the normalized weight of rule i, and $^{5}O^{i}$ is the element i of the fifth layer's output vector, corresponding to the respective local function of rule i.

4. Experimental Results and Discussion

The article presents a set of experiments that include (i) an SIL simulation that aims to identify the open-loop PSA process model for use in the MPC formulations, (ii) offline analyses on the PLC memory effects for different design parameter values, (iii) a comparison of the online processing performance between a closed-loop HIL simulation of the MTSK-MPC algorithm and closed-loop SIL simulations with different MPC algorithms, which also validate the MTSK-MPC embedding in the PLC.

A phenomenological model implemented in gPROMS was used to represent the LSRE plant, as described in Section 2. This phenomenological model comprises a system of non-linear partial differential–algebraic equations (PDAE), which was solved using orthogonal colocation with 20 intervals and second-order polynomials to transform the PDAE system into an ODE system. The ODE system was solved using an implicit Runge–Kutta method.

The central platform for executing the experiments is the MATLAB 2023a software, which exchanges data with gPROMS and other application programs according to the type of simulation. MATLAB and gPROMS communicate to exchange input and output signals through FPI (Foreign Process Interface) events and the gO:MATLAB command. Each software runs on the same personal computer with an Intel Core i5-2400 processor, 3.10 GHz CPU, and 8.00 GB RAM capacity. In the SIL simulations, the MPC algorithms run in MATLAB on the same PC of gPROMS, while in the HIL simulation, the MTSK-MPC runs in the PLC and communicates to MATLAB in the PC through a Modbus protocol. The PLC is a Micro850 general purpose automation hardware with a memory limit of 120 KB for user programs and data.

4.1. Process Model Identification Experiment

As this study case focuses on linear-type MPC approaches, the laboratory PSA runs in the following experiments within a limited deviation range from an operating condition. Table 4 shows the operating point used for the linearized model and the process limits considered in the controlled and manipulated variables of the MPC controller.

Table 4. Summary of the operating point and the process limits considered.

Inputs	Tfeed	Trinse	Outputs	CH_4 Purity	CO_2 Recovery
$\overline{u}$	212.6 s	74.2 s	$\overline{y}$	72.7196%	95.9988%
u_{min}	200 s	70 s	y_{min}	68%	91%
u_{max}	300 s	150 s	y_{max}	100%	100%
Δu_{max}	2 s/cycle	2 s/cycle	-	-	-

The adopted criterion in this case study was to apply simple identification methods since the PSA process model is linearized. Therefore, the experiment to obtain an open-loop process model consisted of excitation procedures based on step responses during an SIL simulation. The procedure involved applying step signals with values 20% above the input values at the nominal operating points and then returning to nominal operation. The use of collected data and the *idproc* function of the MATLAB identification toolbox resulted in the transfer functions models summarized Table 5.

Table 5. Transfer functions identified after step test identification.

Inputs\Outputs	CH_4 Purity (%) y_1	CO_2 Recovery (%) y_2
Tfeed (s) u_1	$\frac{-0.7003}{2.375\times s+1}$	$\frac{-0.3693}{0.3534\times s+1}$
Trinse (s) u_2	$\frac{-1.957}{2.426\times s+1}$	$\frac{-1.028}{0.3788\times s+1}$

After comparing the outputs of the simulated process with the outputs of the transfer functions identified in Table 5, the mean absolute percentage error (MAPE) for CH_4 purity and CO_2 recovery was 2.1618% and 0.6273%, respectively. Note that the estimated transfer function gains are negative. That is, around the operating point used, increasing the feed or rinse time leads to a decrease in CH_4 purity and CO_2 recovery at the process outlet, as shown in Figure 3.

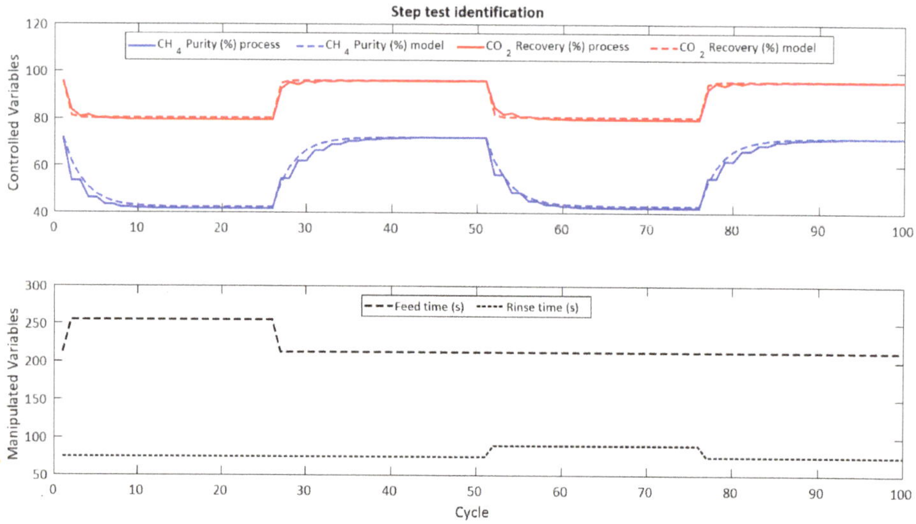

Figure 3. Identification procedure using step input signals. (**Bottom**) The feed and rinse times of the PSA cycle steps. (**Top**) The CH_4 purity and CO_2 recovery responses for these excitations.

The output differences between the simulated nonlinear process in gPROMS and the identified linear models in Table 5 are considered disturbances in the process outputs. Note that, for feed and rinse times longer than those used in the identification step tests, process nonlinearities will increase the modeling errors of these models. Other modeling errors due to nonlinearities and multivariate interactions may arise from variations in the column pressure and inlet total feed flow, which are reported in Table 3, or the molar fraction of gases in this flow. For example, an increase in pressure leads the column adsorption capacity to increase, improving the CH_4 purity and the CO_2 recovery at the process outlet. This behavior will especially impact the transfer function gains, which would increase in this situation. Likewise, an increase in the CH_4 molar fraction in the feed stream will change the process outputs. In this latter case, the transfer function gains will have the values related to CH_4 purity increased and the values related to CO_2 recovery decreased. Finally, the PSA was simulated considering the process as isothermal, so no temperature interference led to modeling errors.

The MPC controller treats the modeled disturbances as extended states, estimated using a Kalman observer, and considering that $\boldsymbol{d}(k+1) = \boldsymbol{d}(k)$. Therefore, after converting the transfer functions in Table 5 to the state-space model with the output disturbances, the complete state-space model in the incremental form of inputs used by the MPC controller is as follows:

$$x(k+1) = \begin{bmatrix} 0.6563 & 0 & 0 & 0 & 0 & 0 & 0.5 & 0 \\ 0 & 0.0590 & 0 & 0 & 0 & 0 & 0.5 & 0 \\ 0 & 0 & 0.6622 & 0 & 0 & 0 & 0 & 1 \\ 0 & 0 & 0 & 0.0714 & 0 & 0 & 0 & 1 \\ 0 & 0 & 0 & 0 & 1 & 0 & 0 & 0 \\ 0 & 0 & 0 & 0 & 0 & 1 & 0 & 0 \\ 0 & 0 & 0 & 0 & 0 & 0 & 1 & 0 \\ 0 & 0 & 0 & 0 & 0 & 0 & 0 & 1 \end{bmatrix} x(k) + \begin{bmatrix} 0.5 & 0 \\ 0.5 & 0 \\ 0 & 1 \\ 0 & 1 \\ 0 & 0 \\ 0 & 0 \\ 1 & 0 \\ 0 & 1 \end{bmatrix} \Delta u(k) \quad (33)$$

$$y(k) = \begin{bmatrix} -0.4814 & 0 & -0.6613 & 0 & 1 & 0 & 0 & 0 \\ 0 & -0.6950 & 0 & -0.9542 & 0 & 1 & 0 & 0 \end{bmatrix} x(k) \quad (34)$$

where the extended states vector, the incremental input vector, and the output vector are, respectively, $x(k) = \left[x_1'(k), x_2'(k), x_3'(k), x_4'(k), d_1'(k), d_2'(k), T_{feed}(k-1), T_{rinse}(k-1)\right]^{\mathrm{T}}, \Delta u(k) = \left[\Delta Tfeed(k) \quad \Delta Trinse(k)\right]^T$, and $y(k) = \left[CH_4purity(k). \quad CO_2recovery(k)\right]^{\mathrm{T}}$, and $x_i'(k)$ are the original states of the system related to the transfer function poles in Table 5, and $d_i'(k)$ are the disturbances in the process outputs.

It should be noted that including disturbances as new states increases the order of the MPC model and affects the optimization dimensions of **Problem P0**, leading to higher computational demands. Nevertheless, this inclusion is necessary for the controller to accurately calculate the output at the artificial steady state, $y_{ss}(k)$, and achieve offset-free control.

The Kalman observer was implemented in the PLC through the following equations:

$$\hat{x}(k+1/k) = A\hat{x}(k/k) + B\Delta u(k) + K_f \begin{bmatrix} y(k) - \hat{y}(k/k-1) \\ u(k-1) - \hat{u}(k-1/k-1) \end{bmatrix} \quad (35)$$

Note that the equation from the observer considers the state correction using the difference between the measured outputs in the current instant, $y(k)$, and from the estimated outputs of the system in the previous instant, $\hat{y}(k/k-1)$. In addition, as the final extended states are the last-instant input signals in the current instant, $\hat{u}(k-1/k)$, this equation also considers the difference between the real last-instant input applied to the system, $u(k-1)$, which is known, from the last-instant input signal estimated in the previous instant, $\hat{u}(k-1/k-1)$. If this difference is not considered, the observer poles would be close to unity, hindering the state estimation. Furthermore, this formulation causes the estimated last-instant input to converge quickly to the last-instant input real value, allowing estimation errors to be treated in the other states. The Kalman gain used in Equation (35) was:

$$K_f = \begin{bmatrix} -0.1110 & -0.0002 & 0.4809 & -0.0064 \\ -0.0004 & -0.0124 & 0.4563 & -0.0010 \\ -0.1659 & -0.0080 & -0.0037 & 0.9569 \\ -0.0042 & -0.0235 & -0.0007 & 0.9123 \\ 0.7015 & -0.0043 & 0.0150 & 0.0413 \\ -0.0146 & 0.5440 & 0.0162 & 0.0444 \\ -0.0010 & -0.0009 & 0.9083 & -0.0001 \\ -0.0029 & -0.0024 & -0.0001 & 0.9080 \end{bmatrix} \quad (36)$$

4.2. PLC Memory Sensitivity Analysis to Design Parameters

The investigation of computational characteristics of the MTSK-MPC algorithm first explored how extending the prediction horizon would impact the Micro850 memory. This study evaluated the MTSK-MPC with three different prediction horizons for each of the single-type constraints from Table 4. The weighting matrices remained constant during all these experiments, as $Q = I$, $S_y = 10^4.I$,$R = \text{diag}(1, 0.1)$. Figure 4 shows the results of this study.

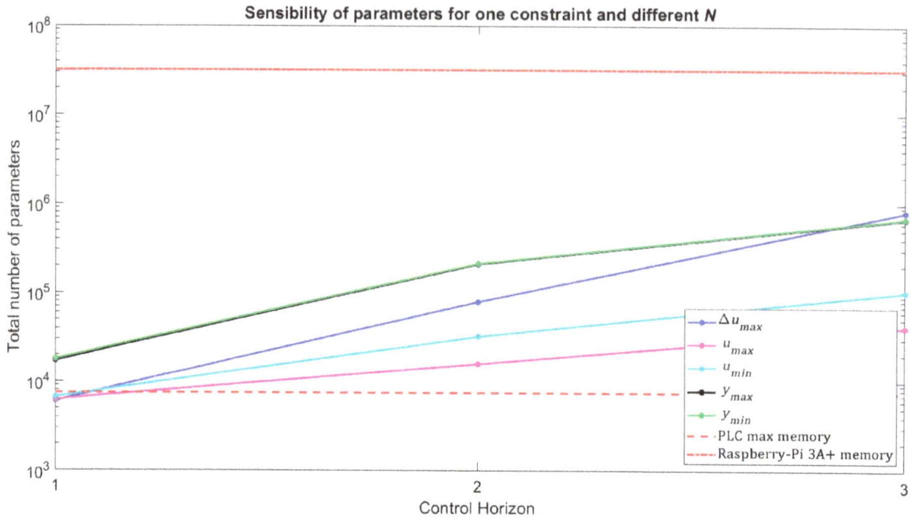

Figure 4. Sensitivity analysis for one constraint and different prediction horizons of the MTSK-MPC algorithm.

Observe in Figure 4 the results from different single-type constraints, which exhibit similar behaviors. As the prediction horizon increases, the MTSK-MPC exponentially increases the required parameter numbers. Additionally, input and input variation constraints result in fewer parameters compared to output constraints. This figure also shows the maximum memory capacity of the Micro850 in the red-dashed line and the memory of a standard Raspberry-Pi® 3 A+ micro-computer in the red-pointed line, assuming that all parameters require 16-bit floating-point variables. Considering such a limit, only three cases are embeddable in this PLC, which involve single input constraints and a unitary control horizon in the MPC formulation.

An investigation of computational characteristics of the MTSK-MPC followed to explore the impact of including and combining various constraints for a unitary control horizon in the Micro850 memory. Figure 5 shows the result of this study.

Note in Figure 5, the increased number of parameters required when considering output constraints in the PSA process. If one includes any of the output constraints in the MPC formulation, individually or in combination with others, the MTSK-MPC implementation in the Micro850 will be unfeasible.

On the other hand, observe that adding new constraints does not necessarily increase the number of parameters and, consequently, the amount of memory required. In fact, in some cases, there is a slight decrease. The reason for this is the high dimensionality of the MPC controller for this PSA unit, which can lead to overlapped critical regions and more critical regions with saturated outputs for different constraint combinations. Taking advantage of this behavior, if one includes the three input constraint types simultaneously and with a unitary control horizon, the MTSK-MPC implementation in the Micro850 will be feasible.

Therefore, considering the laboratory PSA process and the Micro850 of the case study, the main result of these two memory investigations is that embedding the MTSK-MPC into this PLC is feasible under particular conditions. To use larger control horizons and include additional constraints would require changing the Micro850 for other PLC models with more memory and processing capacities.

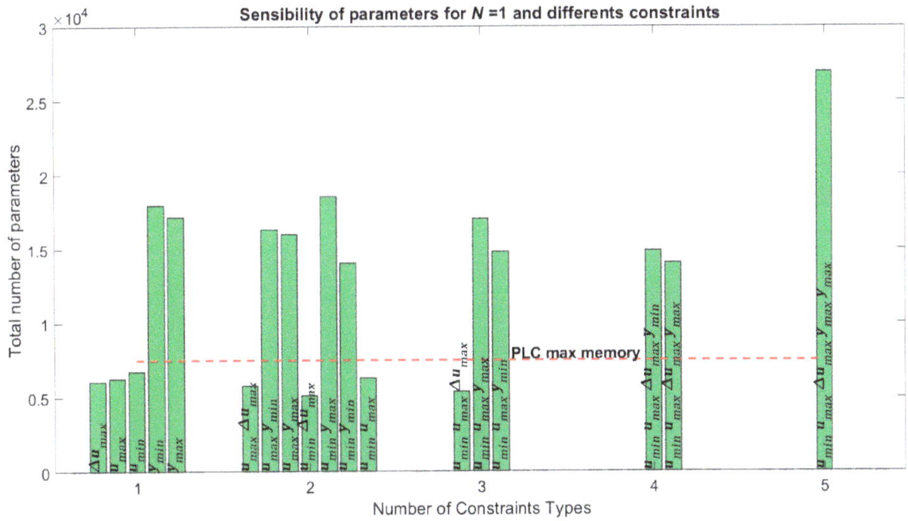

Figure 5. Sensitivity analysis for a fixed prediction horizon and different constraint combinations.

Using the MPC formulation described in Section 3, the values of the input constraints listed in Table 4 with a unitary control horizon, the transfer functions identified in Section 4.1, and the implementation methodology described in Mendes et al. (2022) [29], the MTSK-MPC was successfully embedded into a PLC to control a PSA unit. Figure 6 illustrates the PLC memory used for the program and the data required by this implementation.

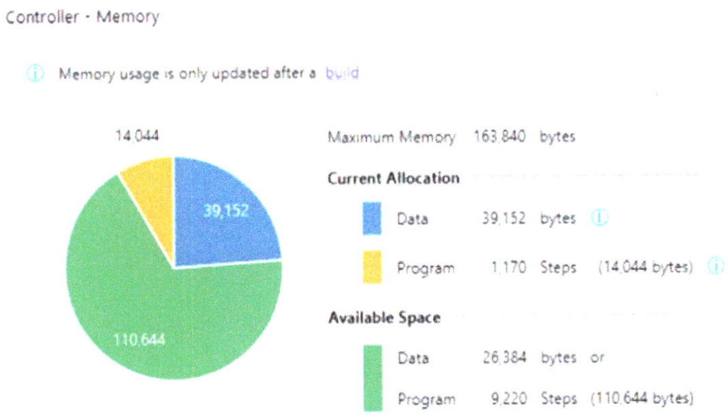

Figure 6. Final memory allocation after MTSK-MPC programmed block and parameter values downloaded to Micro850.

4.3. Comparison of Processing Performance of MPCs in Closed-Loop Simulations

In addition to studying the PLC memory required by the MTSK-MPC implementation, an investigation compared the computational processing of different MPC algorithms and simulations. The comparison evaluated the following algorithms running in PC and SIL simulations: an implicit MPC (iMPC) with interior point optimization from the *quadprog*

MATLAB function, an explicit MPC (eMPC) with multiparametric optimization from the MATLAB MPT toolbox [39], and an MTSK-MPC.

Additionally, the comparison included the MTSK-MPC running in the Micro850 and HIL simulation. The evaluation considered the absolute error difference between these MPC algorithms and their respective processing times. Figure 7 displays the PSA process outputs with the respective setpoints, and the PSA process inputs for the MTSK-MPC in the Micro850 and HIL simulation.

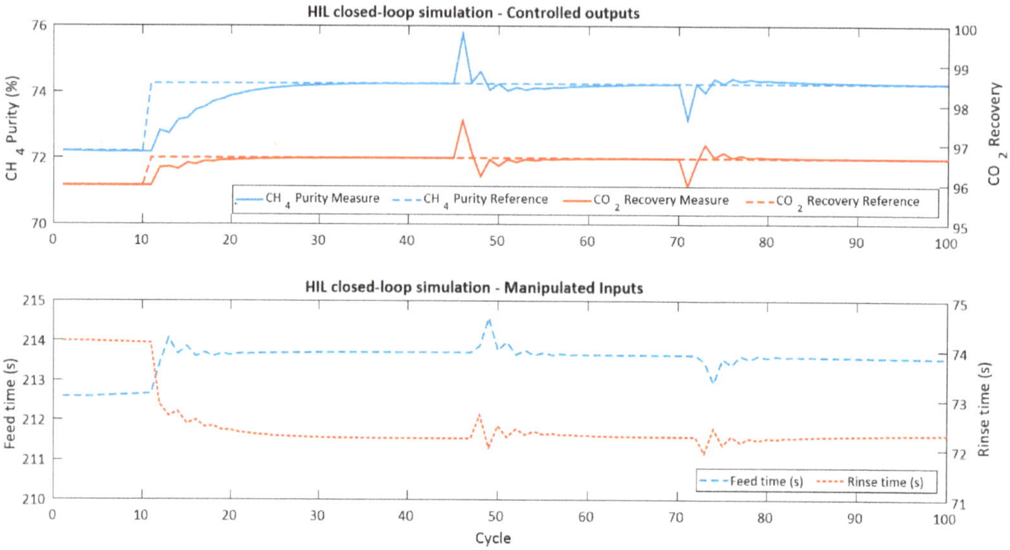

Figure 7. Closed-loop behavior of the PSA unit under MTSK-MPC controller.

Note that a setpoint change occurred at the beginning of the experiment, and the MPC controller successfully tracked this change. Additionally, since PLCs are devices that could present slight oscillations of execution times, intentionally introduced input disturbances perturb the process at cycles 45 and 70 in feed and rinse time, respectively. As expected, the controller rejected these disturbances and restored the system to the desired operating setpoint.

Other results of the comparison are shown in Figure 8, which illustrates the absolute error difference behavior over time, and in Table 6, which summarizes the mean and max absolute error difference between the MPC algorithms. In addition, Tabel 7 presents the mean and max processing execution time of each MPC algorithm with the respective type of simulation and hardware.

Table 6. Absolute error difference between MPCs outputs.

MPC (Hw/Simulation)	iMPC (PC/SIL) – eMPC (PC/SIL)		iMPC (PC/SIL) – MTSK-MPC (PC/SIL)		MTSK-MPC (PC/SIL) – MTSK-MPC (CLP/HIL)	
	u_1	u_2	u_1	u_2	u_1	u_2
Mean (10×10^{-6})	0.086	0.031	0.086	0.031	0.184	0.233
Max (10×10^{-6})	0.009	0.003	0.009	0.003	0.76	0.497

In Figure 8 and Table 6, note that the absolute error differences between the outputs of the MPCs from the SIL simulations are negligible as they are in the eighth decimal place. On the other hand, based on Table 7, it is possible to notice that despite presenting very close output calculations, these MPCs processing times are different. Moreover, the MTSK-MPC algorithm stands out with the lowest value. This result is because implicit MPC, explicit MPC, and MTSK-MPC algorithms are optimization methods with equivalent output responses. The differences between these approaches arise mainly due to the implementation aspects, such as different requirements for memory and processing times [5–7].

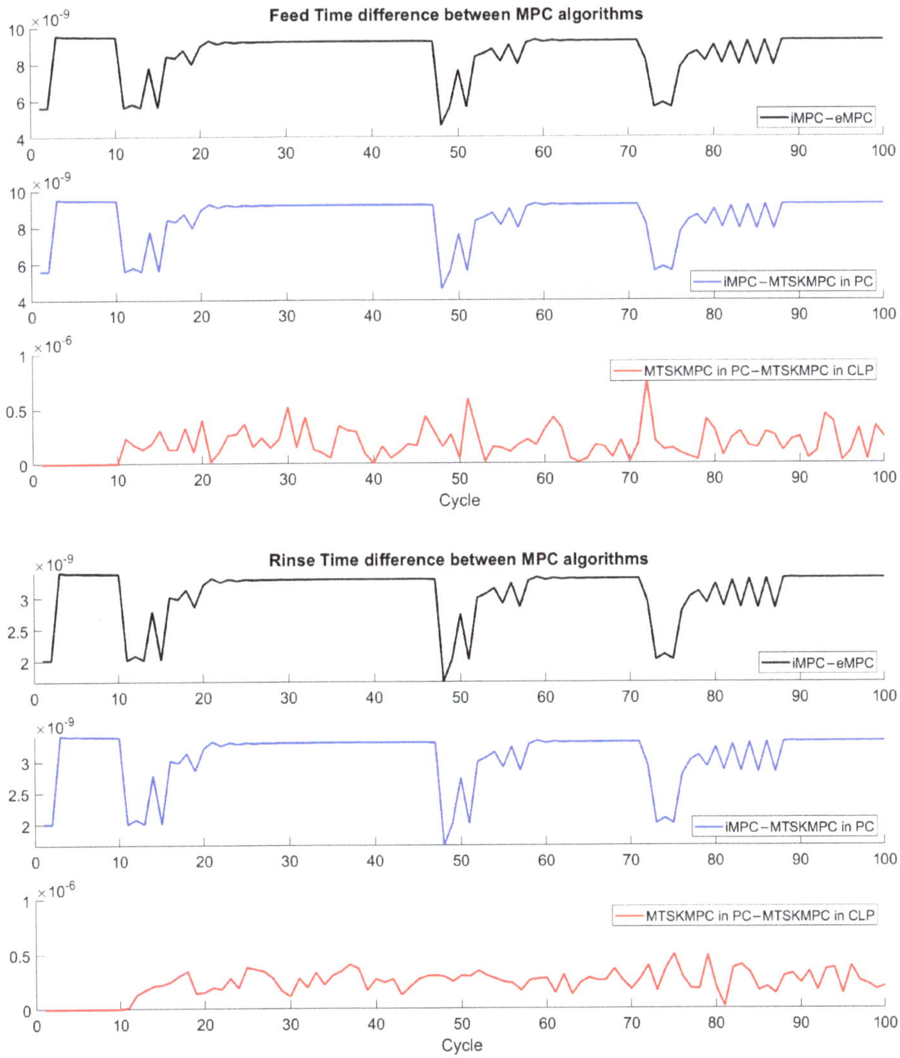

Figure 8. Absolute error difference over time between the MPCs outputs: (**Top**) Feed time and (**Bottom**) rinse time.

Table 7. Processing time of MPCs algorithms.

Simulation Type (Hardware)		SIL (PC)		HIL (CLP)
MPC Algorithm	iMPC	eMPC	MTSK-MPC	MTSK-MPC
Mean Time (ms)	5.329	9.487	0.329	95
Max Time (ms)	61.583	33.259	4.124	96

When comparing the two MTSK-MPC algorithms, the one implemented on the PC in the SIL simulation and the other implemented on the PLC in the HIL simulation, note that the absolute error difference is smaller than in the previous comparison cases. Despite that, it is at the sixth decimal place and still negligible. The reason for the discrepancy is the use of double-precision floating-point variables in the PC implementation, while the PLC has single-precision floating-point variables within the developed program. The single-precision floating-point variables are preferable in the PLC program because they use only 32 bits rather than 64 bits of double precision to save memory.

The MTSK-MPC implemented on the PC in the SIL simulation is approximately 300 times faster than the MTSK-MPC implemented on the PLC in the HIL simulation. That is an expected result, and it reinforces that the challenge of applying MPC to PSA units lies in implementing an algorithm of this type on industrial hardware. Another central result regarding processing time is that the shortest step in the PSA cycle is the rinsing time, which is of the order of tens of seconds. Furthermore, the processing time of the MPC controller in the PLC is about 100 ms. Therefore, as this latter time is faster than the PSA step times, the MPC implementation in the Micro850 successfully handles the timing requirements of the under-studied PSA dynamics.

5. Conclusions

This paper is distinguished from previous literature on PSA control by exploring two unexplored gaps. The first gap handled is the comparison of implicit and explicit MPC algorithms applied to PSA processes, achieved by the presented computational investigation. The second gap addressed is the implementation of MPC in industrial hardware for PSA units, attained by integrating the MTSK-MPC into a Micro850 PLC and testing this controller in a closed-loop using an HIL simulation.

The computational investigation consisted of memory-sensitive analyses and a processing comparative evaluation. Based on the memory analyses, the MTSK-MPC implementation in the Micro850 would be feasible if it only considers the constraints associated with the inputs and with a unitary prediction horizon. This strict result is due to the limited memory capacity of the Micro850. The processing performance comparison showed that MTSK-MPC in the PLC-based HIL simulation could calculate control outputs accurately and with values very close to other standard MPC algorithms in PC-based SIL simulations. Additionally, the MTSK-MPC algorithm exhibited a lower processing time than other algorithms in SIL simulations.

The MTSK-MPC implemented into the Micro850 tracked setpoint changes and rejected input disturbances. In addition, the MTSK-MPC absolute error difference between the PLC-based in the HIL simulation and the PC-based in the SIL simulation was negligible. Finally, the processing time of the MTSK-MPC in the Micro850 is about 100 ms, which is fast enough to control the under-studied PSA dynamics. The successful implementation of the MTSK-MPC into the Micro850 and the results obtained in the HIL simulation form the basis for future work to control the physical laboratory PSA unit of LSRE. In this scenario, future work should consider: (i) the execution time of the identification step tests, since the 100 cycles will result in more than 8 h of experiments, (ii) the PSA process without simplifications assumed in the mathematical model, which may present stronger nonlinear behavior than the simulated process, (iii) the effects of disturbances in key process variables, including feed flow rates, feed flow concentrations, column pressure, and column temperature, among others.

Author Contributions: Conceptualization, T.P.G.M.; methodology, T.P.G.M.; software, T.P.G.M.; validation, I.B.R.N.; formal analysis, T.P.G.M.; investigation, T.P.G.M.; resources, A.M.R.; data curation, T.P.G.M.; writing—original draft preparation, T.P.G.M.; writing—review and editing, L.S. and I.B.R.N.; visualization, T.P.G.M.; supervision, I.B.R.N. All authors have read and agreed to the published version of the manuscript.

Funding: This research was supported by the Federal University of Recôncavo da Bahia and by CAPES, the Brazilian Federal Agency for Post-graduate Education, under financial code (001).

Data Availability Statement: Available from the corresponding author upon request.

Acknowledgments: We would like to thank CAPES under Finance Code 001 and national funds through FCT/MCTES (PIDDAC): LSRE-LCM, UIDB/50020/2020 (DOI: 602 10.54499/UIDB/50020/2020) and UIDP/50020/2020 (DOI: 10.54499/UIDP/50020/2020); and ALiCE, 603 LA/P/0045/2020 (DOI: 10.54499/LA/P/0045/2020) for their financial support.

Conflicts of Interest: The authors declare no conflicts of interest.

References

1. Urich, M.D.; Vemula, R.R.; Kothare, M.V. Implementation of an embedded model predictive controller for a novel medical oxygen concentrator. *Comput. Chem. Eng.* **2022**, *160*, 107706. [CrossRef]
2. Luberti, M.; Ahn, H. Review of Polybed pressure swing adsorption for hydrogen purification. *Int. J. Hydrogen Energy* **2022**, *47*, 10911–10933. [CrossRef]
3. Khajuria, H.; Pistikopoulos, E.N. Dynamic modeling and explicit/multi-parametric MPC control of pressure swing adsorption systems. *J. Process Control* **2011**, *21*, 151–163. [CrossRef]
4. Khajuria, H.; Pistikopoulos, E.N. Optimization and Control of Pressure Swing Adsorption Processes Under Uncertainty. *AIChE J.* **2013**, *59*, 120–131. [CrossRef]
5. Sun, W.; Shen, Y.; Zhang, D.; Yang, H.; Ma, H. A Systematic Simulation and Proposed Optimization of the Pressure Swing Adsorption Process for N_2/CH_4 Separation under External Disturbances. *Ind. Eng. Chem. Res.* **2015**, *54*, 7489–7501. [CrossRef]
6. Oliveira, P.H.M.; Martins, M.A.F.; Rodrigues, A.E.; Loureiro, J.M.; Ribeiro, A.M.; Nogueira, I.B.R. A robust model predictive controller applied to a pressure swing adsorption process: An analysis based on a linear model mismatch. *IFAC-PapersOnLine* **2021**, *54*, 219–224. [CrossRef]
7. Martins, M.A.F.; Rodrigues, A.E.; Loureiro, J.M.; Ribeiro, A.M.; Nogueira, I.B.R. Artificial Intelligence-oriented economic non-linear model predictive control applied to a pressure swing adsorption unit: Syngas purification as a case study. *Sep. Purif. Technol.* **2021**, *276*, 119333. [CrossRef]
8. Martins, M.A.F.; Rodrigues, A.E.; Loureiro, J.M.; Ribeiro, A.M.; Nogueira, I.B.R. Handling model uncertainty in control of a pressure swing adsorption unit for syngas purification: A multi-model zone control scheme-based robust model predictive control strategy. *Sep. Purif. Technol.* **2023**, *306*, 122668. [CrossRef]
9. Peng, H.; Couenne, F.; Le Gorrec, Y. Robust Control of a Pressure Swing Adsorption Process. *IFAC Proc. Vol.* **2011**, *44*, 7310–7315. [CrossRef]
10. Bitzer, M. Model—Based Nonlinear Tracking Control of Pressure Swing Adsorption Plants. In *Control and Observer Design for Nonlinear Finite and Infinite Dimensional Systems*; Springer: Berlin/Heidelberg, Germany, 2005; pp. 403–418.
11. Brizuela-Mendoza, J.A.; Sorcia-Vázquez, F.D.J.; Rumbo-Morales, J.Y.; Ortiz-Torres, G.; Torres-Cantero, C.A.; Juárez, M.A.; Zatarain, O.; Ramos-Martinez, M.; Sarmiento-Bustos, E.; Rodríguez-Cerda, J.C.; et al. Pressure Swing Adsorption Plant for the Recovery and Production of Biohydrogen: Optimization and Control. *Processes* **2023**, *11*, 2997. [CrossRef]
12. Martínez García, M.; Rumbo Morales, J.Y.; Torres, G.O.; Rodríguez Paredes, S.A.; Vázquez Reyes, S.; Sorcia Vázquez, F.d.J.; Pérez Vidal, A.F.; Valdez Martínez, J.S.; Pérez Zúñiga, R.; Renteria Vargas, E.M. Simulation and State Feedback Control of a Pressure Swing Adsorption Process to Produce Hydrogen. *Mathematics* **2022**, *10*, 1762. [CrossRef]
13. Morales, J.Y.R.; López, G.L.; Martínez, V.M.A.; Vázquez, F.d.J.S.; Mendoza, J.A.B.; García, M.M. Parametric study and control of a pressure swing adsorption process to separate the water-ethanol mixture under disturbances. *Sep. Purif. Technol.* **2020**, *236*, 116–214.
14. Han, Z.Y.; Xing, R.; Zhang, D.H.; Shen, Y.H.; Fu, Q.; Ding, Z.Y.; Tian, C.X. Vacuum pressure swing adsorption system for N2/CH4 separation under uncertainty. *Chem. Eng. Res. Des.* **2019**, *142*, 245–256. [CrossRef]
15. Du, W.; Alkebsi, K.A.M. Model predictive control and optimization of vacuum pressure swing adsorption for carbon dioxide capture. In Proceedings of the 6th International Symposium on Advanced Control of Industrial Processes, Taipei, Taiwan, 28–31 May 2017; pp. 412–417. [CrossRef]
16. Ramos-Martinez, M.; Torres-Cantero, C.A.; Ortiz-Torres, G.; Sorcia-Vázquez, F.D.J.; Avila-George, H.; Lozoya-Ponce, R.E.; Vargas-Méndez, R.A.; Renteria-Vargas, E.M.; Rumbo-Morales, J.Y. Control for Bioethanol Production in a Pressure Swing Adsorption Process Using an Artificial Neural Network. *Mathematics* **2023**, *11*, 3967. [CrossRef]
17. Urich, M.D.; Vemula, R.R.; Kothare, M.V. Multivariable model predictive control of a novel rapid pressure swing adsorption system. *AIChE J.* **2018**, *64*, 1234–1245. [CrossRef]
18. Urich, M.D.; Rama Rao, V.; Kothare, M.V. Piecewise linear model predictive control of a rapid pressure swing adsorption system. *AIChE J.* **2020**, *66*, e16998. [CrossRef]
19. Mulholland, M.; Latifi, M.A. Predictive Control Of Pressure Swing Adsorption. In Proceedings of the 2009 International Conference on Process Control, Štrbské Pleso, Slovakia, 9–12 June 2009; Volume 9, pp. 323–331.

20. Nogueira, I.B.R.; Martins, M.A.F.; Rodrigues, A.E.; Loureiro, J.M.; Ribeiro, A.M. Novel Switch Stabilizing Model Predictive Control Strategy Applied in the Control of a Simulated Moving Bed for the Separation of Bi-Naphthol Enantiomers. *Ind. Eng. Chem. Res.* **2020**, *59*, 1979–1988. [CrossRef]
21. Nogueira, I.B.R.; Ribeiro, A.M.; Martins, M.A.F.; Rodrigues, A.E.; Koivisto, H.; Loureiro, J.M. Dynamics of a True Moving Bed separation process: Linear model identification and advanced process control. *J. Chromatogr. A* **2017**, *1504*, 112–123. [CrossRef] [PubMed]
22. Nogueira, I.B.R.; Fontes, R.M.; Ribeiro, A.M.; Pontes, K.V.; Embiruçu, M.; Martins, M.A.F. A robustly model predictive control strategy applied in the control of a simulated industrial polyethylene polymerization process. *Comput. Chem. Eng.* **2020**, *133*, 106664. [CrossRef]
23. Huyck, B.; De Brabanter, J.; De Moor, B.; Van Impe, J.F.; Logist, F. Towards online model predictive control on a programmable logic controller: Practical considerations. *Math. Probl. Eng.* **2012**, *2012*, 912603. [CrossRef]
24. Johansen, T.A. Toward Dependable Embedded Model Predictive Control. *IEEE Syst. J.* **2014**, *11*, 1208–1219. [CrossRef]
25. Qin, S.J.; Badgwell, T.A. A survey of industrial model predictive control technology. *Control Eng. Pract.* **2003**, *11*, 733–764. [CrossRef]
26. Fu, Q.; Yan, H.; Shen, Y.; Qin, Y.; Zhang, D.; Zhou, Y. Optimal design and control of pressure swing adsorption process for N2/CH4 separation. *J. Clean. Prod.* **2018**, *170*, 704–714. [CrossRef]
27. Kvasnica, M. Implicit vs. explicit MPC—Similarities, differences, and a path owards a unified method. In Proceedings of the 2016 European Control Conference (ECC), Aalborg, Denmark, 29 June–1 July 2016; p. 603. [CrossRef]
28. Bemporad; Morari, M.; Dua, V.; Pistikopoulos, E.N. The explicit linear quadratic regulator for constrained systems. *Automatica* **2002**, *38*, 3–20. [CrossRef]
29. Mendes, T.P.G.; Schnitman, L.; dos Reis Nogueira, I.B.; Mafalda Almeida Peixoto Ribeiro, A.; Egídio Rodrigues, A.; Miguel Loureiro, J.; Martins, M.A.F. A new Takagi-Sugeno-Kang model-based stabilizing explicit MPC formulation: An experimental case study with implementation embedded in a PLC. *Expert Syst. Appl.* **2022**, *210*, 118369. [CrossRef]
30. Tiegelkamp, M.; John, K.H. *IEC 61131-3: Programming Industrial Automation Systems*; Springer: Berlin/Heidelberg, Germany, 2010; ISBN 978-3-662-07847-1. [CrossRef]
31. Mergenthal, M.; Tawai, A.; Amornraksa, S.; Roddecha, S.; Chuetor, S. Methane enrichment for biogas purification using pressure swing adsorption techniques. *Mater. Today Proc.* **2023**, *72*, 2915–2920. [CrossRef]
32. Nogueira, I.B.R.; Martins, M.A.F.; Regufe, M.J.; Rodrigues, A.E.; Loureiro, J.M.; Ribeiro, A.M. Big Data-Based Optimization of a Pressure Swing Adsorption Unit for Syngas Purification: On Mapping Uncertainties from a Metaheuristic Technique. *Ind. Eng. Chem. Res.* **2020**, *59*, 14037–14047. [CrossRef]
33. Campos, A.F.; da Silva, N.F.; Pereira, M.G.; Vasconcelos Freitas, M.A. A review of Brazilian natural gas industry: Challenges and strategies. *Renew. Sustain. Energy Rev.* **2017**, *75*, 1207–1216. [CrossRef]
34. Regufe, M.J.; Tamajon, J.; Ribeiro, A.M.; Ferreira, A.; Lee, U.H.; Hwang, Y.K.; Chang, J.S.; Serre, C.; Loureiro, J.M.; Rodrigues, A.E. Syngas Purification by Porous Amino-Functionalized Titanium Terephthalate MIL-125. *Energy Fuels* **2015**, *29*, 4654–4664. [CrossRef]
35. Martins, M.A.F.; Odloak, D. A robustly stabilizing model predictive control strategy of stable and unstable processes. *Automatica* **2016**, *67*, 132–143. [CrossRef]
36. Kacprzyk, J.; Pedrycz, W. *Springer Handbook of Computational Intelligence*; Springer: Berlin/Heidelberg, Germany, 2015. [CrossRef]
37. Sugeno, M.; Kang, G.T. Structure identification of fuzzy model. *Fuzzy Sets Syst.* **1988**, *28*, 15–33. [CrossRef]
38. Sugeno, M.; Takagi, T. Fuzzy identification of systems and its applications to modeling and control. *IEEE Trans. Syst. Man Cybern.* **1985**, *15*, 116–132. [CrossRef]
39. Herceg, M.; Kvasnica, M.; Jones, C.N.; Morari, M. Multi-Parametric Toolbox 3.0. In Proceedings of the 2013 European control Conference (ECC), Zurich, Switzerland, 17–19 July 2013; pp. 502–510. [CrossRef]

processes MDPI

Article

Edge Computing-Based Modular Control System for Industrial Environments

Gonçalo Gouveia [1,†], Jorge Alves [1,2,†], Pedro Sousa [2], Rui Araújo [1] and Jérôme Mendes [3,*]

1 University of Coimbra, Institute of Systems and Robotics, Department of Electrical and Computer Engineering, Pólo II, 3030-290 Coimbra, Portugal; gongou00@gmail.com (G.G.); jorge.alves@isr.uc.pt (J.A.); rui@isr.uc.pt (R.A.)
2 Oncontrol Technologies, Rua Cidade Poitiers, n° 155—1°, 3000-108 Coimbra, Portugal; pedro.sousa@oncontrol-tech.com
3 University of Coimbra, CEMMPRE, ARISE, Department of Mechanical Engineering, Pólo II, 3030-290 Coimbra, Portugal
* Correspondence: jerome.mendes@uc.pt
† These authors contributed equally to this work.

Abstract: This paper presents a modular hardware control system tailored for industrial applications. The system presented is designed with electrical protection, guaranteeing the reliable operation of its modules in the presence of various field noises and external disturbances. The modular architecture comprises a principal module (mP) and dedicated expansion modules (mEXs). The principal module serves as the network administrator and facilitates interaction with production and control processes. The mEXs are equipped with sensors, conditioning circuits, analog-to-digital converters, and digital signal processing capabilities. The mEX's primary function is to acquire local processing field signals and ensure their reliable transmission to the mP. Two specific mEXs were developed for industrial environments: an electrical signal expansion module (mSE) and the vibration signals expansion module (mSV). The EtherCAT protocol serves as a means of communication between the modules, fostering deterministic and real-time interactions while also simplifying the integration and replacement of modules within the modular architecture. The proposed system incorporates local and distributed processing in which data acquisition, processing, and data analysis are carried out closer to where data are generated. Locally processing the acquired data close to the production in the mEX increases the mP availability and network reliability. For the local processing, feature extraction algorithms were developed on the mEX based on a Fast Fourier Transform (FFT) algorithm and a curve-fitting algorithm that accurately represents a given FFT curve by significantly reducing the amount of data that needs to be transmitted over the mP. The proposed system offers a promising solution to use computational intelligence methodologies and meet the growing need for a modular industrial control system with reliable local data processing to reach a smart industry. The case study of acquiring and processing vibration signals from a real cement ball mill showed a good capacity for processing data and reducing the amount of data.

Keywords: industrial modular hardware system; industrial control system; edge computing; Cyber-Physical System; smart industry; EtherCAT

1. Introduction

Recently, there has been a noticeable evolution across the industrial processes, characterized by the need for efficient energy usage, sustainable resource management, and zero defects. These developments have led to a significant increase in the complexity of industrial processes and an urgent need for efficient, flexible, and reliable industrial control systems [1,2]. As a result, the concept of smart industry has emerged [3,4].

An area of notable innovation lies in the evolution of modular hardware systems designed to acquire, process, and transmit signals in alignment with industry requirements [5]. These modular systems introduced greater control over online data acquisition and processing processes. The ability to easily add new modules or replace faulty modules makes these systems more flexible and adaptable to the constantly evolving requirements of the industry [6,7]. Industrial control systems integrate computational and physical components to monitor and control the physical processes of a system [8]. Therefore, combining the physical layer and processes in industrial automation to achieve complex optimization decisions that humans traditionally make increases the demand for flexible and adaptable control systems. Therefore, it has led to essential concepts and architectures in the industry, such as edge computing [9] and fog computing [10], allowing resource's proximity to a lower level, improving the response times (latency and performance), system autonomy, and reliability even on network failures [11]. Therefore, in order to effectively achieve a smart industry, innovative industrial control systems are a current necessity. Some noteworthy examples on the market follow architectures appropriate for industrial applications [12], such as MyPi and Revolution Pi, which are modular industrial computers based on Raspberry Pi with an improved base for industrial applications; National Instruments Corp (NI) offers a data acquisition and control hardware system that can be customized by the user, for example, CompactRio and CompactDAQ with robustness for industry, and there is also a wide range of expansion modules; and Beckhof that offers industrial automation systems with industrial PCs (Portable Computers) and expansion components/modules. However, the development and implementation of expansion modules on the above architectures demand substantial user time and extensive expertise to establish communication, configuration, and programming within modular systems. Also, some architectures integrate proprietary automation software that needs knowledgeable users in the previously explained area. Moreover, these solutions, originally intended for non-industrial sectors, must contend with industry-specific challenges, encompassing factors like dirt, mechanical disturbances, electrical noise, overcurrent and overvoltage spikes, and bad implementations by users. To cover some limitations of the mentioned architectures, [12] proposes a new modular hardware system. It comprises a main processing module and expansion modules that can be added in a Plug&Play way, without third-party dependencies, allowing the easy development and implementation of modules. It also incorporates electrical protection and condition circuits to be robust for industrial environments. However, the hardware architecture developed is not based on the distributed processing concept, including also centralized data processing on the main processing module or outside of the proposed system.

This paper presents an edge computing-based modular hardware control system for industrial environments. The main contribution is the design of an architecture that incorporates the Plug&Play concept by having a main principal module (mP) and expansion modules (mEXs) that can be easily added to a network. The architecture owns local processing features in which data acquisition, processing, and analysis are carried out closer to where data are generated, contributing to the edge computing paradigm. The mP is the central coordinator and controls the entire architecture, performing the external remote control and data transmission, ensuring seamless connectivity and interoperability with external devices and networks. The mEXs are designed to acquire the field signals generated by the sensors, process them locally, and communicate the processed data to the mP. The mEXs are equipped with a System-On-Chip (SoC) capable of executing computational intelligence algorithms to process the data from the sensors or the source. Two specific mEXs were developed, the electrical signal expansion module (mSE) and the vibration signals expansion module (mSV), acquiring and local processing data from electrical signals (current and voltage) and vibration signals, respectively. The mEXs proposed in this paper perform the local processing in two stages: analyzing the field signals in the frequency domain using the FFT algorithm, and then, the FFT signal is approximated by a curve-fitting method to significantly reduce the amount of data that needs to be transmitted and post-processing. The communication between modules is completed by the Ethernet-based EtherCAT bus, thus facilitating the addition of new modules to the control system and deterministic real-time communication framework between the individual modules. The proposed system also incorporates electrical protection to be robust for industrial environments.

The paper is organized as follows. Section 2 presents an overview of the proposed industrial modular hardware control system and the hardware protection techniques employed. The mP is described in Section 3, while Section 4 presents the mEXs. Section 5 presents the results of the proposed system and the local processing algorithms. Finally, the conclusions are drawn in Section 6.

2. Proposed Industrial Hardware Control System

This section presents the global view of the proposed industrial modular hardware control system architecture. Figure 1 depicts the global hardware architecture. It features a main principal module (mP) and expansion modules (mEXs). Together, these modules form the system's backbone, facilitating data acquisition, processing, and transmission. This architecture is designed to provide high performance, scalability, and flexibility, allowing for continuous integration of additional mEX to address new industrial control challenges.

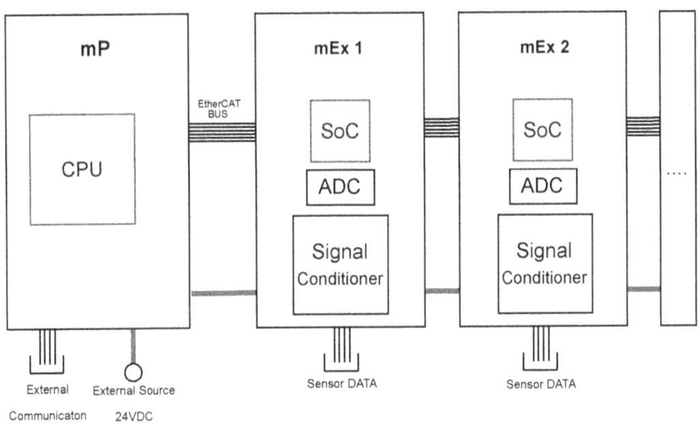

Figure 1. Global view of the proposed architecture for industrial environments.

The principal module, the mP, is the central coordinator of the internal communication network equipped with a Central Processing Unit (CPU) capable of running an operating system such as Linux, allowing remote control of the entire architecture. External remote control and data transmission are accomplished via the Ethernet Transmission Control Protocol/Internet Protocol (TCP/IP), ensuring seamless connectivity and interoperability with external devices and networks. Section 3 presents the mP in more detail.

The mEXs proposed in this paper are designed to capture and process signals generated by the field sensors. Two mEXs were developed in this paper, one to acquire and pre-process data from current and voltage signals, the electrical signal expansion module (mSE), and the other for vibration signals, the vibration signals expansion module (mSV). Section 4 presents the developed mEXs in more detail. In the mEX, signals are acquired from the field and undergo a signal-conditioning step, bringing the signal to the acceptable range of the ADC. Additionally, the module is equipped with a SoC capable of interfacing with the ADC using the Serial Peripheral Interface (SPI) communication and executing computational intelligence algorithms, both in a CPU and in a Field-Programmable Gate Array (FPGA) [13].

The bus presented and developed in this architecture consists of a dual-bus system comprising the power supply bus and the EtherCAT bus. The power supply bus encompasses channels for 24 Volts (V) and 0 V, receiving the 24 V direct current (DC) from an external power supply (represented by the red line in Figure 1). This power supply is received by the principal module and then distributed to the expansion modules via the bus network. EtherCAT functions as a master–slave communication system in which one or more master devices oversee the operation of numerous slave devices. These devices are interlinked via a single Ethernet bus, facilitating data transmission through the communication network. Within this framework, the mP assumes the role of a network master and is furnished with a dedicated Ethernet port tailored to accommodate the EtherCAT protocol. Meanwhile, each expansion module (the generic mEX) operates as an EtherCAT slave, boasting two Ethernet ports that contribute to its role within the system, receiving the EtherCAT frame from the previous module and sending it to the next module. This setup empowers data processing by individual slave devices and network control in the master device. Also, the master device can incorporate data processing features. The EtherCAT protocol offers numerous advantages, including high-speed, deterministic, and real-time communication. Additionally, the EtherCAT's wide acceptance in the market and its ability to support many modules on the communication bus make it suitable for emerging industrial applications.

With this proposed architecture design, it is possible to accommodate additional mEXs, enabling more efficient and swift adaptation to new industrial control challenges. This flexibility and scalability facilitate the seamless integration of new functionalities into the system, empowering users to stay ahead of evolving requirements and emerging trends in industrial control, such as distributed control of processes, the easy addition of new modules to a network, real-time processing, and deterministic communication.

Hardware Protections from External Environment

The proposed system must be designed with the necessary robustness for application in an industrial environment. The industry is aware of various field noises and external disturbances. Therefore, protecting and isolating the modules from possible field disturbances is essential. Thus, the implemented protective measures primarily encompass an assemblage of safeguards, such as Electromagnetic Compatibility (EMC) and Electromagnetic Interference (EMI) filters and radio frequency interference (RFI) filters.

Figure 2 illustrates the 24 V DC input circuit, which originates from an external power supply designated for both the input voltage regulator of the CPU and as a power source for the mEXs module. This circuit incorporates a range of protective elements: a high-current fuse (F2) to safeguard against excessive power supply currents, a diode (D3) to ensure correct input polarity, a varistor (RV3) providing protection against voltage spikes, the integration of a common-mode choke filter (FL3) functioning as an EMI and RFI filter to counteract common-mode noise and capacitors that are used for current noise suppression in differential mode.

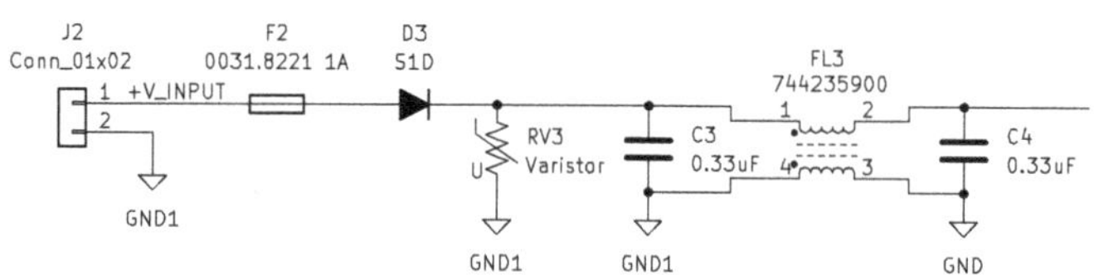

Figure 2. Input protection from External Power Supply (24 V DC).

Figure 3 presents the protection circuit for overvoltages from the 5 V direct current voltage regulator, which serves as a power supply for the Power Management Integrated Circuit (PMIC). The PMIC exerts control over voltage levels utilized by various components as well as the microprocessor and SoC housed within the mP and mEXs, respectively. The protection circuit consists of a reference diode (U1), active when there is a reference voltage above a given threshold proportional to the input voltage. In this circuit, the FET transistor (Q2) will conduct when Q1 is cut off and will not conduct when BJT transistor (Q1) is active, and this holds for an input voltage of less than about 5 V DC. In the first transistor (Q1), a BJT transistor is used since it will be activated by passing the current. In the second case, a FET transistor (Q2) is used since this is activated by applying a voltage and has better characteristics than a BJT transistor. The FET transistor has a higher input impedance and a lower gain in the passage of currents, being more stable with temperature and generating less input noise in the PMIC power supply.

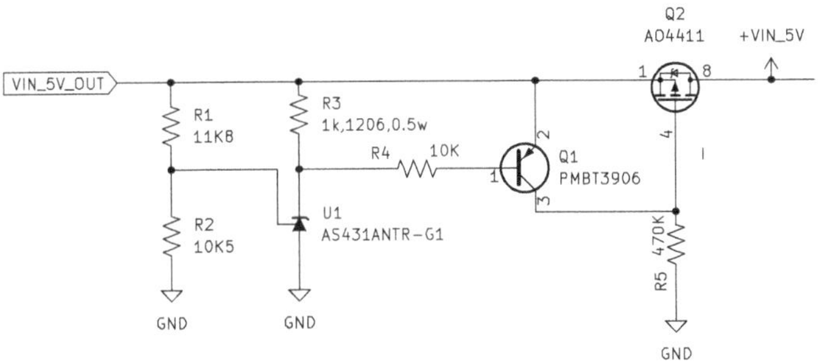

Figure 3. Input protection for overvoltages.

Galvanic isolation systems are also used in modules (both the mP and mEXs) to protect them from external interference. For example, a DC/DC isolator was used to separate the internal and external power supply. Other safeguarding mechanisms were implemented to ensure robust protection, including the strategic use of decoupling capacitors and ferrite beads in the power supplies. These additional measures bolster the system's stability and mitigate potential power fluctuations and electromagnetic interference risks.

3. Principal Module—mP

This section presents the principal module, the mP. It is composed of a CPU (in this case, the MCIMX6D5EY-M10AE, manufactured by NXP Semiconductors/Freescale) featuring two ARM Cortex A9 cores, a 32-bit data bus, and a maximum frequency of 1 GHz. The system can support the following peripherals: Serial Advanced Technology Attachment (SATA) connector, Secure Digital SD card, High-Definition Multimedia Interface (HDMI), Universal Serial Bus connector (USB), Ethernet communication, EtherCAT communication, and Random Access Memory (RAM) memory. The overall architecture of the mP can be seen in Figure 4.

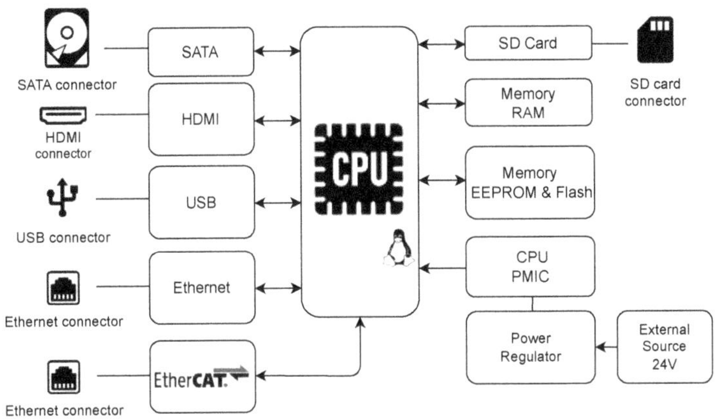

Figure 4. mP architecture overview.

The SATA connection allows for the connection of a Solid-State Drive (SSD) to the module, allowing for the storage of collected data and serving as a backup for the modular system. The SD card allows for the incorporation and initialization of the operating system. The HDMI connection is primarily necessary to facilitate the configuration of the microprocessor and its operating system (Linux), allowing for graphical displays. The USB connection facilitates connections to computers for operating system configurations, as well as the exchange of information with external disks. The USB port is standard in current embedded systems. The Ethernet connections also allow for system configuration, as well as the module's remote control and data transmission to the cloud. Additionally, the Ethernet connector allows the mP to be established as an EtherCAT master, controlling the EtherCAT network, controlling slave devices, and managing the information collected and processed in the mEX. For RAM, the use of 1 GB of memory was dimensioned as sufficient, considering the processes used, the Linux operating system, and the applications used. The 2 Kbits Read-Only Memory (ROM) and 32 Mbits Flash memory are both nonvolatile memories, with the information contained in the ROM memory unable to be updated, while in the Flash memory, there is an updating possibility. The ROM and Flash memories store the BIOS (Basic Input/Output System) or firmware. The module operates on a 24 V DC power supply obtained from an external source located in the electrical panel. To ensure proper functioning and safeguard the components, the module necessitates protection against overvoltage, as previously explained in Section 2. This is achieved by a PMIC routing the supply voltage through a voltage regulator, which adjusts the voltage to an appropriate level for all the components within the module.

4. Expansion Modules

The expansion modules, the mEXs, capture and locally process data directly from the field through sensors. Figure 5 presents an example of a generic mEX architecture composed of two sensors. These sensors convert external stimuli into analog values, which are then prepared for further processing through a signal-conditioning circuit. This circuit ensures that the signals fall within the acceptable range for the ADC (0 to 5 V). More than external disturbances, internal disturbances can also be present due to the components' imprecision in the condition circuit. These components were chosen to have high precision, for example, with impedances with a tolerance of 0.1% and ADCs with high resolution (12 bits of resolution). The error provided by the internal components will be reduced at the acceptable range of output (0–5 VDC). The SPI protocol established communication between the SoC device and the ADC near it. This protocol was favored because of the close distance between devices and the high-speed full-duplex communication capabilities.

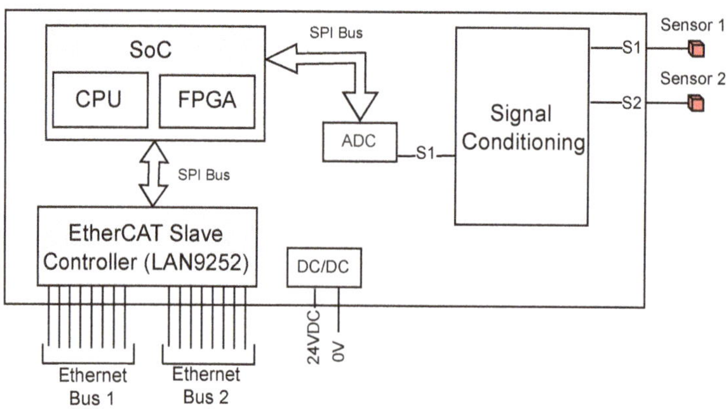

Figure 5. mEX architecture overview.

The SoC incorporated on the mEX has a CPU and an FPGA, offering exceptional flexibility for implementing algorithms. In scenarios where a low latency and real-time application is required, and if the algorithm is simple to implement in hardware, it can be efficiently implemented using the SoC's built-in FPGA. However, if the algorithms to be implemented in the mEX are complex and time-consuming to implement using VHDL or any other Hardware Description Language, or if there is a need for a quick time-to-market, they can be initially implemented in a processor.

Each mEX module is equipped with an EtherCAT slave controller, specifically the LAN9252 IC. The SoC communicates with the LAN9252 through the SPI protocol and allows the module to function effectively as a participant on the EtherCAT bus. The benefits of utilizing the EtherCAT protocol are elaborated upon in Section 2.

In this paper, two mEXs developed are presented: the mEX for electrical signals (Section 4.1) and the mEX for vibration signals (Section 4.2).

4.1. Electrical Signals Expansion Module

An innovative electrical signal expansion module designed specifically for industrial environments has been developed. This module efficiently gathers field electrical signals of a three-phase power supply, processes them, and transmits the processed data to the main module via EtherCAT for further analysis and integration into the industrial system. One of the key features of this mEX is its ability to perform voltage and current readings directly by simultaneously measuring each phase in parallel with the neutral. Analyzing these signals allows for the monitoring of industrial equipment or the prediction of failures. In [12], a case study of "Load Disaggregation Application" can be observed with the respective module, where you could locally process some features through the new module using the methodologies presented below in Section 4.3. This expansion module is updated in relation to the module in [12] by considering local processing.

Figure 6 presents the electrical signal expansion module developed. This expansion module obtains features from a three-phase power supply in the industry using current and voltage sensors. The sensors were dimensioned to detect disturbances that are essential to a precise analysis of the electrical network. Rogowski sensor [14] is used as a current sensor. This sensor is adequate for an industrial environment, dealing with high currents, has an easy application, is isolated from the network (galvanic isolation), and is less perturbed by temperature changes [15]. The current sensor also has an extensive range of reading frequencies and good linearity, which is essential to better analyze current signals. The current sensor was selected to measure a current until to a range of 1 kA corresponding to a 22 mVpk of output from the sensor. This range was considered to analyze the industry's current signals and their respective disturbances. The divider voltage is used as a voltage sensor. It allows the direct reading of the actual voltage signal, deals with high voltages, and has a minor measurement error compared to transformer use. The divider voltage and the conditioning circuit voltage were designed to collect voltages of 350 Vpk, although the typical voltage present in the electrical network is 325/330 Vpk. These 350 Vpk were established to allow the measurement of the disturbances at the three-phase power voltages, which can be essential to the electrical network's analysis.

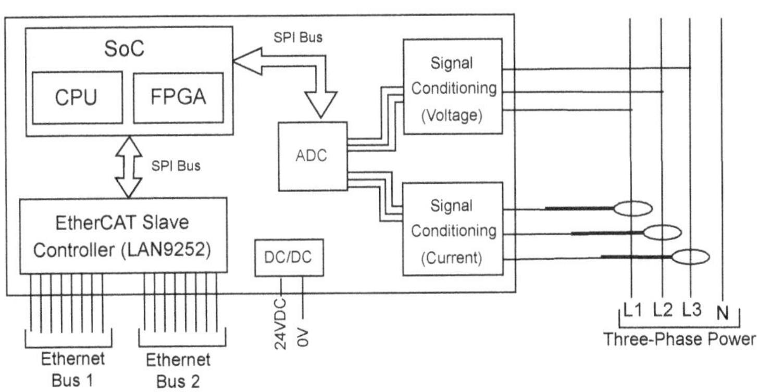

Figure 6. Architecture of the proposed mSE.

As previously presented, the signals from voltage and current sensors need to be conditioned to the range of the ADC. Therefore, two conditioning circuits are presented, one to the current sensor (Figure 7) and the other to the voltage sensor (Figure 8).

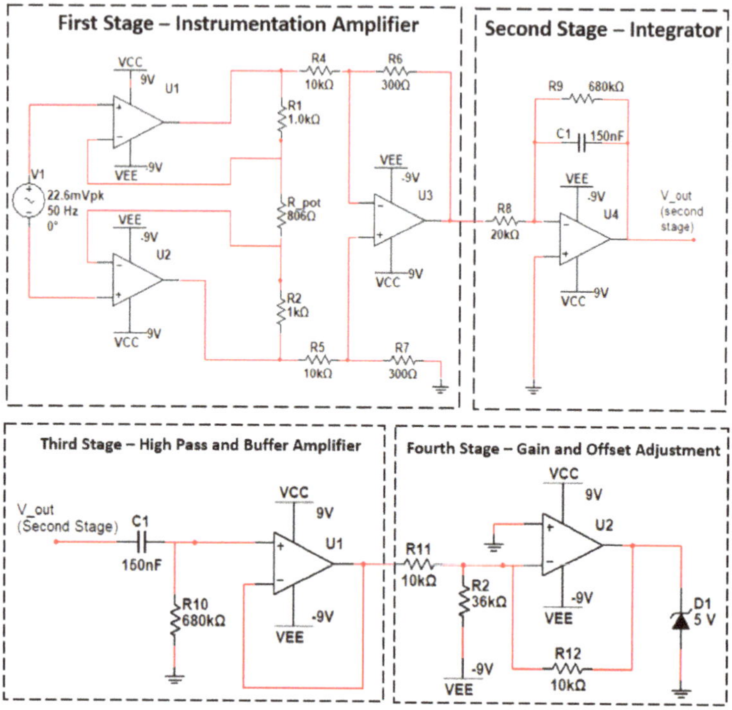

Figure 7. Conditioning circuit of the current signal.

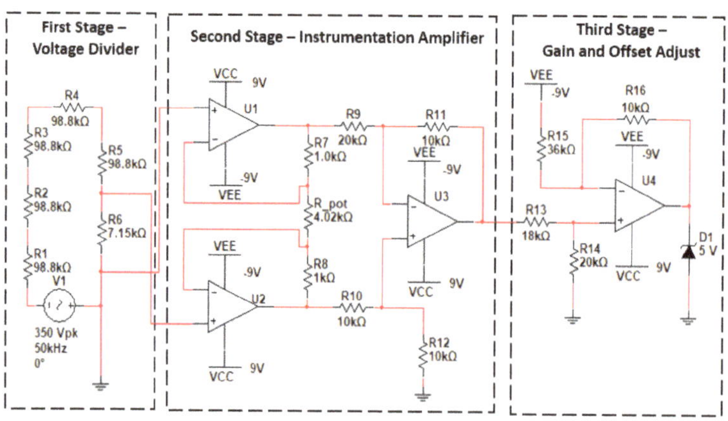

Figure 8. Conditioning circuit of the voltage signals.

Figure 7 presents the conditioning circuitry bridging the current sensor and the ADC. This intricate circuitry is constructed through a sequence of four distinct stages:

- The first stage is an instrumentation amplifier where the main objectives are to amplify the signal to the magnitudes of ADC input and establish a high input impedance on the expansion module as protection from external currents.

- The second stage is an integrator due to the signal current derivative from the current sensor. The stage needs to work as an integrator to the electrical network frequency (50 Hz).
- The third stage is a high pass filter because the integrator works as a high amplifier to the DC components (0 [Hz]). Therefore, a filter was implemented to eliminate the offset voltage. Also, a buffer amplifier was implemented in this stage to isolate the signal to the last stage.
- The last stage has implemented an inverting amplifier to the gain and offset adjustments. In this case, it was adjusted to an output of 0 to 5 V to the ADC. The Zener diode prevents overvoltages above 5 V DC, being an input protection for the ADC.

The circuit was designed not to have any phase shift beyond the integrator of 90 degrees, which is expected to obtain the original current signal due to the derivative signal originated by the current sensor.

In Figure 8, it is observed that the voltage conditioning circuit to the ADC is composed of three stages: (1) the first stage uses a voltage divider typology to reduce the real voltage to one magnitude admissible by the ADC, where various resistors were used to reduce the power dissipated by each resistor; (2) the second stage is an instrumentation amplifier to establish a high input impedance in the circuit to prevent high current flow inside the module; and (3) the third stage intends to adjust the gain and DC offset to obtain an output of 0 to 5 V on the ADC.

After signal conditioning to the range of the ADC, the analog signal is converted to digital. The SPI communication enables transmitting the collected information from the ADC to the processing unit, the SoC, where data can be processed (see Section 4.3) in a CPU or FPGA. In this module, the ADC requires six channels for data collection: three dedicated to gathering information from a voltage and another three dedicated to the current. For this module, it is also important that the ADC follow the following requirements: has simultaneous measuring of voltage and current in the same instant from the three-phase power supply and has a minimum of 10*k* samples per second, following the Nyquist theorem because the motor signature on the industry is typical in frequencies 0 to 5 kHz.

The module features a dual Ethernet port configuration that excels as an EtherCAT slave. This design empowers the module to engage with other dedicated external modules and the mP.

4.2. Vibration Signals Expansion Module

Another mEX was developed, a vibration signal acquisition and processing mEX designed specifically for industrial environments. This mEX was developed for a particular case study of finding the filling level of a cement horizontal ball mill [16]. For this, it was necessary to acquire field signals from vibration signals and extract essential features using frequency analysis. This module serves the purpose of gathering field vibration signals, processing them, and transmitting the processed data to the main module for further analysis. This module can be used in other scenarios, such as analyzing engine and bearing vibrations to detect or predict faults [17], and so on.

In Figure 9 presents the overall architecture of the dedicated mEX for vibration signal analysis applied on a cement mill as a case study. The cement mill (in this case, a horizontal ball mill, the most commonly used in the cement industry) is composed of two chambers [16]: the first chamber, consisting of larger metal balls, corresponds to the raw material inlet chamber for the coarse grinding of the raw material; and the second chamber, with smaller balls, corresponds to the outlet chamber where the fine grinding process takes place. The acquisition of vibration signals is carried out by placing a vibration sensor at the inlet and another at the mill outlet to estimate the filling level of the first and second chambers, respectively.

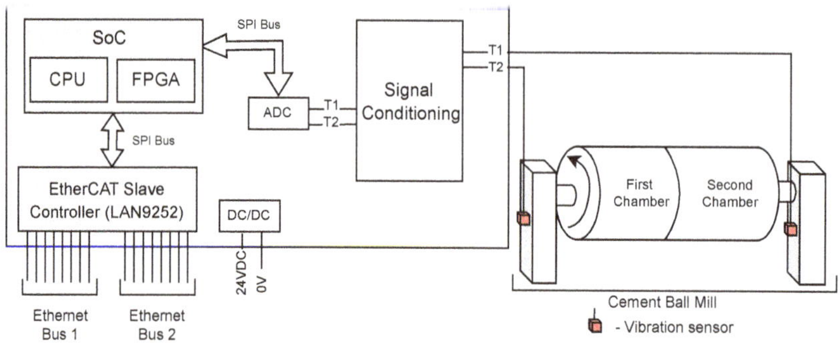

Figure 9. Architecture of the proposed mSV.

A piezoelectric accelerometer is used as an element of the vibration sensor to acquire vibration values. For the case study on the cement mill, the vibration sensor requires an enclosure that facilitates its attachment to the mill, especially in situations involving calibrations or moving it to areas with different vibration levels. Furthermore, ensuring adequate isolation and robust protection for the sensor against external factors is of utmost importance, as it will be subjected to intense disturbances and dust particles.

The vibration sensor is power supplied through the vibration signal module, with a supply voltage (5 V) and 0 V (GND). This results in a bias voltage of $Vcc/2$ at the particular sensor to the case study and an output voltage range corresponding to the vibration readings from the environment, ranging between 0.5 and 4.5 V.

This mEX features a signal-conditioning circuit (see Figure 10) for the vibration sensors' signals (ranging from 0.5 to 4.5 V) to convert them into a voltage between 0 and 5 V. The signal-conditioning circuit for the vibration signal consists of two stages: (1) the first stage is responsible for amplifying the signal to achieve a peak-to-peak voltage of 5 V, and (2) the second stage, adjustments to the DC offset and signal inversion are performed, resulting in an output voltage signal ranging from 0 to 5 V.

The signal circuit consists of two inverter amplifiers providing a final signal without phase shift.

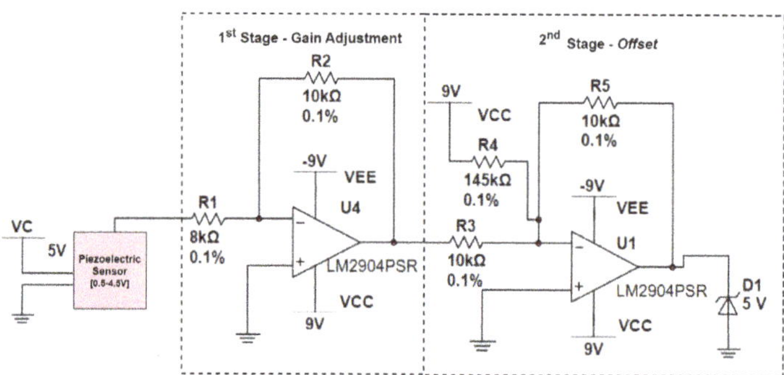

Figure 10. Vibration signal-conditioning circuit.

After signal conditioning to the range of the ADC, the analog signal is converted to digital. The SPI communication enables the transmission of the collected information from the ADC to the processing unit, the SoC, where data can be processed (see Section 4.3) in a CPU or FPGA. The ADC necessitates two distinct channels for data collection, with each channel exclusively designated for acquiring information from a sensor positioned within its corresponding mill chamber. The ADC does not need simultaneous sampling. Instead, the ADC calls for a sampling frequency that ensures accurate sensor readings aligned with the individual sensor's frequency in adherence to the Nyquist theorem.

Similar to the previous module (the mSE), this module also features a dual Ethernet port configuration working as an EtherCAT slave, benefiting from the EtherCAT characteristics.

4.3. Local Processing

Figure 11 presents the data workflow on the mEXs. In the mEXs, the acquired signals can transform in their frequency domain to be later used on intelligent computational algorithms. Thus, the Fast Fourier Transform (FFT) was implemented on the proposed mEXs, more precisely, the Cooley–Tukey Generalized (CTG) algorithm of the FFT. In addition to the FFT, the mEXs incorporate a post-processing step where the transformed data are passed through a curve-fitting algorithm (the Cubic Spline method). This algorithm derives an interpolation function for the transformed dataset, obtaining the coefficients that accurately represent the dataset to be transmitted from the mEX to the mP over the EtherCAT network, reducing the amount of communication data.

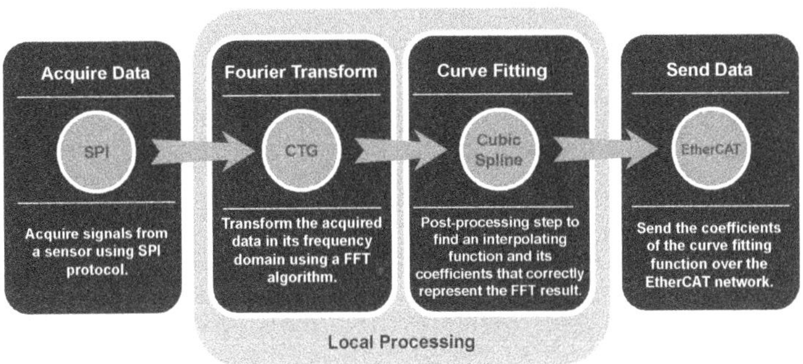

Figure 11. Data workflow within the expansion modules.

4.3.1. Cooley–Tukey Generalized (CTG) Algorithm

The CTG algorithm [18] was preferred to be implemented in the mEXs given its flexibility and ease of implementation. Adopting the CTG algorithm in expansion modules can effectively cover a broader range of case studies involving frequency analysis. This advantage arises from its capability to handle a wider range of input data sizes for computing frequency domain representations.

Recognizing the broader potential of their algorithm, Cooley and Tukey acknowledged its capacity to embrace vectors of diverse dimensions. This flexibility extends to vectors x of length N, particularly when N assumes a composite value. N is composite when $N = N_1 \cdot N_2$, where N_1 and N_2 are integer factors of N. The first step of the Cooley–Tukey algorithm derivation is to substitute the indices in the one-dimensional DFT expression along with the relation $N = N_1 \cdot N_2$ to obtain the following expression:

$$X_{k_1 k_2} = \sum_{n_1=0}^{N_1-1} \sum_{n_2=0}^{N_2-1} W_{N_1 N_2}^{(n_1+n_2 N_1)(k_2+k_1 N_2)} x_{n_1 n_2}, \tag{1}$$

where $n_i, k_i = 0, 1, \ldots, N_i - 1$ ($i = 1, 2$ the dimension of the matrix) and k refers to the index of the frequency component being analyzed in the frequency domain. The twiddle factors matrix W can be expanded as:

$$W_{N_1 N_2}^{(n_1+n_2 N_1)(k_2+k_1 N_2)} = W_{N_1 N_2}^{n_1 k_2} \cdot W_{N_1 N_2}^{n_1 k_1 N_2} \cdot W_{N_1 N_2}^{n_2 k_2 N_1} \cdot W_{N_1 N_2}^{n_2 k_1 N_1 N_2} \tag{2}$$

where $W_{N_1 N_2}^{n_2 k_1 N_1 N_2} = 1$, and $W_{N_j N_i}^{n_i k_i N_j} = W_{N_i}^{n_i k_i}$ [19]. Therefore, the two-dimensional Cooley–Tukey Fourier transform equation is given by:

$$X_{k_1 k_2} = \sum_{n_1=0}^{N_1-1} W_{N_1}^{n_1 k_1} W_{N}^{n_1 k_2} \sum_{n_2=0}^{N_2-1} W_{N_2}^{n_2 k_2} x_{n_1 n_2}. \quad (3)$$

4.3.2. Cubic Spline Algorithm

The curve-fitting process involves selecting a mathematical function to represent the data and finding the coefficients of the function that best fit the function to the data points. The choice of function depends on the nature of the data and the desired level of accuracy. Common functions include linear, polynomial, logistic functions, and spline methods. Although linear and polynomial curve-fitting algorithms are straightforward to develop and computationally efficient, they may be incapable of capturing more complicated patterns in data. Furthermore, high-degree polynomial fits might be prone to overfitting, resulting in poor generalization performance on new data. In light of these considerations, using a Cubic Spline interpolation technique is preferable. Unlike regression-based approaches that approximate data with a single function, Cubic Spline interpolation finds an interpolation function that precisely passes through selected data points. This method maintains the smoothness of the curve while being less prone to overfitting, offering better generalization performance on new data [20,21].

The cubic spline algorithm is a technique that divides the full dataset into smaller intervals and fits a cubic polynomial function to each of the data partitions. These cubic polynomials are selected to pass through every data point and ensure smoothness by maintaining continuous first and second derivatives at the interval intersections [21].

The nodes, which represent local maxima or minima in the dataset, act as the boundaries between these segments. A cubic polynomial interpolates between every two consecutive nodes. In the case of having p knot points $((x_1, y_1), (x_2, y_2) \dots (x_p, y_p))$, there will be $p-1$ segments, each defined by a cubic polynomial $S_i(x)$, as follows:

$$\begin{aligned} S_{p-1}(x) &= y_{p-1} + b_{p-1}(x - x_{p-1}) + c_{p-1}(x - x_{p-1})^2 + d_{p-1}(x - x_{p-1})^3 \\ & \quad x_{p-1} \leq x \leq x_p \end{aligned} \quad (4)$$

Additionally, the first (S') and second (S'') derivatives of all polynomials are identical in the continuity of adjacent polynomials. Thus, coefficients a, b, and c for each polynomial are obtained by solving the following equations that guarantee the continuity and smoothness of the final total spline function.

$$\begin{aligned} S_i(x_i) = \ & y_i \text{ and } S_i(x_{i+1}) = y_{i+1} \text{ for } i = 1, 2, \dots, p-1, \\ & S'_{i-1}(x_i) = S'_i(x_i) \text{ for } i = 1, 2, \dots, p-1, \\ & S''_{i-1}(x_i) = S''_i(x_i) \text{ for } i = 1, 2, \dots, p-1. \end{aligned}$$

The spline system of equations is commonly solved by adding new boundary conditions and using the iterative Jacobi method or the Gauss-Newton method.

5. Results and Discussion

This section discusses the developed modular hardware system (Section 5.1) and the results of the implemented local processing methodologies (Section 5.2).

5.1. Proposed Industrial Control System

As a final result, the Figure 12 illustrates the envisioned hardware components of the developed industrial control system, which consists of the principal module alongside the mSE and the mSV expansion modules. The printed circuit boards (PCB) of modules, including the architectures and the respective hardware protections previously explained, were designed for the enclosure offered by Phoenix Contact, the ICS series for IoT (Internet of Things) applications. This casing is acclaimed for its compactness, robustness, thermal endurance, and facilitated integration of heatsinks [12]. The communication between modules is completed by a shared EtherCAT bus, employed through a dedicated communication rail tailored for industrial settings. This rail also shares a common power supply to all interconnected modules.

Figure 12. Industrial Control System developed with expansion modules coupled.

The system presented in this paper goes towards the objective of creating a modular hardware system with distributed processing specifically designed for industrial environments. The developed modular hardware architecture successfully incorporates the established features for the desired system, enabling the acquisition and processing of field data within the mEXs. Notably, the proposed architecture incorporates Plug&Play functionality [22] by facilitating the addition of new modules to the network by incorporating EtherCAT communication between the mP and the mEX.

The mEXs are equipped with an SoC, housing a CPU and an FPGA, which bestows the module with the versatile capability of executing signal processing algorithms through software and programmable logic. In scenarios where a low-power, low latency, and real-time application is required, and if the algorithm is relatively simple, it can be efficiently implemented using programmable logic, the FPGA. Moreover, the mSV has already been tested in the real cement industry, and the mSE was tested at the laboratory on an electrical network composed of several industrial motors.

5.2. Local Processing

The present system successfully acquires and processes data locally near where it is acquired, incorporating local processing concepts. The mSV case study of acquiring and processing vibration signals from a cement ball mill was tested in a real industrial scenario. First, according to the data workflow described in Figure 11, the data acquired from sensors are transformed and analyzed in its frequency domain by the CTG algorithm presented in Section 4.3.1, and later, it undergoes a post-processing method, the cubic spline algorithm presented in Section 4.3.1, to reduce the amount of communication data.

5.2.1. CTG Implementation

It is essential to transform the incoming data to the frequency domain to implement some computational intelligence algorithms based on the data acquired from the field sensors. As mentioned before, the CTG transform algorithm was selected as the most appropriate choice. The CTG algorithm offered several advantages in being implemented in the mEX. It enables low-latency computation of the Fourier Transform, ensuring timely data processing. Moreover, it exhibits remarkable flexibility in handling different input sizes, surpassing its counterparts in the FFT algorithm family. Figure 13 presents the FFT obtained by the CTG on the proposed mSV. The signals were acquired from a vibration sensor placed in a cement ball mill, and the FFT was successfully computed locally in the mSV.

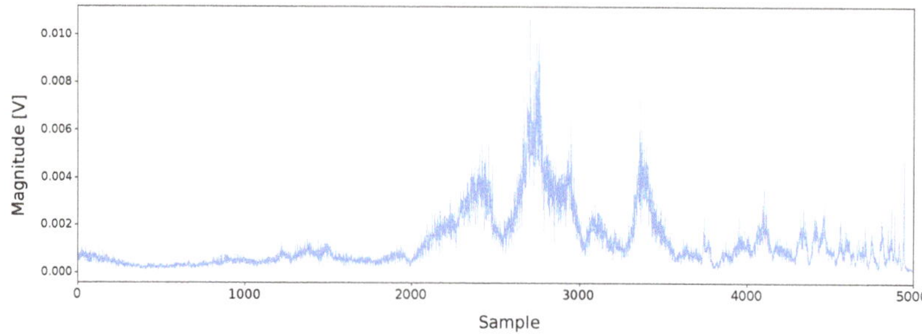

Figure 13. Results of the FFT from a vibration sensor placed in a cement ball mill.

5.2.2. Cubic Spline Implementation

An effective post-processing technique, the Cubic Spline, was successfully employed to enhance network efficiency by reducing data rates and communication latency. This technique was specifically applied to the final result of the FFT CTG algorithm to transmit the essential information without transmitting the entire set of FFT data points. By integrating the Cubic Spline algorithm into the SoC platform, a significant increase in network availability was achieved by not transmitting the complete dataset. This advancement entailed transmitting the pertinent fitted function coefficients solely, accurately capturing the essence of the original data rather than the complete dataset. Figure 14 presents the FFT of acquired field data from a vibration sensor in a cement ball mill and its respective Cubic Spline curve fitting applied to the FFT curve computed locally in the mSV. Compared to the FFT curve, the obtained interpolation function dominates the algorithm's success since the fitted curve represents the original data with satisfactory results. This approach makes it possible to send the computed function coefficients over the network instead of sending all the FFT data, sending less data and enhancing network availability.

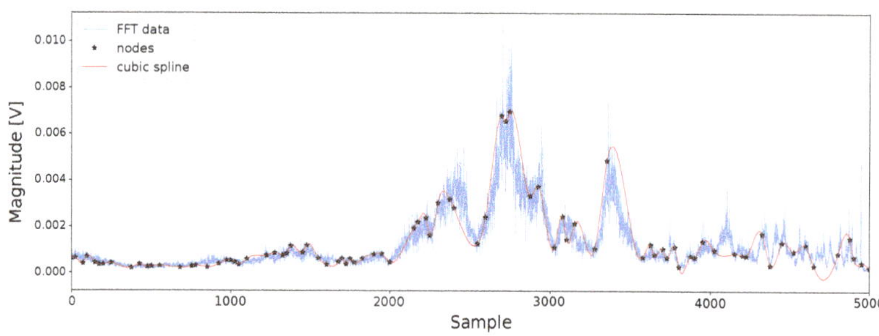

Figure 14. Results of an FFT and Cubic Spline curve fitting from a cement ball mill.

6. Conclusions

The results presented in this paper go towards the objective of creating a modular industrial hardware control system with distributed processing specifically designed for industrial environments.

Initially, the system's architecture was defined as scalable, modular to the new challenges of the industry, and easy for the user to use. The industrial control system also presents essential hardware protections from external perturbations to stable work and to prevent failures in the industrial environment. The system is constituted by a principal module with network control features and high local processing capability. Additionally, two case studies of mEXs (mSE and mSV) incorporated in the industrial control system are presented. The mEXs are essential to the literature for designing boards collecting vibration and electrical signals due to their large range of applications and relevance in the industrial field. Additionally, the mExs incorporate the SoC, allowing for distributed and local processing in the system. This processing allows the data analysis to be carried out closer to where data are generated. Notably, the EtherCAT protocol is employed in the modular architecture, which enables not only deterministic real-time communication among these modules but also incorporates Plug&Play functionality, fostering the easy addition and replacement of modules to the network. The referenced features of the presented architecture increase the efficiency and speed of data processing, making the architecture more adequate for real-time applications.

Finally, the developed modular hardware architecture successfully incorporates feature extraction algorithms for the desired system, incorporating local processing concepts. The industrial control system simplifies the incorporation of different local processing algorithms in the mEX for different case studies and industrial applications. It provides the capability for hardware acceleration for algorithms amenable to parallelization and allows for the deployment of algorithms on a robust processor when it proves more suitable for the specific use case. Two methods of local processing algorithms were used: frequency domain analysis and the curve-fitting process. Frequency domain analysis was used to process sensor data and, consequently, feed intelligent computational methodologies. The cubic curve fitting (Cubic Spline Algorithm) optimizes data transmission over the network by selectively sending only the essential characteristics of the dataset.

As a result, the system becomes highly adaptable, deterministic, reliable, and easily expandable with robust data processing and communication, allowing the versatile implementation of more complex algorithms and models for real-time applications.

As future work, better analysis and new applications domains for the modular control system can be tested. Key performance indicators such as processing speed, dependability, and adaptability need to be evaluated and compared with current control systems. Performance benchmarks can offer verifiable proof of the benefits of our system. Additionally, complex computational analyses, including advanced optimization methods or predictive maintenance models based on machine learning, will illustrate the effectiveness and suitability of the suggested techniques more successfully. The performance analysis through the modular control system, more specifically through expansion modules, requires the system to be tested in different industrial process environments, validating its reliability. The system will be tested in cement, microalgae, biomass, mines, and wastewater treatment plants, among others. More and better analysis and new application domains, as well as an evolution of the system with technical advancements, can be developed with possible technical advancements. A significant technical advancement can be the evolution of architecture to be compatible with distributed processes, which do not have a dependency on central resources to manage the system. The current system is characterized as a centralized architecture, with the expansion modules depending on the principal module. A distributed system will lead to a more reliable architecture and industry process. The approach of new techniques will be essential to achieve a distributed system, like the IEC 61499 standard [23], orchestrations of microservices and the virtualization of services. The introduction of these new technologies also leads to new challenges, such as real-time control in distributed and virtualized processes.

Author Contributions: Conceptualization, methodology, formal analysis: G.G., J.A., P.S. and J.M.; writing—original draft preparation: G.G. and J.A.; validation, writing—review and editing, supervision: P.S., R.A. and J.M. All authors have read and agreed to the published version of the manuscript.

Funding: This research was co-financed by the European Regional Development Fund, through Centro Regional Operational Program 2014/2020 (Centro2020), of Portugal 2020. Project iProMo (CENTRO-01-0247-FEDER-069730). Jorge Alves is supported by Fundação para a Ciência e a Tecnologia (FCT) under the grant ref. 2023.00998.BDANA. This research is partially sponsored by national funds through FCT under projects UIDB/00285/2020 and LA/P/0112/2020.

Data Availability Statement: Data are contained within the article.

Conflicts of Interest: The authors declare no conflicts of interest.

References

1. Mao, W.; Zhao, Z.; Chang, Z.; Min, G.; Gao, W. Energy-Efficient Industrial Internet of Things: Overview and Open Issues. *IEEE Trans. Ind. Inform.* **2021**, *17*, 7225–7237. [CrossRef]
2. de Matos, B.; Salles, R.; Mendes, J.; Gouveia, J.R.; Baptista, A.J.; Moura, P. A Review of Energy and Sustainability KPI-Based Monitoring and Control Methodologies on WWTPs. *Mathematics* **2023**, *11*, 173. [CrossRef]
3. Teixeira, J.E.; Tavares-Lehmann, A.T.C. Industry 4.0 in the European union: Policies and national strategies. *Technol. Forecast. Soc. Chang.* **2022**, *180*, 121664. [CrossRef]
4. Queiroz, R.; Cruz, T.; Mendes, J.; Sousa, P.; Simões, P. Container-Based Virtualization for Real-Time Industrial Systems—A Systematic Review. *ACM Comput. Surv.* **2023**, *56*, 1–38. [CrossRef]
5. Törngren, M.; Thompson, H.; Herzog, E.; Inam, R.; Gross, J.; Dán, G. Industrial Edge-based Cyber-Physical Systems—Application Needs and Concerns for Realization. In Proceedings of the 2021 IEEE/ACM Symposium on Edge Computing (SEC), San Jose, CA, USA, 14–17 December 2021; pp. 409–415. [CrossRef]
6. Ramamurthy, H.; Prabhu, B.S.; Gadh, R.; Madni, A.M. Wireless Industrial Monitoring and Control Using a Smart Sensor Platform. *IEEE Sens. J.* **2007**, *7*, 611–618. [CrossRef]
7. Chi, Q.; Yan, H.; Zhang, C.; Pang, Z.; Xu, L.D. A Reconfigurable Smart Sensor Interface for Industrial WSN in IoT Environment. *IEEE Trans. Ind. Inform.* **2014**, *10*, 1417–1425. [CrossRef]
8. Qiu, T.; Chi, J.; Zhou, X.; Ning, Z.; Atiquzzaman, M.; Wu, D.O. Edge Computing in Industrial Internet of Things: Architecture, Advances and Challenges. *IEEE Commun. Surv. Tutor.* **2020**, *22*, 2462–2488. [CrossRef]
9. Nain, G.; Pattanaik, K.; Sharma, G. Towards edge computing in intelligent manufacturing: Past, present and future. *J. Manuf. Syst.* **2022**, *62*, 588–611. [CrossRef]
10. Kaneko, Y.; Yokoyama, Y.; Monma, N.; Terashima, Y.; Teramoto, K.; Kishimoto, T.; Saito, T. A Microservice-Based Industrial Control System Architecture Using Cloud and MEC. In Proceedings of the Edge Computing—EDGE 2020, Beijing, China, 19–23 October 2020; Katangur, A., Lin, S.C., Wei, J., Yang, S., Zhang, L.J., Eds.; Springer International Publishing: Berlin/Heidelberg, Germany, 2020; pp. 18–32. [CrossRef]
11. Kumari, K.; Sadasivam, G.; Dharani, D.; Niranjanamurthy, M. *Edge Computing: Fundamentals, Advances and Applications;* Advances in industry 4.0 and machine learning; CRC Press: Boca Raton, FL, USA, 2021.
12. Alves, J.; Sousa, P.; Matos, B.; Mendes, J.; Souza, F.; Matias, T. Modular Cyber-Physical System for Smart Industry: A case study on energy load disaggregation. In Proceedings of the 2023 International Conference on Control, Automation and Diagnosis (ICCAD), Rome, Italy, 10–12 May 2023; pp. 1–6. [CrossRef]
13. Crockett, L.H.; Elliot, R.A.; Enderwitz, M.A.; Stewart, R.W. *The Zynq Book: Embedded Processing with the Arm Cortex-A9 on the Xilinx Zynq-7000 All Programmable Soc;* Strathclyde Academic Media: Glasgow, UK, 2014.
14. Xiao, C.; Zhao, L.; Asada, T.; Odendaal, W.; van Wyk, J. An overview of integratable current sensor technologies. In Proceedings of the 38th IAS Annual Meeting on Conference Record of the Industry Applications Conference, Salt Lake City, UT, USA, 12–16 October 2003; Volume 2, pp. 1251–1258. [CrossRef]
15. Shafiq, M.; Stewart, B.G.; Hussain, G.A.; Hassan, W.; Choudhary, M.; Palo, I. Design and applications of Rogowski coil sensors for power system measurements: A review. *Measurement* **2022**, *203*, 112014. [CrossRef]
16. Gonçalves, M.; Sousa, P.; Mendes, J.; Danishvar, M.; Mousavi, A. Real-Time Event-Driven Learning in Highly Volatile Systems: A Case for Embedded Machine Learning for SCADA Systems. *IEEE Access* **2022**, *10*, 50794–50806. [CrossRef]
17. Vanraj; Goyal, D.; Saini, A.; Dhami, S.S.; Pabla, B.S. Intelligent predictive maintenance of dynamic systems using condition monitoring and signal processing techniques—A review. In Proceedings of the 2016 International Conference on Advances in Computing, Communication & Automation (ICACCA), Dehradun, India, 8–9 April 2016; pp. 1–6. [CrossRef]
18. Temperton, C. A Generalized Prime Factor FFT Algorithm. *SIAM J. Sci. Stat. Comput.* **1992**, *13*, 676–686. [CrossRef]
19. Hayes, M. *Schaum's Outline of Digital Signal Processing;* Schaum's Outline Series; McGraw-Hill Companies, Incorporated: New York, NY, USA, 1999.
20. Wold, S. Spline Functions in Data Analysis. *Technometrics* **1974**, *16*, 1–11. [CrossRef]
21. Sun, M.; Lan, L.; Zhu, C.G.; Lei, F. Cubic spline interpolation with optimal end conditions. *J. Comput. Appl. Math.* **2023**, *425*, 115039. [CrossRef]
22. Liu, S.; Guzzo, J.A.; Zhang, L.; Smith, D.W.; Lazos, J.; Grossner, M. Plug-and-play sensor platform for legacy industrial machine monitoring. In Proceedings of the 2016 International Symposium on Flexible Automation (ISFA), Cleveland, OH, USA, 1–3 August 2016; pp. 432–435. [CrossRef]
23. lewis, R. *Modelling Control Systems Using IEC 61499: Applying Function Blocks to Distributed Systems*, 1st ed.; The Institution of Engineering and Technology: London, UK, 2008.

processes

MDPI

Article

Virtual Test Beds for Image-Based Control Simulations Using Blender

Akkarakaran Francis Leonard, Govanni Gjonaj, Minhazur Rahman and Helen E. Durand *

Department of Chemical Engineering and Materials Science, Wayne State University, 42 W Warren Ave, Detroit, MI 48202, USA
* Correspondence: helen.durand@wayne.edu

Abstract: Process systems engineering research often utilizes virtual testbeds consisting of physics-based process models. As machine learning and image processing become more relevant sensing frameworks for control, it becomes important to address how process systems engineers can research the development of control and analysis frameworks that utilize images of physical processes. One method for achieving this is to develop experimental systems; another is to use software that integrates the visualization of systems, as well as modeling of the physics, such as three-dimensional graphics software. The prior work in our group analyzed image-based control for the small-scale example of level in a tank and hinted at some of its potential extensions, using Blender as the graphics software and programming the physics of the tank level via the Python programming interface. The present work focuses on exploring more practical applications of image-based control. Specifically, in this work, we first utilize Blender to demonstrate how a process like zinc flotation, where images of the froth can play a key role in assessing the quality of the process, can be modeled in graphics software through the integration of visualization and programming of the process physics. Then, we demonstrate the use of Blender for testing image-based controllers applied to two other processes: (1) control of the stochastic motion of a nanorod as a precursor simulation toward image-based control of colloidal self-assembly using a virtual testbed; and (2) controller updates based on environment recognition to modify the controller behavior in the presence of different levels of sunlight to reduce the impacts of environmental disturbances on the controller performance. Throughout, we discuss both the setup used in Blender for these systems, as well as some of the features when utilizing Blender for such simulations, including highlighting cases where non-physical parameters of the graphics software would need to be assumed or tuned to the needs of a given process for the testbed simulation. These studies highlight benefits and limitations of this framework as a testbed for image-based controllers and discuss how it can be used to derive insights on image-based control functionality without the development of an experimental testbed.

Keywords: process control; image-based control; digital twin

1. Introduction

Image-based control (IBC) focuses on the use of image sensors to capture state information for the purposes of control. IBC systems are feedback control schemes in which the measurements are obtained from data-intensive image-based sensors. The advances and emergence of affordable camera sensors and image-processing algorithms [1] have contributed to the increased exploration of IBC systems [2]. This includes applications in autonomous driving systems [3], robot manipulators [4], and bioprocess monitoring [5,6]. Images have also been used in drug discovery [7]. The sensors in many process control applications measure a single state of the system, such as temperature or pressure. Images are able to capture significant information in a single observation, and offer the possibility of creating a dense and relatively easy method to capture system states. Over the last few decades, significant advances in image-based sensors and image processing algorithms have been reported. This has led to an increase in studies of IBC systems and their capabilities to enable real-time process control based on camera sensors to perceive the environment and measure variables that would otherwise be impractical or time-expensive to measure (e.g., concentration grade or cell counting). One of the uses of image sensors is object detection. Images have found use in robotics, in applications such as the control of autonomous robots [8,9], drone flight control [10], manipulation of surgical robots [11], and spatial exploration with mobile devices [12]. In addition to classical controllers, vision-based control strategies have also been studied using advanced controllers such as model predictive controllers (MPC's) [13,14]. In particular, image-based visual servoing control using a nonlinear MPC was proposed in Lee et al. [13] as a vision-based obstacle avoidance strategy in a dynamic environment for an unmanned aerial vehicle, where constraints related to actuator and visibility limitations can be added to the control formulation. In the context of visual feedback control, different control architectures integrated with a vision system (e.g., a fixed camera sensor) have been utilized to improve the performance of autonomous systems [4,15]. In Su and Zheng [4], for example, an image-based transpose Jacobian proportional-integral-derivative (PID) control was proposed to address the asymptotic regulation problem of robot manipulators with a vision-based feedback using Lyapunov's direct method and LaSalle's invariance theorem. In chemical engineering, IBC implementation is largely applied to physical models and relies on experimentation to deliver results and test optimal parameters. Image analysis systems were tested, for example, to monitor an industrial boiler system [16] or estimate bubble size at a phosphorus oxide flotation process [17]. The analysis of images for liquid crystal responses for sensing purposes was explored in [18]. The use of convolutional neural networks (CNNs) to analyze images for real time control has been used to develop frameworks for the safe implementation of computer vision in process control [19] and to extract geometrical and pattern information from images [20]. Image sensors have been evaluated for crystal size monitoring [21,22]. Specialized probes and lighting conditions can aid such studies. Drop size distributions monitoring aimed to use image sensors [23]. In the work by Chen et al. [24], image sensors are used to control combustion, using a principle component analysis (PCA)-based model of flame images. Similar to PCA, dynamic mode decomposition (DMD) is also used to create low-dimensional state representations of images that can be used in conjunction with model predictive control (MPC) [25]. Although image-based closed-loop systems have been performed and tested with real systems involving camera sensors [24,26], chemical processes are often large-scale and complex, making it challenging to visually replicate a next-generation manufacturing environment for the process industries without obtaining data from an actual plant. In light of this, it can be more difficult for process systems engineering researchers to test how new next-generation manufacturing concepts such as image-based control algorithms might fare in an actual plant. A simulation-based test environment for these systems would be desirable. Blender is a free and open source 3D modeling and animation software that is managed by the Blender Foundation, with the aim of providing the latest in animation technology to everyone free of charge and with the ability for creators to own the projects that they create with it. It has been used to create open source repositories for models of laboratory equipment [27], plugins for molecular dynamics simulations [28], and toolboxes for the simulation of range scanners [29] and has been used to develop models for packed beds [30].

In the prior work published by our group [31], we proposed the use of Blender for enabling the generation of images as part of a closed-loop image-based control test paradigm. In that work, a simple example of level control in a tank was used to demonstrate the use of Blender for closed-loop image-based control simulations. In this study, we significantly extend this prior work to seek to gain further insight into the potential use cases of Blender for chemical process control applications and also to better understand its benefits and limitations over a wider range of process considerations. To do this, the paper is structured to focus on three primary studies, which are the use of Blender in simulating a precursor to a froth flotation process, a precursor to a controlled self-assembly process, and a reactor for a process with a weather monitoring system. The specific organization is as follows: in the preliminaries, the method that is used in this work to acquire an optimal policy for the self-assembly precursor system (which is a stochastic system), Bellman optimization, and the basics of modeling objects and animation basics in Blender are covered. The sections following the preliminaries then discuss the simulations carried out to illustrate techniques to replicate physical phenomena and IBC systems in Blender. First, the precursor to the froth flotation process is analyzed to focus on how the animation acquired from Blender compares to results from a numerical integration technique to ensure that physical phenomena are represented with an acceptable degree of accuracy. The precursor simulation in this case is the simplified motion of a gas bubble in a liquid. Second, the precursor simulation to controlled self-assembly is presented to discuss the use of images to detect states that may be desired to be monitored using images as part of a closed-loop process. The precursor simulation in this case is the Brownian motion of a nanorod, simulated in a 2-dimensional space, with the aim of developing a control policy that guides the rod to the center of the simulated space. The control policy is optimized offline and the virtual camera in Blender is used to capture images every sampling period, detect the position and orientation of the nanorod, and perform the optimal action. The simulation functions as a precursor to more complex problems, such as those that involve controlling multiple particles in a medium, as in colloidal self-assembly. Finally, an outdoor tank reactor is simulated that is subject to sunlight as a disturbance in ambient conditions. This simulation describes one of the more practical uses of image sensors for disturbances, since weather effects might be identified through their effects on other system parameters. Camera sensors can enhance the responsiveness of a control system by directly utilizing image information to predict probable impacts on the process due to weather. The optimizer used to control this system includes a disturbance model that accounts for the varying sunlight, which correlates information from the image in order to update the control action. The performance of this optimizer is compared to one that does not possess a similar disturbance model.

Our main goal in this work is to analyze the capabilities of Blender for process engineering applications and to explore its benefits and limitations with respect to such applications and also to discuss how simulations that integrate the animation and visualization capabilities of Blender with more typical process engineering considerations (e.g., numerical integration and control) might be developed. The goal is not to imply that Blender is necessarily the best tool for such simulations (though we view its open-source nature as a positive characteristic for research) but rather to discuss how such simulations might be set up and developed and some of the considerations that must be taken into account in building engineering-relevant simulations within it.

2. Preliminaries

2.1. Notation

$\mathbb{R}$ and $\mathbb{Z}^+$ represent the sets of real numbers and non-negative integers, respectively. The variables denoted in bold (e.g., $\mathbf{r}$, $\mathbf{D}_t$) are tensors. The random variables described in this work are normally distributed. For a random variable x, the mean is denoted by $\langle x \rangle$, and the variance is denoted by $\langle x, x \rangle$. For a normally distributed vector $\mathbf{r}$, the variance is denoted with a transpose as $\langle \mathbf{r}, \mathbf{r}^\mathsf{T} \rangle$.

2.2. Stochastic Processes and Their Control

One of the examples of the use of Blender for integrated process modeling and visualization toward image-based control testing that is presented in this work involves stochastic processes. We, therefore, highlight the key background on stochastic processes and their control that comes up in the discussion of that system.

Stochastic processes are defined with random variables, leading to a non-deterministic behavior [32,33]. A process is said to have the Markov property if the probability of a future state is equivalent regardless of whether all the previous states are known or if only the current state is known. A Markov decision process (MDP) is a subset of stochastic processes based on Markov chains, where the probability of a future state of a process is dependent on its current state. MDPs modify this definition by introducing a choice of action to be taken, i.e., the probability of a given future state is dependent on both the current state and the action taken in that state. A MDP is defined as a tuple of 4 sets: S, the state space of the system; A, the set of actions available for control; P, the transition probabilities; and R, the immediate reward of achieving a given state [34]. The reward set R is generally used to define MDPs in the context of robotics, but in the chemical engineering applications in this work, it is recast as a cost set C.

The control of an MDP is carried out with policy $\pi : S \mapsto A$, which returns an action to be executed when the process occupies a given state. The optimal policy π^* is the one that minimizes the infinite horizon objective function $J(\pi)$, defined as:

$$J(\pi^*) = \min \left[E\left\{ \sum_{t=0}^{\infty} \gamma^t C(s_t) \right\} \right] \tag{1}$$

Dynamic programming can be applied to optimize this problem, and in this work, Bellman's optimality condition is used to arrive at an optimal policy. The value function $V : S \mapsto \mathbb{R}$ of state s is a representation of the cost of the current state as well as the cost of any probable future states that can be accessed from the current one. The Bellman optimality condition defines the optimal value function V^* for an MDP recursively, as in Equation (2) [35]:

$$V^*(s) = C(s) + \gamma \min_{a \in A} \sum_{s' \in S} P(s'|s,a) V^*(s') \tag{2}$$

A greedy algorithm can attempt to minimize the cost associated with the next action. The transition probability $P(s'|s,a)$ in Equation (2) captures the non-determinism of the system by weighing the values $V^*(s')$ of future states s' by the probability of acquiring that future state, given the current state s and the action a taken at the time. S represents the set of all possible states of the system. The discount factor γ penalizes the value of future states by adding a fraction of the future states to the current one.

Equation (2) is recast as an update rule to perform value iteration, as in Equation (3), known as the Bellman update rule [35]:

$$V_{i+1}(s) = C(s) + \gamma \min_{a \in A} \sum_{s' \in S} P(s'|s,a) V_i(s') \tag{3}$$

Here, i represents the index of iteration. Through repeated iterations, the value of every state in the system is converged, resulting in an optimal value function $V^*(s)$. This algorithm presented here and used in this work is called value iteration and is one of several algorithms used to optimize MDPs. The mapping of the value function from a previous iteration V_i to the next V_{i+1} is a contraction mapping, referred to as the Bellman operator [35]. The action selected to minimize the value (and by extension the cost) forms the policy of the system. Unlike the policy of deterministic systems that are trajectories through time, the optimal policy $\pi^* : S \mapsto A$ of stochastically controlled systems consists of optimal actions connected to a given state of the system.

2.3. Modeling and Animation in Blender

Blender is a computer graphics software that is able to represent the three-dimensional world spatially, represent it in action (i.e., animated), and interface such actions on the visual world with codes through a Python coding interface that can be used to code typical modeling strategies in chemical engineering (e.g., control laws or process physics). Blender is an open-source program that can be downloaded for free and is licensed under the GNU General Public License. This means that projects created with Blender are the sole property of the user, which makes it of interest to evaluate for potential research purposes. In Blender, three-dimensional worlds can be constructed by adding basic shapes (e.g., cylinders and icospheres) and then, from there, performing operations (e.g., extrusion, scaling, translating) to make them into models of various objects in three-dimensional space. The movement of these objects can then be added, either by adding rigs to achieve allowable motions of the object to give it more complex movements, or through, as is performed in any cases in which animation is used in this work, translations, rotations, and scalings of the original object throughout space. This section goes over the features of Blender that make modeling possible and how they can be leveraged to simulate target processes. Figure 1 details the information flow of the simulations described in this paper. In each of the simulations, the objects in the 3D environment, materials, lighting, and environment are initially designed in Blender and stored in a .blend file. Then, a Python script (.py script) is used to encode the process dynamics, animation of object properties, and image capture using the virtual camera. These images are stored in a local directory, then loaded back into the script using Pillow [36], which is then used to analyze the image data in order to perform state detection. Additionally, parameters of the process such as calibration information, control policies that are optimized offline, or other physical parameters can be loaded from the local directory through the Python script. In this work, the parameters used for the simulation are stored as .mat files and are loaded into the simulation using the `savemat` function, available in the SciPy library. After the script is run, the results are stored in .mat files, and the video of the animation, as well as images from the process, may be rendered from the .blend file.

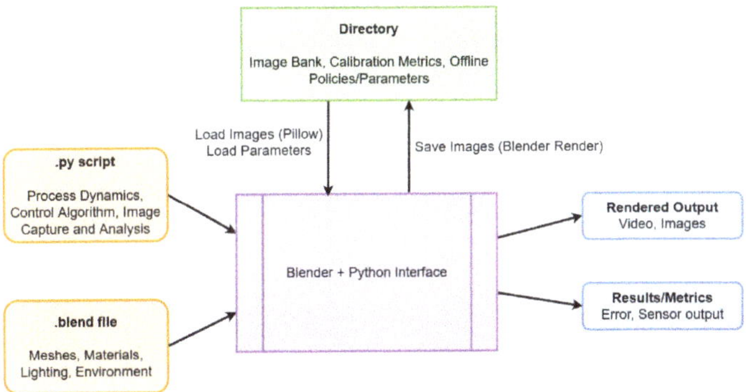

Figure 1. Information flow for IBC simulations.

2.3.1. Object Creation and Sizing

Most three- and two-dimensional objects in Blender are categorized as "meshes" and can be added to a scene. When an object is added, it appears at the location of the cursor occupying the space of a 2 m × 2 m × 2 m cube. The location, size, and orientation of an object can be found by selecting the object in the 3D viewport and navigating to the object properties tab. The location of a new object can be specified relative to the location of the cursor in meters, and its orientation can also be modified. The rotation of objects in Blender is determined in Euler mode by default but can be changed to Quarternion or Axis Angle mode. The scaling of objects in Blender is performed relative to the initial size and, hence, does not use units. For example, to define a cuboid of size 2 m × 20 cm × 4 m, the scale factors in the X, Y, and Z directions are set to 1, 0.1, and 2, respectively, from a base-case cube that has been added to the scene. When implemented, the rendered output would look like the image displayed in Figure 2, with the inset displaying the scale factors. This implies that when setting physical parameters, a scaling factor relative to how objects are scaled in the simulation must be included. By default, a newly created object is not equipped with a material and, hence, appears white in the 3D viewport window when the solid shading option is selected. Selecting an object and navigating to the material properties tab, the object can be equipped with a material, which allows the specification of material properties like color and roughness. These properties specify how an object appears in rendered images and animations and can be previewed by changing the shading in the 3D Viewport to material preview or rendered.

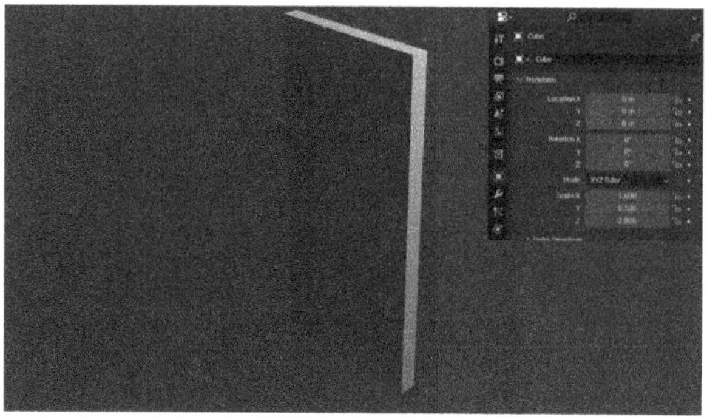

Figure 2. Rendered output of rescaled cube, with inset added to show scale factors.

2.3.2. Object Property Animation Using Blender's Python Interpreter

In order to animate a process, images are rapidly replaced in order to produce the illusion of motion. These images are called frames, and the rate of replacement of frames determines how smooth the animation appears to a viewer, termed as frames per second (FPS). For an object translating from an initial to a final position, each frame captures the intermediate position of the object as it moves. Instead of specifying what intermediate position the object should occupy at each frame, only the initial and final positions are specified at their corresponding frames in Blender. When the position, rotation, or scaling of an object is specified in a frame, the frame is called a keyframe. Blender interpolates the intermediate object properties between keyframes. The number of frames between keyframes is determined by the time interval between keyframes and the FPS. In the projects described in the following sections, keyframes are generated using Python scripts after determining the total runtime for a simulation. Blender includes a version of Python, and many commonly used Python packages can be installed. Script files may be stored within the project file (.blend file), or locally as a Python script file (.py file) that can be loaded into the project and edited outside of Blender. Most modifiable properties in the Blender interface possess a Python data path that can be acquired by enabling Python Tooltips in the Preferences menu in Blender. Using the corresponding data path of a property, Python scripts can be programmed to modify object properties, insert keyframes, acquire data from intermediate frames, and render still images to simulate IBC processes. Using Blender's Python interpreter, any property that can be accessed in the Properties window can be animated in this way.

2.3.3. State Detection Using Images

Animations in Blender are captured and rendered through a virtual camera, which can be manipulated like any other object in the environment (e.g., it can be placed in different locations in the scene or rotated). The camera can also be used to render images during the running of Python scripts used to simulate processes. To load these images during the runtime of the script in order to analyze them, Python libraries that provide this capability can be loaded. In order to carry this out for image-based state detection in this work, a fork of the Python Image Library, Pillow [36], is used to load images for analysis. This library allows us to isolate the different color channels, cast them into arrays, and carry out the necessary algorithms in order to carry out state detection. Setting surroundings and lighting for objects in Blender can aid in state detection from images, as well as bring simulations closer to representing physical processes. Light sources in Blender come in 4 types: Point, Sun, Spotlight, and Area. Depending on the light source being simulated, these 4 types can be modified, for example, by changing color and intensity, to more closely simulate a target process. Material properties also interact with lighting to produce reflections and glare in the simulated process, which may or may not be desirable depending on the use case. With these fundamentals in hand, the following sections detail Blender's capabilities in the context of several pathfinding examples for chemical process simulation.

3. Use of Blender for Modeling and Image-Based Control Tests of Self-Assembly and Environment Recognition

Our prior work [31,37,38] provided a proof of concept that due to the properties of animation, modeling, and coding available in Blender as described in Section 2.3, Blender is a potentially viable option for simulating and testing image-based control laws. However, despite the preliminary analyses of its capabilities as presented in Oyama et al. [31,37,38], many aspects of its performance and interaction with the development of image-based control testbeds remain to be explored and understood for chemical processes to provide a more comprehensive guide to how to successfully develop an image-based control simulation for evaluation purposes in Blender. For example, 3D modeling and animation in Blender abstracts information from the user to make it more friendly for artists, including interpolating movement during the animation process and using non-physical parameters in aspects such as lighting source selection. For the purposes of replicating a chemical process, access to some of this information and determining how to (and when to) manipulate it in a useful manner are key to using Blender as a simulation tool. This section uses three process examples that are precursors to more relevant image-based control simulations than the tank level example in [31,37,38] to explore aspects of how one might work with Blender's capabilities to analyze whether an image-based control simulation has an appropriate setup. Specifically, the following simulations are used for these analyses. First, a simulation of bubbles moving upward (a precursor to modeling their upward movement through a fluid as part of moving toward the modeling of industrial processes such as mineral flotation) is presented. Through this simulation, we analyze the extent to which allowing Blender to interpolate motion between frames where positions and rotations are specified for animation impacts the trajectory of the object compared to hard-coding the information more frequently. The interpolated states are acquired by capturing frame data and are compared with results from the numerical integration of the system model. Second, a nanorod moving stochastically in a two-dimensional plane directed by the actions of an on–off controller that is actuated based on image sensing is presented (considered to be a precursor toward simulations of self-assembly in [39]). In this simulation, we demonstrate how Blender can be utilized to identify challenges in properly sensing objects in image-based control, such as the color of the background being difficult to distinguish from the color of the object such that image processing may need to be more sophisticated to locate an object to which it is desired to apply an optimal control action. Finally, Blender is used to model sunlight shining with different strengths on a reactor, which is then mapped to different heat levels impacting the equipment. Thus, Blender is used to model environmental disturbances to a system, which are then identified through the image-based sensors and subsequently compensated. We discuss how one might integrate Blender's lighting models with non-physical parameters with process modeling and build toward more comprehensive disturbance-handling in control through the use of image sensors that can help to identify causes of changes in the process operation. We compare the responses of the closed-loop system with and without the extra information available from the image capture system to showcase the potential benefits of using Blender for evaluating disturbance-handling techniques. This series of examples provides a deeper look at the potential uses of Blender (and other similar graphics software) toward process systems engineering use cases and how to analyze its benefits and limitations and overcome limitations toward developing closed-loop image-based control tests.

3.1. Animation of Bubble Motion in Blender

This section presents our first demonstration of the use of Blender toward more realistic image-based control applications than were outlined in [31,37,38], where only monitoring and control of tank level using a camera sensor was considered. In those simulations, the tank level was only modeled in two dimensions (i.e., the level was modeled as a plane, similar to drawings that might be made during a controls class but not representative of a physical tank with multiple dimensions), and furthermore, processes such as tank level often have robust measurement techniques through, for example, traditional level sensors that make it less likely for image-based sensing to be used. In contrast, this section focuses on modeling bubble motion in Blender, which is considered to be a precursor to modeling a froth flotation process that could aid in facilitating tests of image-based controllers for a process such as zinc flotation. Image analysis has been previously considered for processes such as zinc flotation and studied using experimental systems [40–42]. In general, image analysis can play an important role in industrial processing [1,43], and this section demonstrates how to model systems in Blender toward representing relevant industrial processes.

Chemical engineering processes are generally described with systems of differential equations that are integrated numerically. The results from numerical integration are a collection of discrete state and time pairs, which can be used in model-based control as predictions of future process conditions and thereby aid in selecting control actions. In Blender, keyframes function as analogues to these discrete state–time pairs, and the software creates intermediate frames by interpolating between the states. In this section, Blender is used to animate the motion of an air bubble through water and determine how much the interpolated states deviate from a numerically integrated result. The equations of motion of a gas bubble in a fluid are modeled using Python to acquire the position of the bubble over time. Some of the resulting state–time pairs are then used to set the keyframes in Blender (i.e., to set the bubble location in these keyframes). The trajectory of the bubbles resulting from the full set of state–time pairs obtained from numerical integration is then compared with the results of using Blender with a subset of those state–time pairs for setting keyframes and the remaining positions interpolated at times between keyframes.

The process of adding animation to objects in Blender involves inserting keyframes corresponding to specific times (frames) in the simulation. A frame rate must be specified in Blender; using that frame rate, one can compute the time corresponding to a given frame so that specific times of visual events can be coded into Blender. Keyframes are differentiated from regular frames by ensuring that a property of an object in Blender (e.g., its position or rotation) meets a designated value when the frame is displayed. Though it is possible to specify properties of an object at every frame, this adds effort to the process of animating the scene due to the high level of specificity required at every frame. Instead, it may be desirable to add keyframes at every several frames instead of every frame. When this is done, Blender interpolates object properties between the keyframes. This creates a smoothing effect between keyframes to give an appearance of smooth motion to a viewer of the animation. However, if the distance between keyframes is large, the interpolation that Blender utilizes may not represent the path that an engineer expects the object to take. Thus, one of the important considerations for developing closed-loop simulations in Blender is determining how often to place keyframes to trade off between specifying the details of the simulation at every frame timescale versus allowing Blender to perform some interpolations. For example, for 24 frames per second, an object's property could be updated every frame (i.e., every 1/24 s) using values of the position from numerical integration with a 1/24 s integration step, though a keyframe might be inserted once every second, i.e., every 24th frame. This would ensure that at the start of every second, the object property would match the value that it would have from integrating the dynamic model of the system. However, over the 24 frames in between the keyframes, the value of the object's properties (e.g., position) would be interpolated if a video of the full scene was rendered from Blender (i.e., they may not be exactly the values expected from the numerical integration). This section uses the bubble modeling process to showcase how the difference between frames and keyframes can lead to some differences in the trajectories compared to the results of using numerical integration.

3.1.1. Equations of Motion of a Bubble

In order to simulate the motion of the bubble, the dynamics of the system are first acquired from the equations of motion [44]. The motion of gas bubbles in a liquid is the result of a number of forces. Since this simulation is intended to function as a pathfinding simulation for the use of graphics toward image-based control applications, the model used in this case is simplified and only accounts for two possible forces on the bubble. Specifically, only the net buoyant force (F_b) and drag force (F_d) are included. The net buoyant force in the upward direction is described in Equation (4), where g is the gravitational acceleration constant, V_{bub} is the volume of the gas bubble, ρ represents the density, and the subscripts liq and bub refer to liquid and bubble properties, respectively:

$$F_b = gV_{bub}(\rho_{liq} - \rho_{bub}) \tag{4}$$

The drag force experienced by a gas bubble is dependent on the flow of the liquid around the gas bubble, characterized by the Reynold's number (Re_{bub}). For low bubble velocities near the initial conditions, the drag force (F_d) is described by Stokes' Law for low Reynold's number, i.e., $Re_{bub} < 1.0$. At higher bubble velocities, the flow around the bubble can become turbulent, which results in a higher Reynold's Number, i.e., $Re_{bub} \geq 1.0$. The drag force in these cases is described by Equation (6), where C_d is the coefficient of drag, which is dependent on the shape and size of the bubble (here, $C_d = 0.2$), r_{bub} is the radius of the spherical bubble, A_{bub} is its surface area, and v_{bub} is the speed of the bubble [45]:

$$Re_{bub} = \frac{2r_{bub}\rho_{liq}v_{bub}}{\eta_{liq}} \tag{5}$$

$$F_d = \begin{cases} -6\pi\eta_{liq}r_{bub}v_{bub} & Re_{bub} < 1.0 \\ -0.5C_dA_{bub}\rho_{bub}v_{bub}^2 & Re_{bub} \geq 1.0 \end{cases} \tag{6}$$

The diffusion of the gas within the bubble to the liquid is assumed to be negligible for this simulation, and hence, the bubble size remains unchanged as it moves through the liquid. With these equations and Newton's third law of motion, the velocity of the bubble is related to the net force it experiences, as in Equation (7), where $v_{bub}(0)$ represents the initial velocity of the bubble and m_{bub} represents its mass:

$$\frac{dv_{bub}}{dt} = \frac{1}{m_{bub}}(F_b + F_d) \quad ; \quad v_{bub}(0) = 0\,\text{m/s} \tag{7}$$

$$x(t + \Delta t) = x(t) + v_{bub}(t) \cdot \Delta t \quad ; \quad x(0) = 0 \tag{8}$$

The position (x) of the bubble is acquired from the numerically integrated velocity, as in Equation (8), where $x(0)$ represents the initial position of the bubble, and Δt represents the integration step using the explicit Euler numerical integration method.

Remark 1. *Cleary et al. [45] describe a number of phenomena involved in more visually accurately modeling bubbles and frothing in graphics. These phenomena include modeling both the fluid and bubbles as separate and unique particles, where the liquid particles carry a certain amount of gas within them, and bubbles are generated whenever enough gas has been concentrated within the liquid particle. The bubbles also interact with both each other and the walls to simulate collisions, and cohesion is accounted for at the liquid surface. Fluid simulation to interact with the bubbles may be carried out using, for example, smoothed particle hydrodynamics principles.*

3.1.2. Animation of Bubble Position in Blender

Equations (7) and (8) are integrated with the parameters listed in Table 1. The radius of the bubbles was chosen to be 2.5 mm. The time and corresponding position values from the numerical integration were stored in two arrays (one for time and one for position). These position and time arrays are imported into Blender in order to animate the motion of the bubble. Three bubbles at different initial positions are placed in a tank. The z coordinates of the bubbles are updated with the integrated displacements, except that the position changes are scaled by a factor of 0.005 to avoid having the bubbles leave the camera view for visualization purposes. In order to animate the bubbles moving, each bubble's location is updated with keyframes, as described in Section 2.3.2. A keyframe is inserted at the initial frame and at every second (i.e., every 24th frame) after until the end of the simulation period. In order to test if the interpolation through intermediate frames is approximately equal to the numerically integrated result, Blender's Python interpreter is used to capture the location of each bubble for all frames of the simulation. This location data is then compared to the position of the bubbles as acquired from integrating Equations (7) and (8). A selection of keyframes from the Blender environment's 3D Viewport is shown in Figure 3.

Table 1. Parameters for bubble motion.

Parameter	Value
Radius of bubble, r_{bub}	2.5×10^{-3} m
Density of liquid, ρ_{liq}	$1000\ \text{kg}/\text{m}^3$
Viscosity of liquid, η_{liq}	10^{-3} Pa · s
Density of bubble, ρ_{bub}	$1.22\ \text{kg}/\text{m}^3$
Integration time step, Δt	1/1440 s
Simulation time	120 s

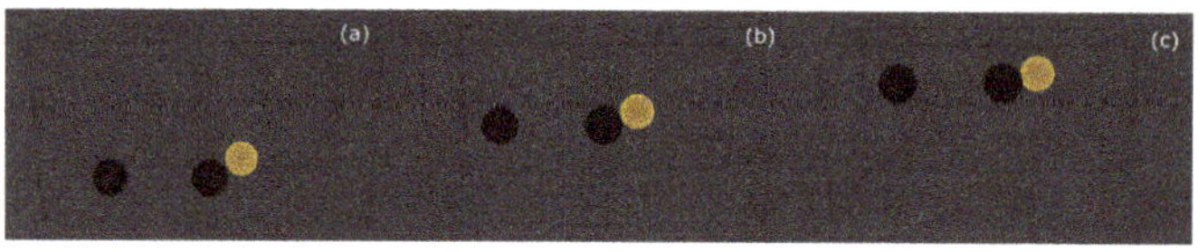

Figure 3. Wireframe captures of the simulation for selected frames: (**a**) frame 0, (**b**) frame 1600, and (**c**) frame 2859. The displacements are scaled by a factor of 0.005. The size of the bubbles are exaggerated for visual identification. The starting positions for each bubble, from left to right, are 1.25885 m, 1.44076 m, 1.30192 m respectively.

3.1.3. Comparing Predictions with Between-Keyframe Interpolations

Because Blender interpolates between keyframes, and since the bubble is given a different position in 3D space at two different keyframes, Blender determines a path between those positions between the two keyframes. This path that it determines may not be equivalent to the numerically integrated path previously described. Therefore, as part of determining how relevant the Blender model is to modeling of the process physics, it is valuable to compare the difference between the numerically integrated path and the path extracted from the simulation including interpolation. To compare the values interpolated by Blender between keyframes and the results of the integration carried out in Python, the displacement of the bubbles in the frames between keyframes is required. This is carried out in Blender by setting up a "for" loop that requests the average displacement (compared to the bubble initial positions, and with the 0.005 scaling applied) from each of the frames divided by the scaling factor of 0.005, and stores it in an array. In this way, the results of the interpolation carried out by Blender between two keyframes can be acquired. These results are stored in .mat files (e.g., with the scipy.io.savemat function) to be compared with the results of the numerical integration. The error between the numerically integrated results and the position of the bubbles acquired from Blender is shown in Figure 4.

The error towards the end of the simulation occurs because Blender assumes that no further motion takes place after the final keyframe. In order to prevent the object from abruptly stopping at the end of an animation, Blender interpolates the object property, in this case, location, to a smooth stop. This results in the object property having a time derivative that approaches 0 near the end of the simulation. This is observed in

Figure 5, where the bubble positions from integration and interpolation between the last two keyframes are plotted. The red line in the plot, which represents the interpolation carried out by Blender, is observed to have a slope of zero towards the 120 s mark, while the blue line, which represents the numerically integrated result, does not, in accordance with the equations of motion. Despite these errors at the end of the simulation, it can be seen that overall, the number of keyframes used in the simulation enabled the results with interpolation from Blender to follow sufficiently well compared to the results from performing the numerical integration. Because the error is greatest toward the beginning and end of the simulation due to the manner in which Blender's interpolation works, it may be useful when performing a physics simulation in Blender to simulate the process in Blender beyond the end time of the simulation to avoid the effect observed in Figures 4 and 5, in which Blender assumes that the object stops moving at the end of the simulation time when that is not implied by the numerical integration result. This also explains a similar error being observed near the start of the simulation, where the objects begin moving from rest. This allows for the results of the animation to replicate the physics more accurately and aids in drawing useful conclusions about the process from the simulation.

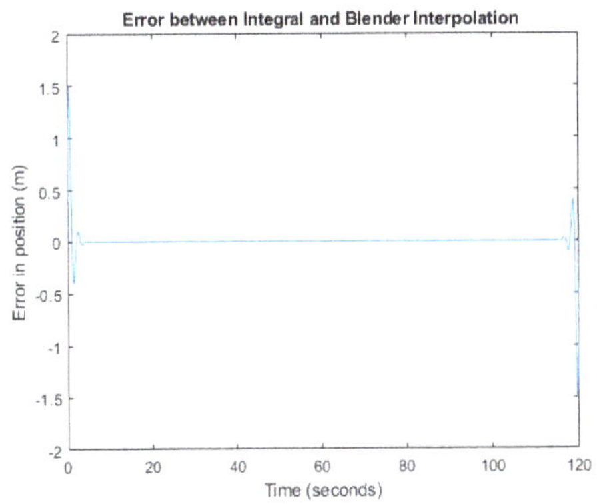

Figure 4. Frame by frame error in bubble position.

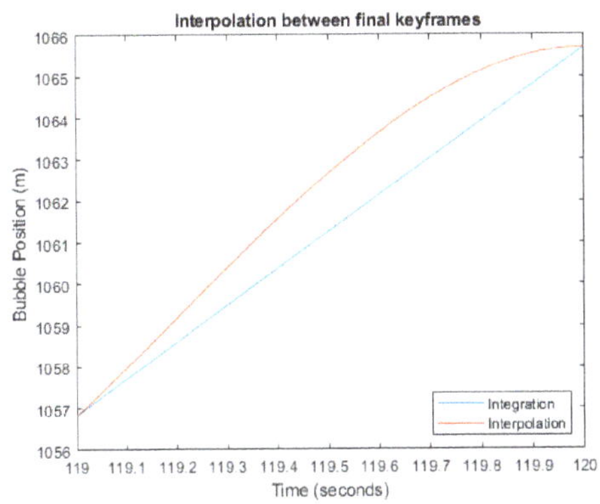

Figure 5. Comparison of numerically integrated and interpolated result for the last second of simulation.

3.1.4. Developing Useful Visuals in Blender

The bubble simulation is considered to be a precursor toward more advanced image-based control studies that could be undertaken, like a full simulation of a flotation process. In this section, we provide several notes on aspects of how to move toward a visually-realistic simulation of froth flotation.

A flotation process involves air introduced into a slurry, which is a mixture of water and treated ore particles [46]. These ore particles are formed by grinding down larger metal ores and treating them with a hydrophobic reagent. The introduction of air agitates the slurry, resulting in the formation of air bubbles. From there, the bubbles rise and come into contact with the hydrophobic ore particles, where they become attached to the bubble surface and rise. This results in the formation of a froth, which is then skimmed off the surface of the slurry and treated to acquire the desired metal from the ore particle. From treating ore particles to reduce their wettability to stabilizing the froth so that it can be skimmed without losing the fine ore particles, every step of the froth flotation process must be carried out carefully. The froth produced in froth flotation cells needs to be chemically analyzed in order to determine which combination of reagent concentration, particle size, and other factors, extract the most ore. However, image sensors can be calibrated to determine these characteristics by correlating representative images against the known results.

Thus, an important characteristic of the development of image-based controllers for a process like froth flotation should be creating models with an appearance that is useful for assessing the performance of an image-based controller through its resemblance to the actual process. The development of a realistic-looking model in Blender requires decisions on the modeling, texture/color, and lighting of a scene. Figure 6, for example, shows a model of foam that might be compared against actual foam from a flotation process to analyze whether it captures key visual characteristics (the set-up of the materials, which controls the "look" of the froth, is shown in Figure 7). In addition, animation in Blender can also be used to capture other physical features of a process, such as the tank shape and also constant movement at a plant, such as an impeller rotating. The ability to modify material properties and lighting in Blender demonstrates its ability to simulate the images captured by a sensor, enabling useful information to be gathered before the control set up is installed in the target process.

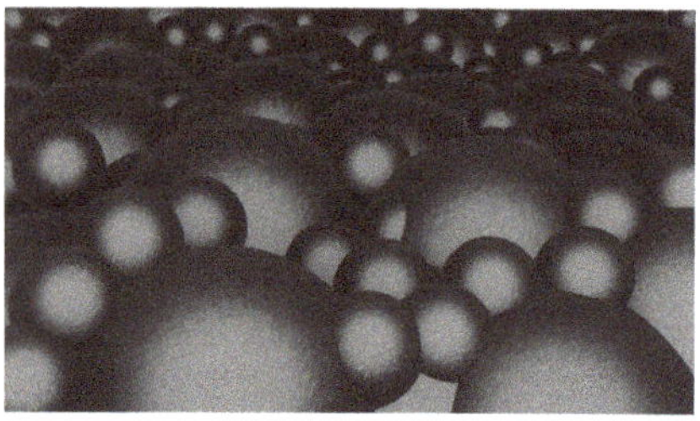

Figure 6. Rendered foam image from Blender.

Figure 7. Material property set up for Figure 6. This is derived from an online tutorial hosted on YouTube [47].

Remark 2. *In general, Blender offers potential to analyze various aspects of image-based control. For example, one might consider multivariate image analysis, wavelet texture analysis, and the gray-level co-occurrence matrix as different image processing methods (or principal component analysis (PCA)) and wish to compare them, specifically for froth flotation applications [40]. Multivariate image analysis (MIA) is a methodology for analyzing multivariate images where the images are analyzed in the form of matrices. MIA can be used to extract features related to the color information in an image by first decomposing the image with the PCA technique, in order to acquire spectral (color-based) features that can be correlated to characteristics of the froth. This is carried out with multivariate image regression (MIR). However, textural features of froth images are better analyzed with gray level co-occurrence matrix and wavelet texture analysis methods. A gray level co-occurrence matrix (GLCM) is a matrix with a number of rows and a number of columns equal to the number of gray levels in the image. The GLCM matrix is then used to extract textural features from the image using statistical methods, such as the variance, correlation, and entropy. It would be expected that Blender may be useful for analyzing not only camera and light positioning but also for comparing different image analysis methods that could be considered. Furthermore, more advanced simulations might be developed by attempting to better correlate images with the physics of the situation. For example, Do [48] describes physics of turbulent flotation, and Cleary et al. [45] describes a detailed graphics modeling strategy for bubbles in liquid. Through more detailed modeling strategies in the future, it may be possible to utilize Blender more for analyzing and modeling image-based control processes in greater detail. However, it should be noted that in general, a traditional process model may not directly translate to an ability to model the process in Blender.*

3.2. Evaluating Identification Techniques for a Controlled System under Image-Based Control in Blender

The prior section described the use of Blender in the modeling of a process that could be placed under image-based control and some of the considerations for checking whether the modeling efforts are sufficient to represent a system. In this section, we focus on a process under image-based control in Blender and discuss how Blender helps to reveal characteristics required by either the image processing algorithm or process design to enable the image-based control to be effectively carried out. The process under stochastic control in this section is a nanorod that was used in [39] as a precursor simulation toward considering control of directed self-assembly (though [39] did not perform image-based control manipulations with Blender). The control development in this section follows that in [39], but we present the details of our exact implementation for completeness and replicability of our work. After discussing the simulation setup, we describe the image-based control framework and how it aids with understanding some of the potential pitfalls of an image-based controller for the process, as well as ideas for overcoming them.

Directed self-assembly (DSA), the motivation behind simulating the optical control of the nanorod, is a manufacturing technique that has been gaining research interest in recent years. The technique has its roots in self-assembly, where instead of external control actions, molecules are assembled by modifying them such that the desired configuration can be acquired stably, or by similarly modifying a surface on which the self-assembly takes place, commonly referred to as a substrate. This idea has been explored in the field of biology specifically by using the DNA [49,50] and in the polymer sciences through experiments on block copolymerization (BCP) [51,52]. The modification of a base substrate to direct multiple species to bind and form a desired configuration is generally seen as the first attempt at directing the self-assembly of particles, though this idea overlaps with those of regular self-assembly. The improvement of BCP for directed self-assembly demonstrates the capability of this method, by forming highly precise nanostructures on a pretreated substrate [53,54] or using forces of the medium to achieve similar results [55]. External forces can also be used in the absence of substrates, sometimes to improve existing self-assembly [56] or to drive particles into positions to facilitate further interaction [57]. The ability to use image-based control in self-assembly applications has been explored in the work of Tang et al. [58]. In this work, images are processed and used to form the feedback of a control loop, where electric fields are used to manipulate the free energy landscape and ensure the colloidal particles form the desired morphology. In that example, the arrangement of the individual particles in the medium, the stability of their configuration, and speed of assembly can be observed directly through imaging techniques. A virtual model of these systems is what an IBC testbed would hope to emulate, simulating ideal conditions for optimum self-assembly, as well as an image-based observation of the process.

Motivated by the potential utility of image-based control for self-assembly, as demonstrated in this prior work, this section focuses on the self-assembly precursor simulation from [39], as discussed above, of a nanorod undergoing Brownian motion in a two-dimensional space. Stochastic control is used to carry out navigation of the rod to the center of the control space. Image-based control is simulated in Blender with the virtual camera that captures images of the rod, which are then used to determine the state of the rod and carry out the optimal action. The optimal policy (consisting of on/off-type control actions) is acquired by performing stochastic optimization, as detailed in Section 2.2.

3.2.1. Equations of Motion for a Self-Propelled Rod

Since the rod simulation is used in the PhD dissertation by Yang [39] as a precursor to self-assembly studies, the rod's motion in free space is modeled using 2D Brownian motion, according to Equations (9) and (10):

$$\begin{aligned} \mathbf{r}(t+\Delta t) = {} & \mathbf{r}(t) + \tfrac{\mathbf{D}_t}{kT} \cdot \mathbf{F}\Delta t + \Delta \mathbf{r}^B \\ & + v\cos(\phi)\Delta t\, \mathbf{e}_1 + v\sin(\phi)\Delta t\, \mathbf{e}_2 \\ \phi(t+\Delta t) = {} & \phi(t) + \tfrac{D_r}{kT}\mathbf{T} \cdot \mathbf{e}_3\Delta t + \Delta\phi^B \\ \mathbf{D}_t = {} & \begin{pmatrix} D_{xx} & D_{xy} \\ D_{yx} & D_{yy} \end{pmatrix} \end{aligned} \tag{9}$$

$$\begin{aligned} \langle \Delta \mathbf{r}^B \rangle = 0 \quad & \langle \Delta \mathbf{r}^B, (\Delta \mathbf{r}^B)^\intercal \rangle = 2\mathbf{D}_t \Delta t \\ \langle \Delta\phi^B \rangle = 0 \quad & \langle \Delta\phi^B, \Delta\phi^B \rangle = 2D_r\Delta t \end{aligned} \tag{10}$$

In these equations, $\Delta\mathbf{r}^B$ and $\Delta\phi^B$ represent the changes to the 2-dimensional position and the orientation of the rod (i.e., the angle between the line along the axis of the cylinder and the positive x axis) under Brownian motion and are normally distributed random variables. $\mathbf{D}_t$ (with components D_{xx}, D_{xy}, D_{yx}, and D_{yy}) and D_r represent translational and rotational "diffusivities", which characterize the random motion of the rod due to particles of the medium. The velocity of the rod under the control action is represented by v and is only non-zero when the control action is on. Since no force or torque fields act on the rod (hence, we refer to it as "self-propelled"), the terms $\mathbf{F}$ and $\mathbf{T}$ are set to 0. Vectors $\mathbf{e}_1$ and $\mathbf{e}_2$ represent unit vectors along each of the two dimensions, and $\mathbf{e}_3$ represents the unit vector normal to the 2D plane. The temperature of the system is represented by T, and k represents the Boltzmann constant. From the definition of the translational diffusivity tensor $\mathbf{D}_t$, the components of $\Delta\mathbf{r}^B$ are coupled. For a cylinder, $\mathbf{D}_t$ is parameterized with $D_\parallel$ and $D_\perp$ [59], diffusivities along and perpendicular to the cylinder axis, respectively. The following equation (Equation (11)) simplifies the tensor to a single parameter by assuming $D_\parallel \approx D_\perp$ as follows:

$$\begin{aligned} & \mathbf{D}_t = \mathbf{n}\mathbf{n}^\intercal D_\parallel + (\mathbf{I} - \mathbf{n}\mathbf{n}^\intercal)D_\perp \\ & \text{where } \mathbf{n} = \begin{pmatrix} \cos(\phi) \\ \sin(\phi) \end{pmatrix} \\ & \text{If } D_\parallel \approx D_\perp \\ & \mathbf{D}_t = \mathbf{I} \cdot D_\perp \\ & \quad = \begin{pmatrix} D_\perp & 0 \\ 0 & D_\perp \end{pmatrix} \\ & \implies D_{xx} = D_{yy} = D_\perp = \bar{D}_t \end{aligned} \tag{11}$$

where $\mathbf{I}$ is the identity matrix.

This decouples the movement in the orthogonal directions, as well as transition probabilities for location and orientation. The equation of motion used for the simulation then simplifies to Equation (12) as follows, where the velocity v is either zero or a constant, depending on the choice of the control action, i.e., on or off:

$$\begin{aligned} x(t+\Delta t) &= x(t) + \Delta x^B + v\cos\phi\Delta t \\ y(t+\Delta t) &= y(t) + \Delta y^B + v\sin\phi\Delta t \\ \phi(t+\Delta t) &= \phi(t) + \Delta\phi^B \\ \langle \Delta x^B \rangle = 0 \quad & \langle \Delta x^B, \Delta x^B \rangle = 2\bar{D}_t\Delta t \\ \langle \Delta y^B \rangle = 0 \quad & \langle \Delta y^B, \Delta y^B \rangle = 2\bar{D}_t\Delta t \\ \langle \Delta\phi^B \rangle = 0 \quad & \langle \Delta\phi^B, \Delta\phi^B \rangle = 2D_r\Delta t \end{aligned} \tag{12}$$

3.2.2. State Space and Transition Probabilities

The optimal policy generation method from Section 2.2, which we wish to employ to control the rod, uses a discrete state-space. In general, the space in which a rod moves is a continuous space; it is, therefore, necessary to determine how to translate the continuous state space into a discrete state space for the purpose of control. To do this, we select a portion of the free space in an area 101 μm × 101 μm around the target and set it as the neighborhood around which the policy is generated. To create the discretized state space, this space is first discretized into 1 μm × 1 μm zones, and the rod's position can take values at the center of each square. The orientation is discretized as sectors of $\pi/4$ radians each, centered on each of the cardinal and subcardinal directions, as in Figure 8. The set of all possible discrete states can then be defined as a three-dimensional array, where the first two indices represent the position of the nanorod in the space, and the third index represents the orientation sector that it occupies. This set is defined in the following equation (Equation (13)), where the intersection with $\mathbb{Z}^+$ indicates that x, y, and ϕ are integers within the specified intervals:

$$S = \{(x, y, \phi) : x, y \in [0, 100] \cap \mathbb{Z}^+, \phi \in [0, 7] \cap \mathbb{Z}^+\} \tag{13}$$

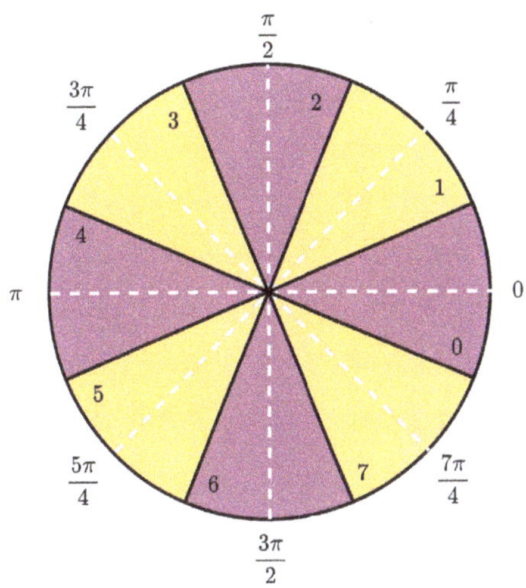

Figure 8. Discretization of orientation into 8 sectors. The numbers in each sector denote its index. The dashed white lines indicate the central angles for each sector, the purple sectors are centered on the cardinal directions, and the yellow ones are centered on the sub-cardinal directions.

The dynamics of the rod as presented in Equation (12) are presented in discrete time and are used to generate the transition probabilities associated with each state. The transition probabilities map the probability associated from one state to another for each action, i.e., $P : A \times S \times S \mapsto (0,1)$, and based on the discretization of the state space in Equation (13), the total number of transition probabilities would be $2 \times (101 \times 101 \times 8)^2$. The normal distributions of Δx^B and Δy^B imply that the probability of moving to a new position tends towards zero outside of a small neighborhood around an initial position. Hence, a neighborhood of a 19 μm × 19 μm grid around the initial position of the rod is set as the maximum possible displacement in a single time step. The transition probabilities are then acquired through Monté Carlo trials. Monté Carlo trials are particularly useful in extracting equivalent discrete probabilities from continuous random variables. First, a large number of values for the random variables (trials) are generated, according to their known distributions. Then, the probability associated with each outcome is found by counting the number of times the outcome is observed and dividing it by the total number of trials. Three sets of random data are generated: the position Δx^B and Δy^B and the orientation $\Delta\phi^B$. Table 2 details the parameters in Equation (12) used to generate the trials. The probability associated with how far the nanorod travels after one time step is termed as positional probability. This is calculated by first dividing the neighborhood of a 19 μm × 19 μm grid around the initial position of the rod into 1 μm × 1 μm zones and determining which zone the nanorod ends up in for each trial (based on the center of the zones). Dividing the number of times the nanorod ends up in each zone by the total number of trials yields the positional probability. This positional probability is then multiplied by the discrete probabilities associated with the orientation. Similar to how the positional probability is acquired, orientation probabilities are acquired by counting the number of times the nanorod ends up in each orientation sector, as described in Figure 8, given an initial orientation sector. This operation is repeated for each control action, where additional components of the velocity $v \cos \phi \Delta t$ and $v \sin \phi \Delta t$ are added to Δx^B and Δy^B, respectively, for the on control action, as described in Equation (12). Hence, the total number of transition probabilities reduces to $2 \times 8 \times 8 \times 19 \times 19$ for a given initial position, where the first index represents the chosen control action, the second and third indicate the initial and final orientation of the rod, and the fourth and fifth indicate the final position of the nanorod relative to the center of the neighborhood. These transition probabilities are stored in a matrix P. The transition probability associated with the rod, starting in orientation sector 0, moving 8.5–9.5 μm towards the positive end of the x-axis, with a final orientation in orientation sector 0, and moving under Brownian motion alone (i.e., when the chosen control action is off), would be stored at $P(0,0,0,9,18)$. Figure 9 displays the transition probabilities for the rod with orientation sector 0, associated with final positions for both on and off control. Each square represents the probability of the rod acquiring that position after Δt time has passed, given its initial position at the center of the heatmap. The colorbar represents the values of the probability of the nanorod occupying a specific position once it starts at the origin. Figure 10 displays how changing orientation affects the transition probability. Since the change in orientation is also normally distributed around 0, the observed transition probabilities when orientations change (compared to the initial orientation) is lower than those where the orientation remains the same as the initial orientation.

Table 2. Probability Generation Parameters.

Parameter	Value
Translational diffusivity, $\bar{D}_t$	0.5 $\mu m^2/s$
Rotational diffusivity, D_r	0.548 rad^2/s
Sample time, Δt	1 s
Discretization length, $\Delta x = \Delta y$	1 μm
Velocity of rod under control, v	2.5 $\mu m/s$

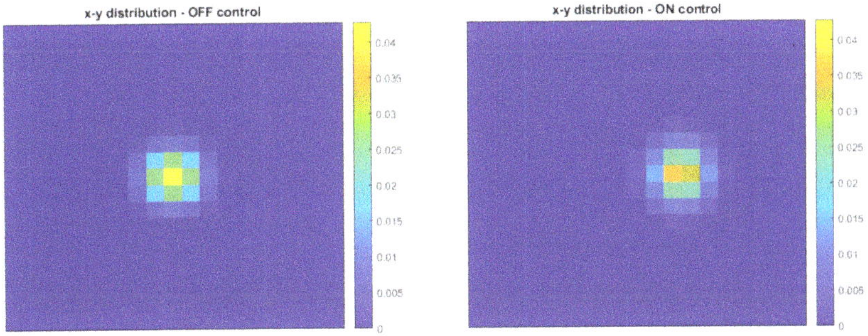

Figure 9. Positional probability distribution for the nanorod in sector 0 of orientation. The figure on the left shows the probability distribution when the control action is off. In this case, the rod starts at the center of the figure and has the highest probability to stay at the center, with lower probabilities moving away from the center. The figure on the right shows the probability distribution when the control action is on. In this case, the rod starts at the center of the figure and has the highest probability to move slightly to the right of the center, with lower probabilities moving away from the right-shifted region of highest probability.

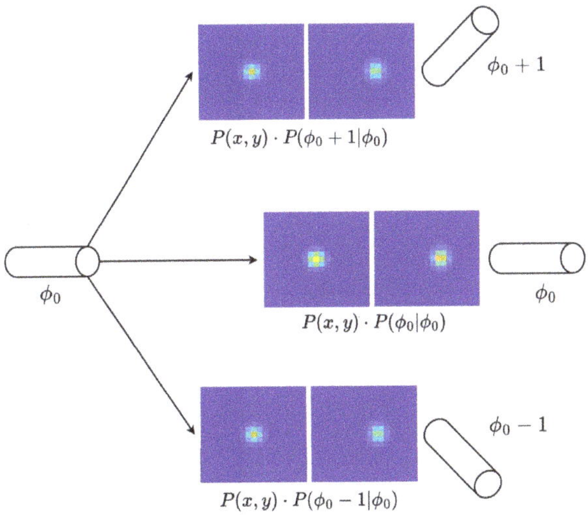

Figure 10. Transition probability distributions for 3 adjacent orientations. The cylinders represent the orientations of the nanorod, with the one on the left representing the initial orientation and the three on the right representing three possible final orientations. The associated transition probability heatmaps show that since the rod is most likely to keep its orientation, the heatmaps above and below the center line show lower probabilities when compared to the center. The colorbar used to indicate the relative magnitude of the transition probabilities shares the same scale and is identical to that in Figure 9. $P(x, y)$ indicates the probability of a certain (x, y) location in space given an initial state at the center, and $P(\phi_j|\phi_0)$, where ϕ_j is either $\phi_0 + 1$, ϕ_0, or $\phi_0 - 1$, indicates the probability of achieving certain orientations given an initial orientation of ϕ_0.

These reduced probability matrices (i.e., considering probabilities within only a fraction of the total state space) are then translated to the corresponding window in order to carry out the value iteration algorithm, as shown in Figure 11. With the state space discretized and the transition probabilities calculated, the policy can now be generated in accordance with Bellman's optimality condition.

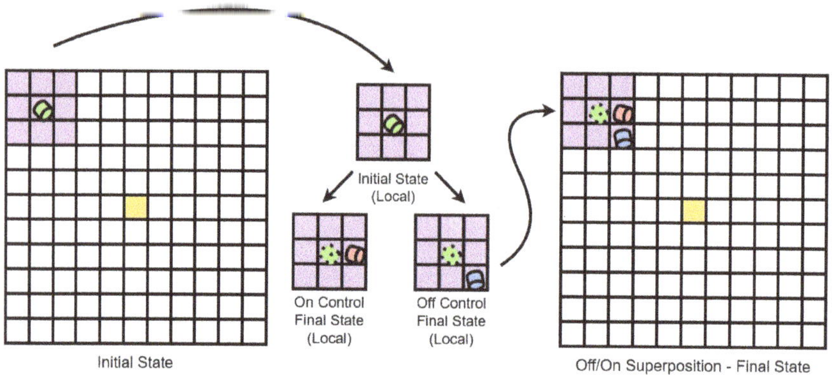

Figure 11. Using reduced transition probabilities to calculate overall transition probabilities. The left figure starts with defining where the neighborhood of possible positions is (indicated in purple), centered on the initial position of the nanorod, depicted as a green cylinder. The middle figures demonstrate how transition probabilities are used to determine possible states, where the red cylinder shows a potential final state under the on control action, and the blue cylinder under the off control action. Furthermore, the figure on the right shows how the reduced transition probabilities relate to the transition probabilities of the entire state space.

3.2.3. Policy Generation

From Equation (2), the optimal value function is described as a weighted sum of the value functions of possible future states and the cost function at the current state. Generally speaking, the cost function is a representation of the cost to maintain a certain state. However, since the final state (considered to be navigating the rod to the center of the portion of free space being simulated) is known in this simulation, the cost function is designed such that its value at the center of the free space is the minimum. From Equation (3), the value of every state beyond the boundary of the 101 μm × 101 μm region must be the sum of its cost as well as the value of neighboring states. From the discussion in Section 3.2.2, the set of possible future states is restricted to a small neighborhood of states surrounding the initial position. If a portion of the future states is outside the boundary of the modeled space, and since states outside the boundary cannot be iterated, it is necessary to have a placeholder value function instead, as shown in Figure 12.

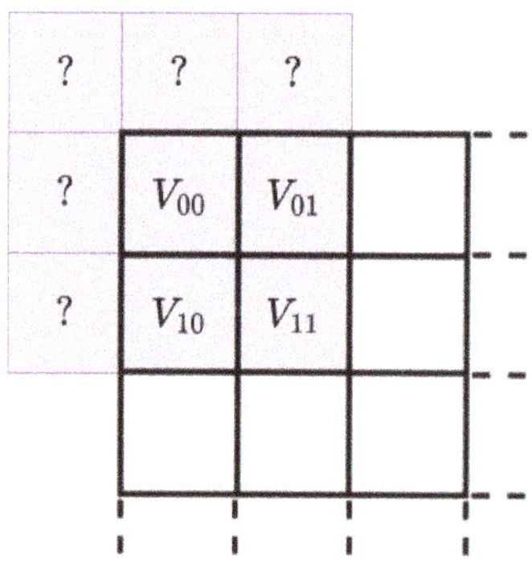

Figure 12. Edge effects in 2D presentation. When determining the neighborhood of possible positions (purple) around the initial nanorod position, if the rod lies close enough to the boundary, the neighborhood can include positions that lie outside the boundary (i.e., where the question marks are located in the figure). The values V_{pq}, $pq = 00, 01, 10,$ and 11 indicate the value function values at different positions.

Since the value function is a function of cost, and the optimization procedure involves the minimization of cost, we can assume that states outside the boundary have higher value function mappings than those within. For states with locations infinitely far away from the target, the result of the value function should be maximum and approximately equal to the value of the neighboring states. In order to meet this requirement, and following the description of the state space from Equation (13), the cost function is described in Equation (14):

$$C(s) = \begin{cases} 0 & \text{If } s \notin S_{\text{target}} \\ -1 & \text{If } s \in S_{\text{target}} \end{cases} \tag{14}$$

$$S_{\text{target}} = \{(x, y, \phi) : x, y = (50, 50), (x, y, \phi) \in S\} \tag{15}$$

S_{target} is chosen such that the target area lies at the center of the modeled space. Since the value function is recursively defined as linear combinations of the cost function, it is bounded by the maximum of 0. This is selected as the placeholder value function result outside the boundary of the simulation, as displayed in Figure 12. With this feature in place, the Bellman update rule (Equation (3)) is carried out for every state. Figure 13 displays the converged policy with $\gamma = 0.8$ (iterations stopped if the sum of the value function results for every state between two consecutive iterations was below a threshold of 10^{-5}).

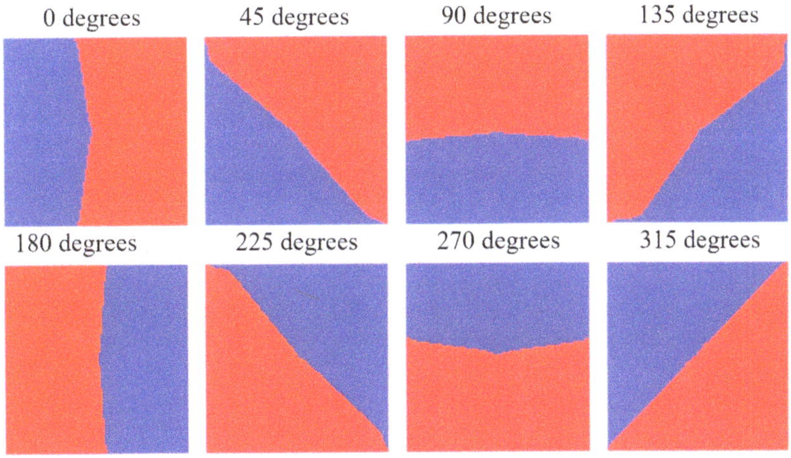

Figure 13. Optimal control policy based on location and orientation with $\gamma = 0.8$. Blue regions are regions where the control action is on; red regions are regions where the control action is off.

3.2.4. Blender Implementation of Rod Dynamics

Blender is an open-source computer graphics program with capabilities in modeling, animation, and rendering, along with a Python programming interface, which facilitates visualization of the control policy described in Section 2.2. The nanorod is modeled as a cylinder (given initial values in Blender with scale = [0.02,0.02,0.10], rotation_euler = [0,$\frac{\pi}{2}$,0], and location = [−0.5,0.5,0.01], and a seed for random number generation of 252), and its motion is restricted to the x-y plane, as in Figure 14. As discussed in Section 3.2.1, the control action moves the rod along its axis at a fixed velocity.

Because both the position and orientation of the rod are important for selecting the control action to be applied in this study, a method for distinguishing the rod's orientation from the image must be obtained. A challenge in this direction is that the rod is a cylinder and, thus, symmetric, unless distinguishing marks are placed on it. To enable the orientation to be ascertained, the direction of the rod is determined by color coding the "front" end of the rod white and the "back" end black. The body of the rod is colored blue, and a contrasting backplate is modeled as a red plane. These various modifications indicate several learnings from the Blender simulation that would aid with designing an image-based controller for an actual process. First, they elucidate that without the color coding, it may be difficult to accurately determine the rod's orientation without a more complex image processing strategy (e.g., a type of edge detection). This helps to indicate some of the options available to a process or control designer for attempting to set up such a system successfully in a real plant before the time to build it. Furthermore, the need to ensure that there is sufficient contrast between the background and the color of the rod (and its ends) is also highlighted, again showcasing some of the features of the system design that are important for achieving the desired control goals.

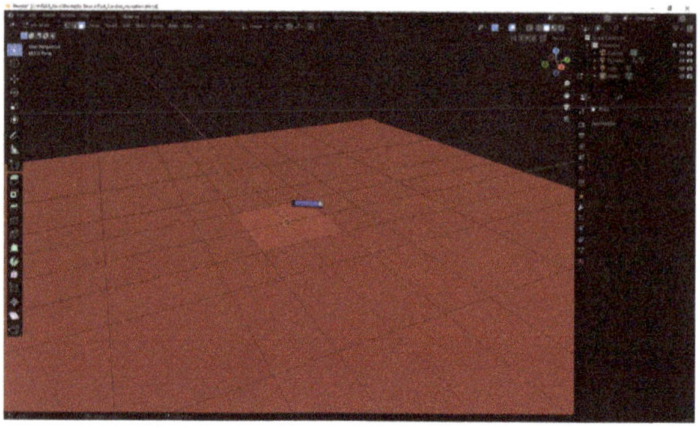

Figure 14. Models of rod and contrasting backplate in Blender.

As discussed in Section 3.2.1, Brownian motion is simulated as a random walk in discrete time, and the motion is characterized in continuous time as the Weiner process [60]. The distance unit of the parameters used to describe the Brownian motion of the rod and the control action is scaled up, such that 1 m in the Blender model corresponds to 50.5 μm. The simulation is run at 24 frames per simulation second, with the position and orientation updated for each frame. The updates are carried out according to Equation (12), with $\Delta t = 1/24$ s of simulation time. With the rod dynamics now in place, Blender's image capture technology can now be explained.

3.2.5. Image Capture and State Measurement

Three-dimensional animation software suites generally employ a virtual camera to render animations with desired position and lighting. In order to capture the state accurately, the camera in Blender is positioned above the contrasting backplate with its line of sight perpendicular to the plane. An area light source is set above the backplate to illuminate the control area and the nanorod uniformly. Each end of the nanorod being colored differently allows the detection of the position and orientation of the rod. In the generation of the optimal policy, the rod is assumed to be a point object with respect to the discretization of the location. The choice of black and white ends is due to the representation of color in the image capture. Digital image capture uses the additive color primaries: red, green, and blue. White pixels have the highest combination of these primaries, and black pixels have the lowest. In order to minimize the possible effects due to lighting conditions, sums of the color channels around points of interest are used to determine color. This is done by determining the sum of the RGB channels for all pixels in a 3×3 window centered on the pixel being inspected, and taking the sum of them to determine the color (i.e., the window with the smallest sum is considered to identify the black end and that with the largest sum is considered to identify the white end). This also informs our choices for the rod and contrasting backplates, to prevent the body of the rod and the medium from being detected by the image capture algorithm. The locations of the ends of the nanorod are used to determine the discretized location by averaging the coordinates of each end to get an approximate location of the mid-point of the rod. Additionally, the tangent inverse of the negative of the slope of the line through each end is used to determine the orientation. Since digital array indices are numbered from the top left corner, i.e., the index (0, 0) is at the top left corner, and the row index corresponding to the y coordinate increases in the downward direction, the tangent inverse of the negative slope is used to obtain the orientation with respect to having the zero orientation at the standard location of zero degrees in polar coordinates (i.e., on the right). Figure 15 details the operations that take place at each image capture. In general, an image-based control testbed can help with analyzing effects such as lighting effects in advance of attempting to implement and design an area instrumented with image-based control.

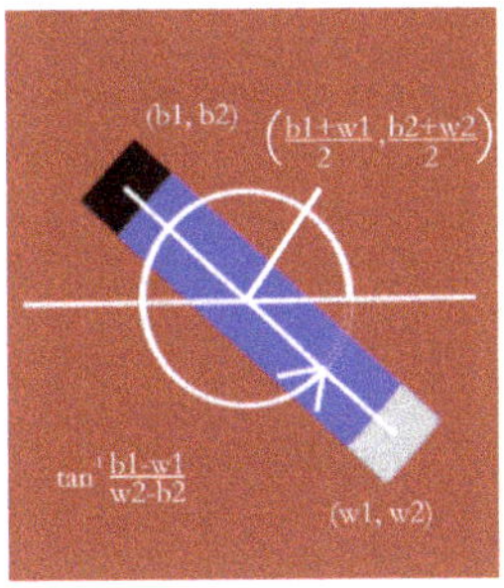

Figure 15. State determination from image information. (b_1, b_2) signifies the array index of the black end of the rod from the upper left, and (w_1, w_2) signifies the array index of the white end. The average of these points is listed and the argument of the inverse tangent, used in the computations of the orientation, is the slope between the points in the Cartesian coordinate system.

These images are generated with render functions that come built-in with the Blender Python library. The policies generated as per Section 3.2.3 have a much lower discretization than the resolution of the image returned in the image capture algorithm. The images are of the size 1920 × 1920, while the location component of the state space, as described in Equation (13), is of the size 101 × 101. These images are compressed in order to match the scale of discretization of the policy. Specifically, the number of pixels that are used to represent the image is changed because the image captured by Blender has a resolution of 1920 × 1920, and in order to identify what the discretized location of the rod is, it is necessary to make this size match that of the policy (i.e., 101 × 101). The state determination and image compression are carried out using the `resize` function in Pillow. At every state observation (which takes place at every sample time), an image is explicitly rendered from the current frame and stored locally. This image is then imported into the running Python script with Pillow, compressed, and processed for state determination. The position and orientation of the rod updated with Brownian motion dynamics are compared with the values acquired from state determination in Table 3. Specifically, the L^1 norm of the error between the actual position and the position obtained by the image processing algorithm is reported in Table 3 (and is calculated as the sum of the absolute values of the differences between the first position index and the second position index between the actual values ("Position Blender" in the table) and the values obtained from the image processing algorithm ("Position Image" in the table)). As shown in the fourth column of Table 3, the maximum L^1 norm error computed in this way is 2, indicating that the position values for both position indices, as obtained from the image processing method, were close to the correct values. In addition, the absolute error between the sector identified for the orientation using the image processing ("Orientation Image" in the table) and the actual sector ("Orientation Blender" in the table) is reported in Table 3. In the full data set for the 1200 s simulation, there are cases where an error was recorded in the sector determination from the images when the rod was actually in Sector 7 (since there are 8 sectors, this can also be considered to be the sector -1) but it was stated by the image processing strategy to be in Sector 0. The absolute error (considered to be the absolute value between the difference between the 0 and -1 sector values) in that case is 1, suggesting once again that the image processing algorithm performed well in many cases in determining the orientation of the rod correctly. The results reflect that the image analysis often produced correct orientations and that the position error was relatively low, suggesting that this control policy may perform well but that there is also potential for improvement of, for example, the position sensing if desired. State determination from the rendered image is one of two core functions of the Python script. The other is rendering the animation, and it is discussed in the following section.

Table 3. Deviation and error of state determination for selected frames of a 1200 s simulation.

Time Seconds	Position Blender	Position Image	L1 Norm Error	Orientation Blender (Sector)	Orientation Image (Sector)	Absolute Error
80	(28, 30)	(29, 31)	2	0	0	0
160	(32, 32)	(33, 33)	2	4	4	0
240	(32, 32)	(32, 33)	1	0	0	0
320	(32, 34)	(32, 34)	0	1	1	0
400	(37, 37)	(38, 38)	2	3	3	0
480	(36, 36)	(37, 37)	2	3	3	0
560	(36, 38)	(36, 38)	0	7	7	0
640	(45, 44)	(45, 44)	0	0	0	0
720	(48, 49)	(49, 49)	1	5	5	0
800	(50, 48)	(50, 47)	1	4	4	0

3.2.6. Animation and Control Execution

In Section 3.2.4, it is explained that the position and orientation are updated for every frame of the animation. According to Section 2.3.2, since keyframes update object properties in the simulation, they are used to update the position and orientation of the rod in the animation. In this work, however, the keyframes help simulate the interplay between the smaller timescales of the rod dynamics and the image-based control. Specifically, the sampling time for the controller as listed in Table 2 is 1 s, which is also when each keyframe is inserted. The Brownian motion deviation variables are updated every 1/24 s in the Python programming interface, as detailed in Section 3.2.4. Since the deviation variables $\Delta \mathbf{r}^B$ and $\Delta \phi^B$ are generated with different time scales than what is used to determine the policy, the simulation demonstrates sample-and-hold control of a Brownian motion process. A keyframe is inserted at every sampling time in this experiment, during which the image is captured, the state is identified, and the control action is determined. The control is actuated in a sample-and-hold fashion, keeping the velocity constant regardless of changes in state until the next sampling time occurs. Figure 16 is a compilation of locally stored images used for the state determination, which are also from keyframes.

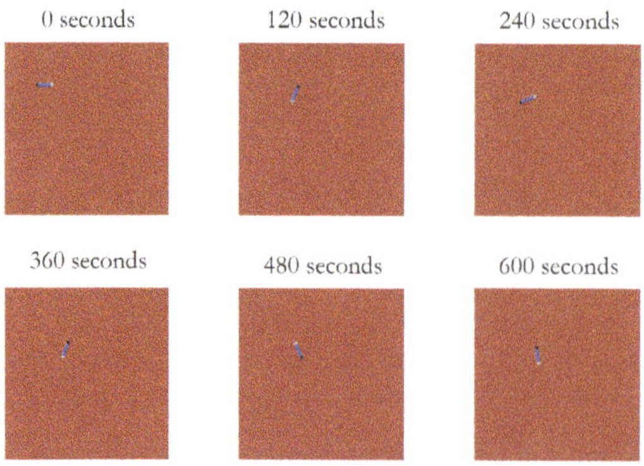

Figure 16. Rendered keyframe images for state determination at different simulation times.

The Blender script is run for 1200 s of total animation time; however, due to the non-determinism of the system, the agent can enter the target space within any amount of time that, on average, scales with the distance of the initial position to the target for a given initial orientation. This series of frames indicates that though the rod starts off the target and takes some time in moving toward it under its stochastic dynamics, it eventually reaches the target, suggesting success of the image-based control strategy for this example.

3.2.7. Opportunities for Image-Based Control Testing in Blender Based on the Stochastic Nanorod

The sections above demonstrate that Blender is successfully used to replicate Brownian motion dynamics of the nanorod in 2-dimensional space, and the optimal stochastic policy is implemented using a virtual image-based controller. By updating the position and orientation of the rod every frame ($\Delta t = 1/24$ s) and picking an optimal action at every sampling time (where keyframes are also inserted), the simulated process attempts to show how the sample-and-hold control policy performs on a process that is updated on a smaller timescale. One aspect of setting keyframes for the simulation of the rod, compared with the bubble of Section 3.1, is that the bubble was taking a straight path at a constant velocity such that, except near the beginning and end of the simulation where the starting and stopping behavior of the bubble in Blender created a mismatch between the numerical integration result and the Blender simulation result, no error was observed between the interpolated position values (between keyframes) by Blender and the values obtained from numerical integration. However, in this simulation involving the rod, due to the stochastic behavior, we would not expect that the results with applying keyframes every frame versus every several frames would cause the position and orientation of the rod in Blender to be the same in both cases. This indicates that the frequency with which keyframes are set can cause different accuracy for different dynamic systems.

The color information is summed within a 3×3 window to detect the rod and extract information about its state features. Image capture by the software and additional image processing with the Python Pillow library is demonstrated, and the errors between the states of the system returned from the virtual environment and those extracted from the image are compared.

Since Blender can render animations, it provides the ability to visually demonstrate and inspect the execution of an image-based control algorithm. When attempting to visualize the nanorod approaching its destination using a rendered video of the animation, the target can be highlighted in the video. In such a case, a section of the contrasting backplate around the target could be colored differently, for example, when a video of the simulation is rendered to aid in showing that the rod approached its target with time. However, care must be executed when attempting to improve visualization of aspects of the simulation such as this to ensure that the visualization aids do not interfere with the behavior of the image-based control algorithm. For example, attempts were made to aid with visualizing the rod moving toward the target by making the target area visible (i.e., a different color than the rest of the backplate) during the execution of the image-based control policy. However, this resulted in failure of the color detection strategy, in part due to the summing of the color channel values in a small neighborhood around the nanorod, which would include portions of the contrasting backplate as the rod moved close to the target. In the simulations performed, a yellow colored target section registered as the white end of the nanorod, since windows of color summing would place the sum total of the color channels in the target section higher than the actual front end of the nanorod due to the backplate being included in windows around the front. A darker color than the backplate would result in similar misidentification of the black end of the nanorod. These issues might be addressed with the use of image convolutions, such as edge detection, in order to further improve state identification. Alternatively, Blender allows for the editing of object properties after keyframes are inserted, which allows the target area to be colored in after the simulation is run, in order to display it in rendered videos. While this allows for highlighting of the simulation features, it is not representative of the images used by the virtual camera for state identification. These studies highlight that making the simulation results more easily understood to a viewer by updating the simulation to include new visualization aspects may affect how true the simulation results remain to what they would be without the additional visualization tools.

Another consideration that could be tested in Blender is cybersecurity or privacy strategies for image-based control systems. For example, one might wish to make the policy between the states and the optimal control actions, as well as the images of the process, unclear to an eavesdropper, even if the policy was to be stored on the Cloud and the images needed to be transferred back and forth between the process and the Cloud for processing for state determination and control action selection. One idea for attempting to obscure, for example, the series of images in Figure 13 is to attempt to utilize a policy inspired by homomorphic encryption being combined with control [61]. In particular, we consider the case that we do not wish to send the actual image to the optimal control look-up strategy, but we wish to process the image locally and then send the state to the Cloud to retrieve the optimal control action from the policy stored there. Since the policy is determined from each state (a 3-dimensional quantity) being mapped to an optimal action, one way of attempting to make the state unclear to an eavesdropper on the Cloud would be to shuffle the full look-up table according to a mapping from the discrete state space to a new value, where that new value would be associated with some control action that could then be returned from the Cloud. This prevents the need to send the actual value of the state to the Cloud, and instead, only the mapped state value is sent. A similar shuffling/mapping idea can also be applied to the optimal action, if it is desired to make the control action returned obscure to an eavesdropper as well. With a strategy for generating the images of the process using Blender, one could simulate this full technique or other cybersecurity and privacy strategies.

The use of Blender opens the possibility of virtually testing various image processing techniques in order to rapidly test novel image-based control algorithms. These various insights regarding image-based control of this process are some of the benefits of utilizing an image-based control testbed like Blender. Additional insights such as camera position, suitable lighting, and minimum resolution for state detection can be simulated in Blender in order to test configurations to test advantages.

3.3. Modeling Sunlight Effects on Outdoor Reactors in Blender

Weather can play a role in the effectiveness of chemical process control policies where equipment is not housed in a controlled environment. The impact of weather on large outdoor reactors can be measured through a variety of methods, but attempting to assess weather changes could aid with reducing plant/model mismatch in model-based control formulations and aid with potentially forming more physics-based models by avoiding treating weather variations as a type of unknown process disturbance affecting a measured state and causing it to deviate from standard operating conditions. In the work of Zavala et al. [62], weather forecasting is used to determine economically optimal actions. With the advent of solar technology, the determination of weather phenomena becomes more relevant, allowing the prediction of energy availability and aiding in ensuring demands can be met by combining a variety of power sources. Going a step further, in the work of Rowe et al. [63], a hybrid solar–electric reactor is designed and then investigated for optimal model predictive control design. The results determined a control setup that accounts for varying sunlight in order to potentially minimize thermal fatigue of the reactor. The ability to simulate the visual recognition of different weather conditions could prove advantageous to determine control parameter adjustments online as weather changes, potentially before its effects show up as measured deviations from normal operation.

An important question in how to set up an advanced disturbance-handling framework, however, that takes advantage of images for identifying different weather patterns and then adjusting control laws as required, is that it is not clear how to adjust the controllers, or whether different visual scenarios might correspond to very different needs in terms of how to update the controller parameters. An image-based control testbed in Blender could help in moving toward addressing some of these questions. However, modeling of weather in Blender can involve, for example, changing lighting model parameters. Lighting models in Blender have non-physical parameters to typically aid in the artistry of lighting for animation. For an engineering testbed application, ideas must be proposed regarding how to make images that correspond to different weather conditions and then translate these into data for a controller as precursors toward more advanced simulation concepts that might be more rigorously compared with more traditional disturbance-handling methods like offset-free model predictive control [64] or online triggering of data-driven model updates [65].

To move toward addressing the potential of Blender for aiding in testing ideas for image-based disturbance-handling methods for model-based control, in this section, a tank reactor that is exposed to outdoor sunlight is simulated in Blender. Using the virtual camera setup in Section 3.2.5, images are captured in order to determine the ambient heat effects on the temperature of the reactor. A model of the tank reactor that takes into account ambient heat due to sunlight is used to simulate the process. The controller uses detected sunlight in order to predict states of the reactor, which are acquired by calibrating the images from the virtual camera with lighting strength in the simulation.

3.3.1. Model Equations for Tank Reactor

The reactor considered in this section is a continuously stirred tank reactor, simulated with a system of ordinary differential equations, which are integrated numerically using the explicit Euler method. The reactor model is detailed in Equation (16), as follows:

$$\begin{aligned} \frac{dC_A}{dt} &= \frac{F}{V}(C_{A0} - C_A) - k_0 C_A^n \exp\left(-\frac{E_A}{RT}\right) \\ \frac{dT}{dt} &= \frac{F}{V}(T_{A0} - T_A) - \frac{k_0 C_A^n (\Delta H_R)}{\rho c_p} \exp\left(-\frac{E_A}{RT}\right) + \frac{Q}{\rho c_p V} \end{aligned} \tag{16}$$

The reactant concentration C_A and the temperature T are the state variables, while the inlet concentration C_{A0} and the heat supply Q are the manipulated variables. The design, reaction, and simulation parameters and their respective values are listed in Table 4.

Table 4. Reaction Parameters.

Parameters	Value
Rate constant k_0	$8.46 \times 10^6\ (\mathrm{m^3 kmol^{-1}})^{n-1}\mathrm{h^{-1}}$
Activation energy E_A	5×10^4 kJ/kmol
Density of reaction medium ρ	10^3 kg/m^3
Molar heat of reaction ΔH_R	-1.15×10^4 kJ/kmol
Specific heat capacity of reaction medium c_p	0.231 kJ kg^{-1}K^{-1}
Order of reaction n	2
Flow rate F	5 m^3/h
Reactor volume V	1 m^3
Integration time step Δt	1/24 s
Feed temperature T_{A0}	300 K

The tank reactor is controlled using an optimization-based controller that ensures that the reactant concentration and temperature reach pre-determined steady-state values, $C_{AS} = 2$ kmol/m^3, $T_S = 400$ K. Equation (16) is used to predict future states with the explicit Euler method, and an optimization problem is set up to minimize the sum of the squared errors for temperature and concentration with respect to the steady-state temperature and concentration of the reactor. The problem is solved using the SciPy [66] `minimize` function with the L-BFGS-B method over a prediction horizon of 10 sampling times. Each initial guess provided to the optimizer is $C_{A0S} = 4$ kmol/m^3, $Q_S = 0$ kJ/h, and the control actions are bounded, with $0.5\ \text{kmol/m}^3 \leq C_{A0} \leq 7.5\ \text{kmol/m}^3$, and $-10^5\ \text{kJ/h} \leq Q \leq 10^5\ \text{kJ/h}$.

3.3.2. CSTR Model with Sunlight Factor

In this final investigation of the potential uses of Blender as an image-based testbed for process systems engineering applications, incident sunlight on a reactor that is not insulated is assumed to affect the process primarily by adding heat to the system. It is assumed that the model of the tank reactor is identified at a particular sunlight strength, and hence, when the ambient sunlight is below or above this reference level, heat is removed or added to the system, respectively. In this simulation, the detected sunlight is considered to have an immediate impact on the rate of heat transferred to the system (i.e., potential time lags associated with heat transfer through the material of the tank or mixing within the tank are not modeled). Specifically, Equation (16) is modified by replacing the heat supply variable Q_c with $\hat{Q}_c$, the apparent heat supply, as in Equation (17):

$$\hat{Q}_c = Q_c + \Delta S \epsilon \tag{17}$$

The deviation from the reference sunlight strength is characterized by ΔS and is scaled to the heat supply by a factor ϵ. For this experiment, the reference lighting strength parameter in Blender is chosen to be 2, and the heating factor is set to 5×10^3 kJ/h. While the process model includes the lighting strength directly, the controller utilizes the detected images by correlating the RGB sums to the lighting strength, as detailed in the following section.

3.3.3. Ambient Light Simulation and Detection

In this simulation, images of a reactor and its surroundings are captured using a virtual camera. The intensity of the lighting in the simulation (i.e., the lighting strength) is modified in order to consider changing amounts of sunlight during the day. The model of the reactor is modified so that the heat extracted from the reactor is impacted by the incident light. One of the questions for creating a simulation involving the environment in Blender is how best to set up the lighting source. One method for setting up a lighting source is to use a high dynamic range image (HDRI). An HDRI is used in this simulation to attempt to generate images that resemble outdoor camera sensors. HDRIs are images used for the backgrounds in 3D modeling that wrap around objects in order to provide lighting and environment. The HDRI used in this work, `city.exr`, was obtained from Blender.

Another key aspect of the weather detection simulation is to conceptualize how to detect the sunlight strength from images so that the controller can use the data to update its model. In this example, we consider that the sum of the values of the red, green, and blue channels (RGB sums) are related to the set strength of the lighting (the relationship used in this simulation is calibrated using data from Blender). These RGB sums are calculated by importing the images with Pillow [36] and extracting the color channel values. A set of reference images between the minimum and maximum lighting strengths for the simulation are selected, and then the relation between the natural logarithm of the lighting strength and the RGB sums is plotted, as in Figure 17. RGB sums from the captured images are then used to acquire the lighting strength by interpolating between the known points. Equation (18) details the relationship obtained from the data, where r refers to the RGB sum, S is the lighting strength, and subscripts 1 and 2 refer to the two nearest data points used to interpolate the detected lighting strength:

$$S = S_1 \cdot \left(\frac{S_2}{S_1}\right)^{\frac{r - r_1}{r_2 - r_1}} \tag{18}$$

This calibration curve is specific to the chosen background, the resolution of the camera sensor, and the material of the object used to model the CSTR in Blender. Higher image resolutions correlate to higher sums, and an average of the sums could be used to mitigate these effects, subject to the accuracy required. The material of the objects affect the RGB sums by way of their reflectivity, with coarser and finer textures responding differently to the incident light.

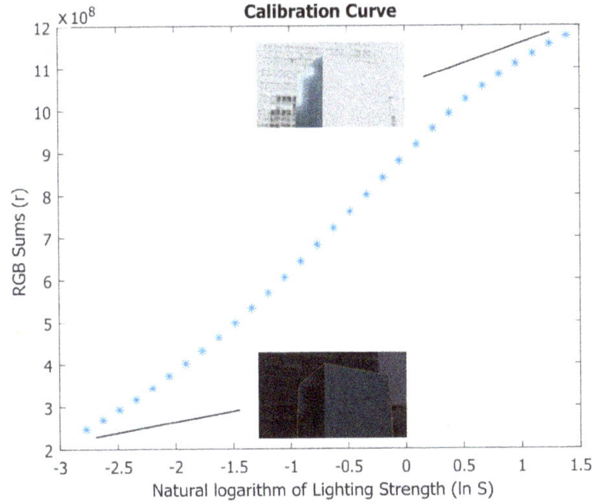

Figure 17. Lighting strength vs RGB sums plot. The yellow outline around the cube in the bottom left reflects that the cube is selected in Blender.

Remark 3. *The lighting strength parameter of HDRIs in Blender allows changes to the lighting model but is more of an artistic rather than a physical parameter. Thus, selecting the lighting strength for an actual application would likely require a comparison of images from an actual plant compared to the virtual model to attempt to find lighting strength values in Blender that match the observed plant behavior. However, even without a data-driven selection of lighting strength, simulations of the type in this work can still be used to test overall controller modification concepts in response to weather variations, knowing that they may need to be fine-tuned once actual process data is available to aid with updating the model. In practice, the lighting strength is replaced by a measurable quantity that is used to calibrate the image sensor. It may also be possible to determine the reference exclusively in terms of images by comparing the sum of the RGB channels of a reference image where the model behaves as designed and assigning a factor that scales the difference in RGB sums between the reference and images captured online to the apparent heat supply* Q_c.

3.3.4. Animation and Control Execution

To investigate the impact of the use of the image-based weather change detection algorithm on the effectiveness of the control policy, simulations were run for a total of 300 s, with each period of constant lighting strength lasting 75 s. In this simulation, lighting strength changes are implemented as step changes to observe the robustness of the controller when weather effects are accounted for. In practice, sunlight strength may vary on a time scale much slower than the controller sampling period. The sunlight was varied across three intensities, i.e., 0.5, 2.0, and 4.0, with time in this simulation, and the specific sunlight variation used in the simulation is detailed in a graph in Figure 18, along with example images captured by the virtual camera.

The differential equations in Equation (16) are used to simulate the state of the CSTR using the explicit Euler method with an integration step size of 1/24 s. The model used to update the reactor includes the lighting strength directly used in Blender. The factor of 1/24 is chosen because the animation runs at 24 frames per second, and the sampling and control is applied every second. The model built into the control optimizer differs from the one used to simulate the process by including the calibrator described in Section 3.3.4.

The sums of the RGB channels of the image are measured by capturing an image from the virtual camera in Blender and then loading it back into the script with the Python Pillow library [36]. The sum calculated is then interpolated to acquire the lighting strength from Equation (18) and then is used in the predictions of the optimizer to acquire an optimized input for the prediction horizon.

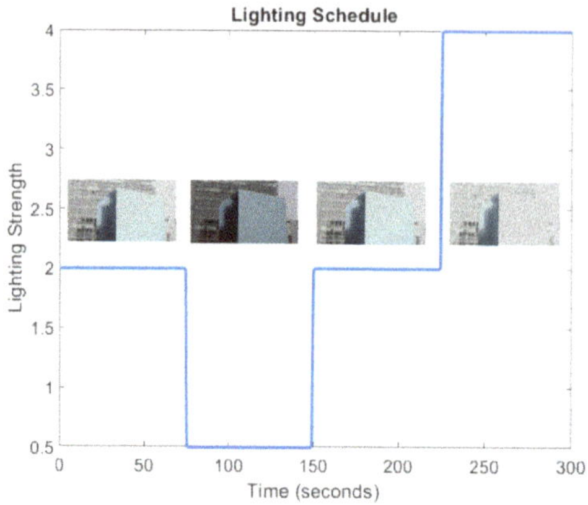

Figure 18. Schedule of lighting strength during simulation, with example images.

3.3.5. Control Effectiveness with and without Weather Detection

In order to compare the effects of light detection on the overall controller performance, the closed-loop responses of the system are compared for the cases where light is accounted for and where it is not accounted for. This is carried out by using Equation (16) for both cases. Figure 19 shows the differing control actions applied in each scenario. The major difference between them is the tracking of reactor temperature. When light detection is active, there is a small disturbance as the optimizer adjusts for a new lighting strength, while when it is inactive, there is a small step change. This also corresponds to the lighting strength in the schedule, with the reactor temperature dipping when the lighting strength is low and rising when it is high. The concentration profiles are largely similar, with differences being extremely small. For the case described, this Blender simulation would demonstrate that an image sensor to detect disturbances in sunlight does not significantly improve the performance of the outdoor reactor on the timescale simulated. This could be useful in determining whether to add such a sensor or not before building one. However, to more comprehensively make such a conclusion, the simulation should be run over a longer time period. Furthermore, longer periods of different sunlight strengths could affect the result.

Table 5 details the lighting strengths set in the simulation and the values detected from the calibration curve using RGB sums of the captured images. Since images with constant lighting strength are identical, errors are displayed for the three lighting strengths in the schedule. This demonstrates that the calibration curve is highly accurate if enough data points are provided. However, this is also subject to disturbances in real world scenarios, where image artifacts and the presence of uncalibrated objects, e.g., operators working in front of a camera sensor, can affect the calibration and interpolation in different ways. However, if a possible disturbance can be modeled with an object in Blender, it can be used to design a calibration curve that meets the requirement of the process. In order to function as a test bed, the calibration of the image sensor requires a more rigorous approach than that demonstrated in this example. Since Blender is built primarily for animation purposes, the lighting strength is an arbitrary unit. Hence, in order to more closely mimic the real world, experimental data that measures deviations from the heat supply and images of the reactor captured by the sensor during that time may be acquired to develop a calibration metric.

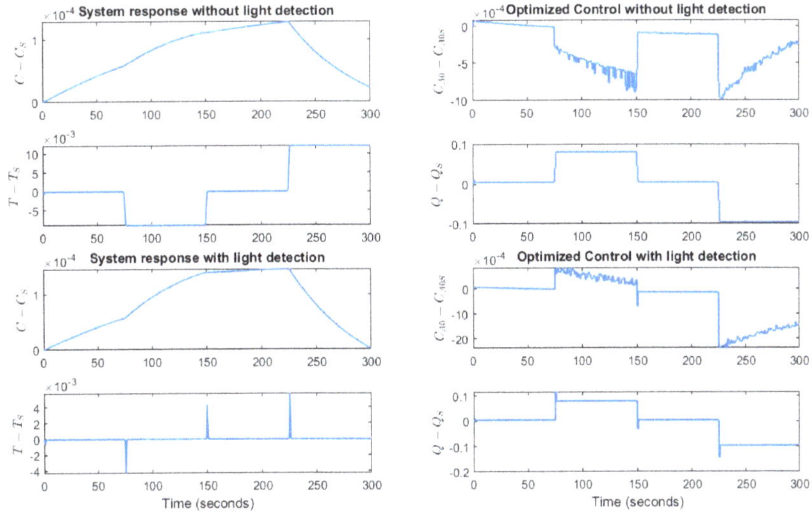

Figure 19. Comparison of system response and optimal control actions when light detection is active and inactive. All variables are plotted as deviations from steady-state values. Units: concentration (C, C_{A0})—kmol/m^3, temperature (T)—K, heat (Q)—kJ/h.

Table 5. Errors in detecting scheduled lighting strengths (see Remark 3 for a discussion of lighting strength in Blender).

Simulation Lighting Strength	Detected Lighting Strength
0.5	0.5001
2.0	2.0025
4.0	4.0000

Remark 4. *The general concept of utilizing Blender to test different concepts for weather detection for updating optimization-based control laws can be applied to other weather phenomena as well (e.g., rain). Though this work has discussed a number of potential challenges and opportunities with respect to using Blender for image-based control tests, many other considerations remain to be investigated, including how to make more photo-realistic models to ensure that the types of tests performed are representative of what might be seen in the field. In general, the concept that auxiliary data that is not typically used in updating controller models might be able to aid in knowing when to update them (i.e., revealing that a physics-based model is incomplete (e.g., it may not account for the impacts of weather variations on the process dynamics) can also be potentially applied with other data (e.g., there is potential that if anomalies are detected in the state data [67–69] that causes for these might be embedded in data such as auxiliary process numerical/textual information [70–72] (e.g., weather or maintenance logs) to seek to identify conditions that may have caused the modeling discrepancies and then to develop methods for detecting when such conditions are about to occur (e.g., utilizing pattern recognition [73,74] to note trends in the process data similar to those leading up to the prior occurrence of the anomaly).*

4. Conclusions

In this work, we demonstrate the potential of using Blender as a simulation test bed for image-based control with a few example processes. First, physical phenomena are replicated in Blender utilizing limited data and relying on the interpolation of object properties as a substitute to numerical integration. The motion of a bubble moving through a liquid is simulated with a select number of forces, and the position over time is acquired and animated in Blender using keyframe information. The position values acquired from Blender between keyframes where they are explicitly defined are compared with the results of numerical integration carried out at a finer timescale. Second, Blender is successfully used to replicate Brownian motion dynamics of the nanorod in two-dimensional space, and the optimal stochastic policy is implemented using a virtual image-based controller. Keyframe features are leveraged to demonstrate the scaling of time between the simulation and the control algorithm, where Brownian motion dynamics were updated frame by frame, but keyframes were inserted and images were captured every second to enact control. A virtual camera and light source are used to simulate the image-based control policy. The color information is summed and is used to detect the rod and extract information about its state features. Image capture by the software and additional image processing with the Python Pillow library are demonstrated, and the error between the state of the system from Blender and states detected in the image-based control algorithm is ascertained. Finally, a simulation of sunlight as a disturbance in processes that can be affected by ambient conditions is carried out in Blender. Utilizing the HDRI lighting settings, an outdoor environment is simulated, and the strength of the lighting is varied to simulate changes in daylight. These settings are then calibrated with the sum of the red, blue, and green channels of the images captured by the virtual camera in Blender to approximate lighting strength based on a given image. The process is then simulated with a controller that optimizes for steady-state behavior while also accounting for varying sunlight. A comparison of the responses to the process based on whether the controller accounts for the changes in sunlight is illustrated. With the framework currently in place, it opens the possibility of virtually testing these and other image processing techniques in order to rapidly test novel image-based control algorithms in a variety of processes. Future studies could include further investigations into the potential of Blender for use in developing image-based control and safety systems.

Author Contributions: A.F.L.: conceptualization, methodology, software, validation G.G.: software, methodology M.R.: conceptualization, methodology H.E.D.: supervision, funding acquisition, validation. All authors have read and agreed to the published version of the manuscript.

Funding: Financial support from the Air Force Office of Scientific Research (award number FA9550-19-1-0059), National Science Foundation CNS-1932026 and CBET-1839675, and Wayne State University is gratefully acknowledged.

Data Availability Statement: The codes associated with this article will be posted at: https://durand.eng.wayne.edu/PublicationResources/index.html accessed on 2 September 2023.

Acknowledgments: We would like to thank Michael Williamson, Keshav Kasturi Rangan, and Henrique Oyama for the discussions, and Jacob Noll for finding reference [39]. Figures in this publication have been created with Blender versions 3.6 and 4.0, MATLAB R2023a, GNU Image Manipulation Program 2.10.36, and draw.io v22.1.8.

Conflicts of Interest: The authors declare no conflict of interest.

References

1. Prats-Montalbán, J.M.; de Juan, A.; Ferrer, A. Multivariate image analysis: A review with applications. *Chemom. Intell. Lab. Syst.* **2011**, *107*, 1–23. [CrossRef]
2. De, S.; Mohamed, S.; Bimpisidis, K.; Goswami, D.; Basten, T.; Corporaal, H. Approximation trade offs in an image-based control system. In Proceedings of the 2020 Design, Automation & Test in Europe Conference & Exhibition (DATE), Grenoble, France, 9–13 March 2020; IEEE: Piscataway, NJ, USA, 2020; pp. 1680–1685.
3. Bengler, K.; Dietmayer, K.; Farber, B.; Maurer, M.; Stiller, C.; Winner, H. Three decades of driver assistance systems: Review and future perspectives. *IEEE Intell. Transp. Syst. Mag.* **2014**, *6*, 6–22. [CrossRef]
4. Su, Y.; Zheng, C. A simple PID control for asymptotic visual regulation of robot manipulators. *Int. J. Robust Nonlinear Control* **2011**, *21*, 1525–1540. [CrossRef]
5. Bluma, A.; Höpfner, T.; Lindner, P.; Rehbock, C.; Beutel, S.; Riechers, D.; Hitzmann, B.; Scheper, T. In-situ imaging sensors for bioprocess monitoring: State of the art. *Anal. Bioanal. Chem.* **2010**, *398*, 2429–2438. [CrossRef]
6. Höpfner, T.; Bluma, A.; Rudolph, G.; Lindner, P.; Scheper, T. A review of non-invasive optical-based image analysis systems for continuous bioprocess monitoring. *Bioprocess Biosyst. Eng.* **2010**, *33*, 247–256. [CrossRef]
7. Chandrasekaran, S.N.; Ceulemans, H.; Boyd, J.D.; Carpenter, A.E. Image-based profiling for drug discovery: Due for a machine-learning upgrade? *Nat. Rev. Drug Discov.* **2021**, *20*, 145–159. [CrossRef]

8. Mebarki, R.; Lippiello, V. Image-based control for aerial manipulation. *Asian J. Control* **2014**, *16*, 646–656. [CrossRef]
9. Collewet, C.; Chaumette, F. A contour approach for image-based control on objects with complex shape. In Proceedings of the 2000 IEEE/RSJ International Conference on Intelligent Robots and Systems (IROS 2000) (Cat. No. 00CH37113), Takamatsu, Japan, 31 October–5 November 2000; IEEE: Piscataway, NJ, USA, 2000; Volume 1, pp. 751–756.
10. Zheng, D.; Wang, H.; Wang, J.; Chen, S.; Chen, W.; Liang, X. Image-based visual servoing of a quadrotor using virtual camera approach. *IEEE/ASME Trans. Mechatron.* **2016**, *22*, 972–982. [CrossRef]
11. Becker, B.C.; MacLachlan, R.A.; Lobes, L.A.; Hager, G.D.; Riviere, C.N. Vision-based control of a handheld surgical micromanipulator with virtual fixtures. *IEEE Trans. Robot.* **2013**, *29*, 674–683. [CrossRef] [PubMed]
12. Pazzi, R.W.N.; Boukerche, A.; Huang, T. Implementation, measurement, and analysis of an image-based virtual environment streaming protocol for wireless mobile devices. *IEEE Trans. Instrum. Meas.* **2008**, *57*, 1894–1907. [CrossRef]
13. Lee, D.; Lim, H.; Kim, H.J. Obstacle avoidance using image-based visual servoing integrated with nonlinear model predictive control. In Proceedings of the 2011 50th IEEE Conference on Decision and Control and European Control Conference, Orlando, FL, USA, 12–15 December 2011; IEEE: Piscataway, NJ, USA, 2011; pp. 5689–5694.
14. Hajiloo, A.; Keshmiri, M.; Xie, W.F.; Wang, T.T. Robust online model predictive control for a constrained image-based visual servoing. *IEEE Trans. Ind. Electron.* **2015**, *63*, 2242–2250.
15. Lopez-Franco, C.; Gomez-Avila, J.; Alanis, A.Y.; Arana-Daniel, N.; Villaseñor, C. Visual servoing for an autonomous hexarotor using a neural network based PID controller. *Sensors* **2017**, *17*, 1865. [CrossRef]
16. Yu, H.; MacGregor, J.F. Monitoring flames in an industrial boiler using multivariate image analysis. *AIChE J.* **2004**, *50*, 1474–1483. [CrossRef]
17. Lin, B.; Recke, B.; Knudsen, J.K.; Jørgensen, S.B. Bubble size estimation for flotation processes. *Miner. Eng.* **2008**, *21*, 539–548. [CrossRef]
18. Cao, Y.; Yu, H.; Abbott, N.L.; Zavala, V.M. Machine learning algorithms for liquid crystal-based sensors. *ACS Sens.* **2018**, *3*, 2237–2245. [CrossRef] [PubMed]
19. Pulsipher, J.L.; Coutinho, L.D.J.; Soderstrom, T.A.; Zavala, V.M. SAFE-OCC: A novelty detection framework for Convolutional Neural Network sensors and its application in process control. *J. Process Control* **2022**, *117*, 78–97. [CrossRef]
20. Jiang, S.; Qin, S.; Pulsipher, J.L.; Zavala, V.M. Convolutional Neural Networks: Basic Concepts and Applications in Manufacturing. *arXiv* **2022**, arXiv:2210.07848.
21. Wang, X.Z.; Roberts, K.J.; Ma, C. Crystal growth measurement using 2D and 3D imaging and the perspectives for shape control. *Chem. Eng. Sci.* **2008**, *63*, 1173–1184. [CrossRef]
22. Larsen, P.A.; Rawlings, J.B.; Ferrier, N.J. An algorithm for analyzing noisy, in situ images of high-aspect-ratio crystals to monitor particle size distribution. *Chem. Eng. Sci.* **2006**, *61*, 5236–5248. [CrossRef]
23. Maaß, S.; Rojahn, J.; Hänsch, R.; Kraume, M. Automated drop detection using image analysis for online particle size monitoring in multiphase systems. *Comput. Chem. Eng.* **2012**, *45*, 27–37. [CrossRef]
24. Chen, J.; Chang, Y.H.; Cheng, Y.C.; Hsu, C.K. Design of image-based control loops for industrial combustion processes. *Appl. Energy* **2012**, *94*, 13–21. [CrossRef]
25. Lu, Q.; Zavala, V.M. Image-based model predictive control via dynamic mode decomposition. *J. Process Control* **2021**, *104*, 146–157. [CrossRef]
26. Pearson, T.; Brabec, D.; Haley, S. Color image based sorter for separating red and white wheat. *Sens. Instrum. Food Qual. Saf.* **2008**, *2*, 280–288. [CrossRef]
27. Dere, S.; Sahasrabudhe, S.; Iyer, S. Creating open source repository of 3D models of laboratory equipments using Blender. In Proceedings of the 2010 International Conference on Technology for Education, Mumbai, India, 1–3 July 2010; pp. 149–156.
28. Rajendiran, N.; Durrant, J.D. Pyrite: A Blender plugin for visualizing molecular dynamics simulations using industry-standard rendering techniques. *J. Comput. Chem.* **2018**, *39*, 748–755. [CrossRef]
29. Gschwandtner, M.; Kwitt, R.; Uhl, A.; Pree, W. BlenSor: Blender sensor simulation toolbox. In *Proceedings of the Advances in Visual Computing: 7th International Symposium, ISVC 2011, Las Vegas, NV, USA, 26–28 September 2011*; Proceedings, Part II 7; Springer: Berlin/Heidelberg, Germany, 2011; pp. 199–208.
30. Flaischen, S.; Wehinger, G.D. Synthetic packed-bed generation for CFD simulations: Blender vs. STAR-CCM+. *ChemEngineering* **2019**, *3*, 52. [CrossRef]
31. Oyama, H.; Leonard, A.F.; Rahman, M.; Gjonaj, G.; Williamson, M.; Durand, H. On-line process physics tests via Lyapunov-based economic model predictive control and simulation-based testing of image-based process control. In Proceedings of the 2022 American Control Conference (ACC), Atlanta, GA, USA, 8–10 June 2022; IEEE: Piscataway, NJ, USA, 2022; pp. 2479–2484.
32. Parzen, E. *Stochastic Processes*; SIAM: Philadelphia, PA, USA, 1999.
33. Øksendal, B. Stochastic differential equations . In *Stochastic Differential Equations*; Springer: Berlin/Heidelberg, Germany, 2003; pp. 65–84.
34. Wrobel, A. On Markovian decision models with a finite skeleton. *Z. Für Oper. Res.* **1984**, *28*, 17–27. [CrossRef]
35. Gangwani, T.; Li, D.; Ye, Z. Lecture 16: Value Iteration, Policy Iteration and Policy Gradient. Available online: https://yuanz.web.illinois.edu/teaching/IE498fa19/lec_16.pdf (accessed on 4 September 2023).
36. Clark, A. Pillow (PIL Fork) Documentation. ReadTheDocs. 2015. Available online: https://buildmedia.readthedocs.org/media/pdf/pillow/latest/pillow.pdf (accessed on 4 September 2023).

37. Oyama, H.; Messina, D.; O'Neill, R.; Cherney, S.; Rahman, M.; Rangan, K.K.; Gjonaj, G.; Durand, H. Test Methods for Image-Based Information in Next-Generation Manufacturing. *IFAC-PapersOnLine* **2022**, *55*, 73–78. [CrossRef]
38. Oyama, H.; Messina, D.; Rangan, K.K.; Leonard, A.F.; Nieman, K.; Durand, H.; Tyrrell, K.; Hinzman, K.; Williamson, M. Development of directed randomization for discussing a minimal security architecture. *Digit. Chem. Eng.* **2023**, *6*, 100065. [CrossRef]
39. Yang, Y. Stochastic Modeling and Optimal Control for Colloidal Organization, Navigation, and Machines. Ph.D. Thesis, The Johns Hopkins University, Baltimore, MD, USA, 2017.
40. Bartolacci, G.; Pelletier, P.; Tessier, J.; Duchesne, C.; Bossé, P.A.; Fournier, J. Application of numerical image analysis to process diagnosis and physical parameter measurement in mineral processes—Part I: Flotation control based on froth textural characteristics. *Miner. Eng.* **2006**, *19*, 734–747. [CrossRef]
41. Kaartinen, J. ; Hätönen, J.; Hyötyniemi, H.; Miettunen, J. Machine-vision-based control of zinc flotation—A case study. *Control Eng. Pract.* **2006**, *14*, 1455–1466. [CrossRef]
42. Liu, J.J.; MacGregor, J.F.; Duchesne, C.; Bartolacci, G. Flotation froth monitoring using multiresolutional multivariate image analysis. *Miner. Eng.* **2005**, *18*, 65–76. [CrossRef]
43. Bharati, M.H.; MacGregor, J.F. Multivariate image analysis for real-time process monitoring and control. *Ind. Eng. Chem. Res.* **1998**, *37*, 4715–4724. [CrossRef]
44. Bird, R.B.; Stewart, W.E.; Lightfoot, E.N. *Transport Phenomena*; John Wiley & Sons: Hoboken, NJ, USA, 1961.
45. Cleary, P.W.; Pyo, S.H.; Prakash, M.; Koo, B.K. Bubbling and frothing liquids. In *ACM SIGGRAPH 2007 Papers*; ACM: New York, NY, USA, 2007; p. 97-es.
46. Dewitt, C.C. Froth flotation concentration. *Ind. Eng. Chem.* **1940**, *32*, 652–658. [CrossRef]
47. Cartesian Caramel. How to Make Oil Spill Materials in Blender! Video, Uploaded to Youtube. 20 February 2022. Available online: https://youtu.be/xcx_LfXuuX4 (accessed on 31 October 2023).
48. Do, H. Development of a Turbulent Flotation Model from First Principles. Ph.D. Thesis, Virginia Tech, Blacksburg, VA, USA, 2010.
49. Lin, C.; Liu, Y.; Rinker, S.; Yan, H. DNA tile based self-assembly: Building complex nanoarchitectures. *ChemPhysChem* **2006**, *7*, 1641–1647. [CrossRef]
50. Douglas, S.M.; Dietz, H.; Liedl, T.; Högberg, B.; Graf, F.; Shih, W.M. Self-assembly of DNA into nanoscale three-dimensional shapes. *Nature* **2009**, *459*, 414–418. [CrossRef]
51. Wang, X.S.; Winnik, M.A.; Manners, I. Synthesis and aqueous self-assembly of a polyferrocenylsilane-block-poly (aminoalkyl methacrylate) diblock copolymer. *Macromol. Rapid Commun.* **2002**, *23*, 210–213. [CrossRef]
52. Rider, D.A.; Manners, I. Synthesis, Self-Assembly, and Applications of Polyferrocenylsilane Block Copolymers. *Polym. Rev.* **2007**, *47*, 165–195. [CrossRef]
53. Liu, R.; Huang, H.; Sun, Z.; Alexander-Katz, A.; Ross, C.A. Metallic nanomeshes fabricated by multimechanism directed self-assembly. *ACS Nano* **2021**, *15*, 16266–16276. [CrossRef]
54. Lane, A.P.; Yang, X.; Maher, M.J.; Blachut, G.; Asano, Y.; Someya, Y.; Mallavarapu, A.; Sirard, S.M.; Ellison, C.J.; Willson, C.G. Directed self-assembly and pattern transfer of five nanometer block copolymer lamellae. *ACS Nano* **2017**, *11*, 7656–7665. [CrossRef]
55. Duan, H.; Berggren, K.K. Directed self-assembly at the 10 nm scale by using capillary force-induced nanocohesion. *Nano Lett.* **2010**, *10*, 3710–3716. [CrossRef]
56. Yu, J.; Wang, Q.; Zhang, X. Effects of external force fields on peptide self-assembly and biomimetic silica synthesis. *Appl. Surf. Sci.* **2014**, *311*, 799–807. [CrossRef]
57. Motornov, M.; Malynych, S.Z.; Pippalla, D.S.; Zdyrko, B.; Royter, H.; Roiter, Y.; Kahabka, M.; Tokarev, A.; Tokarev, I.; Zhulina, E.; et al. Field-directed self-assembly with locking nanoparticles. *Nano Lett.* **2012**, *12*, 3814–3820. [CrossRef] [PubMed]
58. Tang, X.; Rupp, B.; Yang, Y.; Edwards, T.D.; Grover, M.A.; Bevan, M.A. Optimal feedback controlled assembly of perfect crystals. *ACS Nano* **2016**, *10*, 6791–6798. [CrossRef] [PubMed]
59. Issa, H.; Natale, G.; Ausias, G.; Férec, J. Modeling and numerical simulations of Brownian rodlike particles with anisotropic translational diffusion. *Phys. Rev. Fluids* **2023**, *8*, 033302. [CrossRef]
60. Durrett, R. *Probability: Theory and Examples*; Cambridge Series in Statistical and Probabilistic Mathematics; Cambridge University Press: Cambridge, UK, 2019.
61. Alexandru, A.B.; Morari, M.; Pappas, G.J. Cloud-based MPC with encrypted data. In Proceedings of the 2018 IEEE Conference on Decision and Control (CDC), Miami, FL, USA, 17–19 December 2018; pp. 5014–5019.
62. Zavala, V.M.; Constantinescu, E.M.; Krause, T.; Anitescu, M. On-line economic optimization of energy systems using weather forecast information. *J. Process Control* **2009**, *19*, 1725–1736. [CrossRef]
63. Rowe, S.C.; Hischier, I.; Palumbo, A.W.; Chubukov, B.A.; Wallace, M.A.; Viger, R.; Lewandowski, A.; Clough, D.E.; Weimer, A.W. Nowcasting, predictive control, and feedback control for temperature regulation in a novel hybrid solar-electric reactor for continuous solar-thermal chemical processing. *Sol. Energy* **2018**, *174*, 474–488. [CrossRef]
64. Das, B.; Mhaskar, P. Lyapunov-based offset-free model predictive control of nonlinear process systems. *Can. J. Chem. Eng.* **2015**, *93*, 471–478. [CrossRef]
65. Alanqar, A.; Durand, H.; Christofides, P.D. Fault-tolerant economic model predictive control using error-triggered online model identification. *Ind. Eng. Chem. Res.* **2017**, *56*, 5652–5667. [CrossRef]

66. Virtanen, P.; Gommers, R.; Oliphant, T.E.; Haberland, M.; Reddy, T.; Cournapeau, D.; Burovski, E.; Peterson, P.; Weckesser, W.; Bright, J.; et al. SciPy 1.0: Fundamental Algorithms for Scientific Computing in Python. *Nat. Methods* **2020**, *17*, 261–272. [CrossRef]
67. Lane, T.; Brodley, C.E. An application of machine learning to anomaly detection. In Proceedings of the 20th National Information Systems Security Conference, Baltimore, MD, USA, 7–10 October 1997.
68. Shon, T.; Moon, J. A hybrid machine learning approach to network anomaly detection. *Inf. Sci.* **2007**, *177*, 3799–3821. [CrossRef]
69. Vernekar, S.; Nari, S.; Vijaysenan, D.; Ranjan, R. A novel approach for classification of normal/abnormal phonocardiogram recordings using temporal signal analysis and machine learning. In Proceedings of the 2016 Computing in Cardiology Conference, Vancouver, BC, Canada, 11–14 September 2016; Volume 63, pp. 1141–1144.
70. Turney, P.D.; Pantel, P. From frequency to meaning: Vector space models of semantics. *J. Artif. Intell. Res.* **2010**, *37*, 141–188. [CrossRef]
71. Zhang, Y.; Jiang, R.; Petzold, L. Survival Topic Models for Predicting Outcomes for Trauma Patients. In Proceedings of the IEEE 33rd International Conference on Data Engineering, San Diego, CA, USA, 19–22 April 2017; pp. 1497–1504.
72. Nguyen, N.T.H.; Miwa, M.; Tsuruoka, Y.; Tojo, S. Identifying synonymy between relational phrases using word embeddings. *J. Biomed. Inform.* **2015**, *56*, 94–102. [CrossRef] [PubMed]
73. Lopes, A.T.; de Aguiar, E.; De Souza, A.F.; Oliveira-Santos, T. Facial expression recognition with convolutional neural networks: Coping with few data and the training sample order. *Pattern Recognit.* **2017**, *61*, 610–628. [CrossRef]
74. Uhlmann, E.; Pontes, R.P.; Laghmouchi, A.; Bergmann, A. Intelligent pattern recognition of a SLM machine process and sensor data. *Procedia CIRP* **2017**, *62*, 464–469. [CrossRef]

processes

MDPI

Article

Embedded MPC Strategies for ESP-Lifted Oil Wells: Hardware-in-the-Loop Performance Analysis of Nonlinear and Robust Techniques

Bruno A. Santana [1,2,*], Victor S. Matos [1], Daniel D. Santana [3] and Márcio A. F. Martins [1,3,*]

[1] Mechatronics Program, Polytechnic School, Federal University of Bahia, Salvador 40170-110, Brazil; victorsm@ufba.br
[2] Department of Engineering and Computing, State University of Santa Cruz, Ilhéus 45662-900, Brazil
[3] Department of Chemical Engineering, Polytechnic School, Federal University of Bahia, Salvador 40170-110, Brazil; daniel.diniz@ufba.br
* Correspondence: bruno.aguiar@ufba.br (B.A.S.); marciomartins@ufba.br (M.A.F.M.)

Abstract: This paper proposes embedded model predictive control strategies for oil-production processes equipped with electric submersible pump (ESP) installations. The novelty of this paper is the robustness and computational performance analysis of the Robust Infinite-Horizon Model Predictive Controller (RIHMPC) and Nonlinear Model Predictive Controller (NMPC) strategies, which have not yet been documented by the oil and gas exploration and production literature. The proposed method to embed the control laws is flexible with different hardware and is based on automatic code generation, which facilitates the project workflow. Hardware-in-the-loop simulation cases were used to compare the performance of both control strategies embedded in the Teensy 4.1 microcontroller, using key indices for real applications. The results showed that the RIHMPC strategy is a very promising alternative for real-time operation in ESP-lifted oil wells, with overall performance similar to the NMPC controller, even in noisy and plant–model mismatch scenarios, and using only linear models in its formulation.

Keywords: artificial lift; electrical submersible pump; zone control; robust model predictive control; nonlinear model predictive control; embedded control

1. Introduction

1.1. Overview of ESP-Lifted Oil-Well System

The ESP is the second most widely used artificial lift method for offshore and onshore oil wells and the first in volume of oil produced [1,2]. The economic success of oil wells is heavily dependent on the availability of their artificial lift equipment, which is related mainly to the frequency of failures in ESP-equipped oil wells due to unfavorable operating conditions during the production process. As outlined in Takacs [3], ensuring the safe and stable operation of ESP-lifted oil wells involves adhering to its nonlinearities and a set of physical and safety constraints. Such ESP systems' constrained multivariable characteristics demand therefore automation and control solutions to ensure proper and efficient operation.

Although advanced process control techniques, e.g., model predictive control (MPC), can systematically deal with the particularities of ESP-lifted oil wells, providing an optimal and safer approach, it is a common practice to adopt a manual operation in fields equipped with this artificial lift method, as highlighted by Krishnamoorthy et al. [4]. In the oil and gas exploration and production (E&P) field, implementing model predictive control (MPC) strategies in an actual ESP facility is a scarce subject, with a few examples treated in the literature, such as Pavlov et al. [5] and Patel et al. [6].

On the other hand, some authors are embedding their techniques in relevant hardware for real-time applications in this field, as Binder et al. [7], Santana et al. [8] and Matos et al. [9]. These formulations are usually based on linear models and adaptive schemes without incorporating stabilizing properties. To work around this, the contribution of this work is to investigate the embedded application of two control strategies for oil-production processes equipped with ESP installations, namely: (i) a robust stabilizing MPC based on a multi-linear model description of uncertainty; (ii) an MPC strategy based on a nonlinear phenomenological model (NMPC). Additionally, aiming at practical applications, these strategies incorporate the ESP operational envelope-type constraints in an implementable and softening zone-control scheme.

1.2. State of the Art and Related Works

Regarding MPC-based strategies for ESP-lifted oil wells, Pavlov et al. [5] proposed an MPC controller for a laboratory full-scale ESP-lifted oil-well facility. This controller aimed to track the ESP intake pressure and minimize power consumption by explicitly incorporating a choke valve opening target in the objective function of the controller. A step-response-based linear model was identified from a third-order nonlinear model considering homogeneous fluid under isothermal conditions and choke valve opening around 0–20%. Their formulation considered constraints as a compact convex set: wellhead pressure, ESP intake pressure, ESP motor current, bounds on the ESP rotational speed, and its rate of change. They also included ESP operating envelope-type constraints, limiting the down-thrust and up-thrust regions. Subsequently, Krishnamoorthy et al. [4] proposed another MPC application for such a system, based on the same control objectives of Pavlov et al. [5], but with a data-driven linearized model from a high-fidelity ESP simulator that considers multiphase flow and viscous fluid associated with gas and water under non-isothermal condition. They considered tracking the ESP's reference flow rate instead of explicitly imposing (down and up)-thrust force constraints, resulting in a less computationally intensive optimization problem, even with performance loss. Their results proved to be robust to deal with realistic fluid viscosity changes. Binder et al. [10] included feedforward actions related to the measured disturbance of the plant, such as reservoir pressure, into their MPC formulation to improve the control performance in an ESP-lifted oil-well system.

Binder et al. [7] proposed the first embedded MPC in a programmable logic controller (PLC) for ESP-equipped systems through hardware-in-the-loop (HiL) simulations, in which a nonlinear phenomenological model represented the oil-production process to be controlled. The control law was based on the step-response model identified in a region near 90–100% choke valve opening. They included reference flow rate constraints to indirectly minimize ESP power consumption by regulating the motor current. Furthermore, the first practical MPC large-scale implementation in a real onshore oilfield (more than 100 wells), with multiple ESP-lifted oil wells, was presented by Patel et al. [6]. Their formulation applied a data-driven step-response model and used the rotational speed, the choke valve opening, and the ESP voltage as manipulated variables to track the reference flow rate setpoint, considering practical ESP constraints, achieving a significant power saving (10% to 20%).

Despite ongoing advances, the strategies mentioned so far do not consider either guarantee of stability or feasibility for MPC control laws. Fontes et al. [11] was the first to propose a stabilizing MPC strategy with guaranteed feasibility for such systems. This strategy is based on an infinite-horizon MPC (IHMPC) controller that explicitly includes the down-thrust and up-thrust constraints such as a softening target zone scheme. The stabilizing properties are guaranteed by slacked terminal equality constraints, which also enforces the feasibility of the resulting optimization problem.

Regarding robustness, some developments with adaptive MPC schemes have been observed to extend the operating region and accommodate the typical nonlinearities and uncertainties of the ESP-equipped oilfields. In this scenario, Delou et al. [12] introduced an adaptive MPC control law that uses two step-response linear models, one for a low-gain region with a large choke valve opening and the other for a high-gain region with a small choke valve opening. However, to enhance the system's robustness and incorporate more dynamic information, Delou et al. [13] proposed a control law that employs multiple linear models, including an updated state estimator, to create a robust approach that can handle missing state variable measurements, as previously suggested by Binder et al. [14]. However, the authors do not explicitly consider down-thrust and up-thrust constraints in their control laws.

To overcome the limitations from [12,13], Santana et al. [15] proposed an adaptive and feasible zone IHMPC scheme based on Fontes et al. [11], applying successive linearization of a nonlinear phenomenological model to accommodate the complexities and uncertainties present in the ESP-equipped systems and directly dealing with the pump envelope constraints. However, this strategy does not have robustly stabilizing properties, despite the performance improvement compared to the linear model-based IHMPC [11]. Matos et al. [9] extended that approach, treating the adaptation of the prediction model by a fuzzy structure, which combines data-based local linear models as a linear parameter-varying model. This technique can be directly embedded in ESP installations since it needs minimum available measured data without requiring phenomenological knowledge or flow rate estimation.

Another promising class of MPC controllers associated with ESP-lifted oil-well systems is the nonlinear MPC (NMPC) strategies. Jordanou et al. [16] designed a nonlinear control law using a data-based echo-state neural network as the prediction model. However, their approach does not explicitly consider ESP operational envelope constraints in their controller formulation. Moreover, the average execution time to solve the NMPC law in a simulation environment is more than five times the adopted sampling time, limiting real-time embedded implementation.

In contrast, Santana et al. [17] presented a zone NMPC formulation that can handle these limitations for ESP-lifted oil-well systems. Their approach explicitly deals with ESP operational envelope constraints via an implementable zone-control scheme. Additionally, the method can accommodate the lack of state measurements by coupling a nonlinear state estimator. The most significant advantage of their approach is its real-time applicability for embedded implementation since the computation time is in the order of milliseconds on the simulation reported, following Allamaa et al. [18], which presented recently an NMPC strategy formulated and validated for a real-time application that suggests its implementation on an embedded system. The results showed a low-level controller working at 25 Hz in a constrained and disturbance application.

With recent advances in industrial MPC implementations, automatic code generation techniques have been increasingly improved, aiming to easily provide a solver implementation that is compact, self-contained, efficient, and possibly customized to the embedded hardware on which it needs to run. In [19], a GRAMPC library was used for the low-level integration of an NMPC in assisted driving and obstacle avoidance. The controller was implemented in an ARM-based microcontroller using entirely the C/C++ language, and validated through Processor-in-the-Loop tests. The author [20] presented the real-time implementation of NMPC on embedded systems and its performance evaluation using ACADO toolkit code generation, and had as a case study the position control of hovercraft. Other code generation tools such as Casadi [21] and IPOPT [22] are also excellent for PC-based environments but not as suitable and ready to port for embedded implementation. Moreover, the ACADO toolkit generates highly efficient C/C++ code but it is generally library dependent.

1.3. Contributions

The main contribution of this work is the proposition and systematic analysis of an embedded robust MPC strategy for ESP-lifted oil-well systems from the preceding discussion, not yet documented in the E&P literature. Through a hardware-in-the-loop simulation, its performance is compared against the NMPC scheme proposed by [17]. Both strategies encapsulate the zone-control approach to deal directly with the ESP operational envelope constraints and incorporate economic targets to improve their performance. As a further contribution, the proposed MPC implementation method makes use of the Model-Based Development approach, in which the automatic C/C++ code generation tool facilitates the project workflow and has no dependency on external libraries.

2. Robust and Nonlinear MPC Schemes for Embedded Applications on ESP-Equipped Oil-Well Systems

This section presents the challenges of controlling the ESP-lifted oil wells and the proposed MPC techniques designed for this system: (i) the nonlinear MPC approach and (ii) the robust infinite-horizon MPC (RIHMPC) approach.

2.1. ESP-Lifted Oil-Well System

Figure 1 depicts a typical ESP-lifted oil-well system. The ESP operates by adjusting the rotational frequency of the pump, f, and the production choke valve opening, z_c, to apply additional pressure to the oil and bring it to the surface through the production column and manifold. The fluid properties, such as flow type, the average flow rate in the production column, q_p, and ESP head dynamics, H, defines the operational envelope constraint, which consists of time-varying up-thrust and down-thrust constraints. A typical operational envelope curve is shown in Figure 1, usually available from pump manufacturer datasheets. In this study, the average flow rate of the production column was treated as a measured variable. At each sample interval, the maximum head value, H_{max}, and minimum head value, H_{min}, were established for the current q_p value on the operational envelope curve.

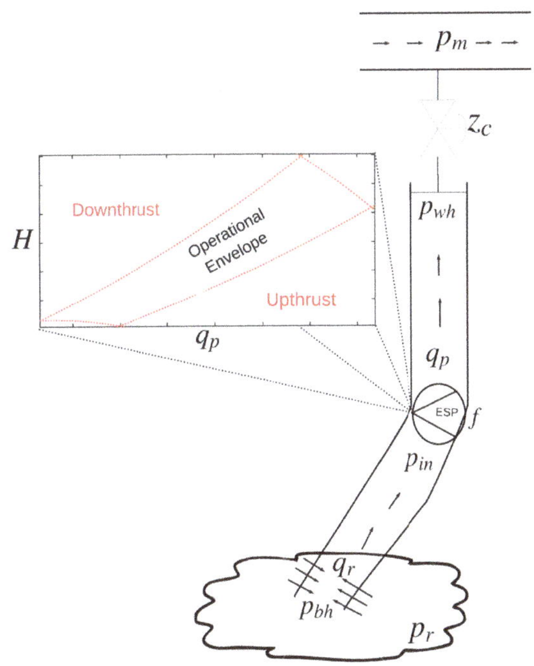

Figure 1. Typical scheme of an ESP-lifted oil-production system [11].

A typical lumped-element dynamic model for an ESP-lifted oil well was used for simulating the dynamic behavior of the ESP-lifted oil well system. It is represented by a set of differential-algebraic equations as follows [5]:

$$\begin{cases} \dfrac{dp_{wh}}{dt} = 1.54 \times 10^8 (q_p(t) - q_c(t)) \\ \dfrac{dp_{bh}}{dt} = 0.8584(p_r(t) - p_{bh}(t) - 3.7 \times 10^8 q_p \\ \dfrac{dq_p}{dt} = 5.02 \times 10^{-9} [p_{bh}(t) - p_{wh}(t) - 6.30 \times 10^8 q_p(t)^{1.75} \\ \quad +9.32 \times 10^3 (H(t) - 1 \times 10^3)] \end{cases} \tag{1}$$

$$\begin{cases} p_{in}(t) = p_{bh}(t) - 1.85 \times 10^8 q_p^{1.75}(t) - 1.9 \times 10^6 \\ H(t) = 0.2664 f(t)^2 + 133.09 f(t) q_p(t) - 1.41 \times 10^6 q_p^2(t) \end{cases} \tag{2}$$

$$q_c(t) = 2 \times 10^{-5} z_c(t) \sqrt{p_{wh}(t) - p_m(t)} \tag{3}$$

$$P(t) = 4.77 \times 10^{-1} f^3(t) + 1.41 \times 10^3 q_p(t) f^2(t) - 3.74 \times 10^5 q_p^2(t) f(t) - 3.12 \times 10^9 q_p^3(t) \tag{4}$$

where p_{wh}, p_{bh}, p_{in}, p_m and p_r are the wellhead, bottom hole, intake, manifold, and reservoir pressures in Pa, respectively; q_c is the flow rate in the production choke in m^3/s; and P is the ESP power in W.

The ESP's commonly manipulated variables are its rotational frequency and choke valve opening. The control problem involves adjusting them to reach a desired production flow while respecting the safe and physical constraints, especially those defined by the operational envelope. To reach this objective, the intake pressure is taken as a controlled variable that must track a setpoint. Additionally, the head can be treated as another controlled variable, requiring the zone-control approach to softly define its optimal setpoint based on the limits defined by the operational envelope [11].

2.2. Nonlinear MPC

This section presents the NMPC approach for ESP systems proposed by Santana et al. [17]: a zone control approach incorporating the model (1)–(4), resulting in a soft optimization problem. In this sense, the constraints of the zone map the operational envelope, as presented in Figure 2, which avoids infeasible solutions for the optimization problem. The optimization problem is given by [17]:

Problem 1

$$\min_{\Delta \mathbf{u}_k, \mathbf{y}_{sp,k}} V_k = \sum_{j=1}^{p} \left\| \hat{\mathbf{y}}(k+j|k) - \mathbf{y}_{sp,k} + \epsilon(k|k) \right\|_{\mathbf{Q_y}}^2 + \sum_{j=0}^{m-1} \|\Delta \mathbf{u}(k+j|k)\|_{\mathbf{R}}^2 + \sum_{j=0}^{m-1} \|\mathbf{u}(k+j|k) - \mathbf{u}_{tg}\|_{\mathbf{Q_u}}^2, \tag{5}$$

subject to:

$$\Delta \mathbf{u}_k \in \mathcal{U}, \tag{6}$$

$$\mathcal{U} = \begin{Bmatrix} \mathbf{u}_{\min} \le \mathbf{u}(k-1) + \sum_{i=0}^{j} \Delta \mathbf{u}(k+i|k) \le \mathbf{u}_{\max} \\ -\Delta \mathbf{u}_{\max} \le \Delta \mathbf{u}(k+j|k) \le \Delta \mathbf{u}_{\max} \\ \Delta \mathbf{u}(k+j|k) = 0, \ \forall j \ge m, \end{Bmatrix} \tag{7}$$

$$\hat{\mathbf{x}}(k+j|k) = \mathbf{G}(\hat{\mathbf{x}}(k+j-1|k), \mathbf{u}(k+j-1|k), \mathbf{d}(k|k)) \tag{8}$$

$$\hat{\mathbf{y}}(k+j|k) = \mathbf{F}(\hat{\mathbf{x}}(k+j|k)) \tag{9}$$

$$\mathbf{y}_{\min}(k|k) \le \mathbf{y}_{sp,k} \le \mathbf{y}_{\max}(k|k) \tag{10}$$

where $\hat{\mathbf{y}}(k+j|K)$ and $\hat{\mathbf{x}}(k+j)$ are output and state predictions at time step $k+j$ from the plant measurement at time step k ($\mathbf{y}(k)$), respectively; $\mathbf{u}$ and $\mathbf{d}$ are the manipulated and disturbance variables, respectively; $\Delta\mathbf{u}_k$ is the sequence of control actions calculated along the control horizon, namely $\Delta\mathbf{u}_k = \left[\Delta\mathbf{u}(k|k)^\top, \ldots, \Delta\mathbf{u}(k+m-1|k)^\top\right]^\top$; $\epsilon(k|k) = \mathbf{y}(k|k) - \hat{\mathbf{y}}(k|k)$ is a constant output-type disturbance model which results in an offset-free control law and is indicated by the $\mathbf{y}(k|k)$ feedback dotted-line in Figure 2. $\mathbf{y}_{sp,k}$ is the artificial setpoint within the proposed zone-control scheme and is a decision variable in the *Problem 1*; $\mathbf{u}_{tg}$ is the optimizing target of the manipulated variables; $\mathbf{u}_{\max}$, $\mathbf{u}_{\min}$, $\Delta\mathbf{u}_{\max}$, are the constraints on manipulated variables and their increments, respectively. $\mathbf{y}_{\max}(k|k)$ and $\mathbf{y}_{\min}(k|k)$ are the upper and lower bounds of the zone control included in constraint (10), evaluated from the operational envelope according to [17]. p and m are the prediction and control horizons, respectively; $\mathbf{Q_y}$, $\mathbf{R}$ and $\mathbf{Q_u}$ are weighting matrices on controlled, control actions and manipulated variables, respectively. Equation (8) represents the discretized nonlinear state equations, Equation (9) the discretized nonlinear output equations. The economic input target ($\mathbf{u}_{tg}$) is also addressed in this control law, without the need for a related constraint.

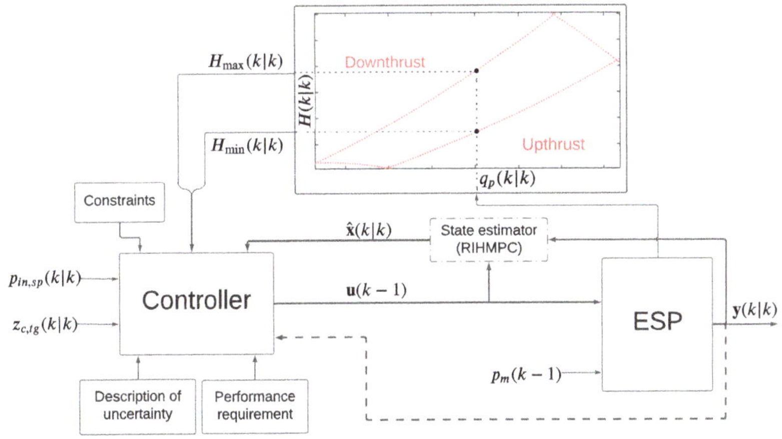

Figure 2. Control scheme elucidating the zone-control strategy.

In this work, the state and other variables needed to evaluate the operational envelope were considered measured, and state estimation was not necessary—presented as a block with a dot–dashed line in Figure 2. In a general scenario, an Extended Kalman Filter (EKF) can be applied to estimate unmeasured variables, for example, the flow rate, as described by Santana et al. [17].

The Figure 2 presents the zone-control scheme for the proposed controllers, where the zone is defined as $\mathbf{y}_{\min}(k) = [p_{in,sp}(k|k), H_{min}(k|k)]$ and $\mathbf{y}_{\max}(k) = [p_{in,sp}(k|k), H_{max}(k|k)]$, the MVs $\mathbf{u}(k-1) = [f(k-1), z_c(k-1)]$, the controlled variables $\mathbf{y}(k|k) = [p_{in}(k|k), H(k|k)]$, and the economic target $\mathbf{u}_{tg}$ is only applied for choke valve opening $z_{c,tg}(k|k)$.

One challenging aspect of NMPC techniques is accommodating a nonlinear model to describe the process. Such a task is not trivial in some cases since it requires a mathematical description of phenomenological phenomena. Additionally, the complexity of these models can increase the computational cost and limit real-time applications, so alternatives such as surrogate models or robust MPC strategies must be employed. In this sense, Section 2.3 presents the robust MPC approach investigated in this work.

2.3. Robust Infinite-Horizon MPC

An implementable RIHMPC strategy was designed for ESP-lifted oil-well systems applying a multi-model uncertainty description. This implementable RIHMPC was first introduced by Odloak [23] and extended by Gonzalez et al. [24] to the zone-control approach. It imposes non-increasing cost constraints for each model in the set, with only the selected nominal model having its cost function minimized [24]. The RIHMPC strategy attains robustly stabilizing properties and the feasibility of the optimization problem (implementable feature) by incorporating slack variables [23]. Introducing slack variables into controller formulation is a desired property for practical applications since it can provide a feasible solution to the optimization problem at each time step.

Considering a set of L linear models of the ESP selected for specific operational points, the operational envelope mapped by setpoint constraints $(\mathbf{y}_{\min,k}, \mathbf{y}_{\max,k})$, and input targets $(\mathbf{u}_{tg,k})$ to track economic performance objectives, the RIHMPC with zone control is presented as the following optimization problem at each time step k:

Problem 2

$$\min_{\Delta\mathbf{u}_k, \mathbf{y}_{sp,k}(\boldsymbol{\Theta}_n=1,\cdots,L), \boldsymbol{\delta}_{y,k}(\boldsymbol{\Theta}_n=1,\cdots,L), \boldsymbol{\delta}_{u,k}} V_k(\boldsymbol{\Theta}_N) \quad (11)$$

$$\begin{aligned} V_k(\boldsymbol{\Theta}_N) = & \sum_{j=0}^{m} \left\| \mathbf{y}(k+j|k) - \mathbf{y}_{sp,k}(\boldsymbol{\Theta}_N) - \boldsymbol{\delta}_{y,k}(\boldsymbol{\Theta}_N) \right\|^2_{\mathbf{Q}_y} + \\ & + \left\| \mathbf{x}^{st}_n(k+m|k) \right\|^2_{\bar{\mathbf{Q}}(\boldsymbol{\Theta}_N)} + \sum_{j=0}^{m-1} \left\| \mathbf{u}(k+j|k) - \mathbf{u}_{tg,k} - \boldsymbol{\delta}_{u,k} \right\|^2_{\mathbf{Q}_u} + \\ & + \sum_{j=0}^{m-1} \left\| \Delta\mathbf{u}(k+j|k) \right\|^2_{\mathbf{R}} + \left\| \boldsymbol{\delta}_{y,k}(\boldsymbol{\Theta}_N) \right\|^2_{\mathbf{S}_y} + \left\| \boldsymbol{\delta}_{u,k} \right\|^2_{\mathbf{S}_u}, \end{aligned} \quad (12)$$

subject to:

$$-\Delta\mathbf{u}_{\max} \leq \Delta\mathbf{u}(k+j|k) \leq \Delta\mathbf{u}_{\max}, \quad j = 0,\ldots,m-1, \quad (13)$$

$$\mathbf{u}_{\min} \leq \mathbf{u}(k+j|k) \leq \mathbf{u}_{\max}, \quad j = 0,\ldots,m-1, \quad (14)$$

$$\mathbf{u}(k+m-1|k) - \mathbf{u}_{tg,k} - \boldsymbol{\delta}_{u,k} = \mathbf{0}, \quad (15)$$

and, for each $n = 1,\cdots,L$,

$$\mathbf{y}_{\min}(k|k) \leq \mathbf{y}_{sp,k}(\boldsymbol{\Theta}_n) \leq \mathbf{y}_{\max}(k|k), \quad (16)$$

$$\mathbf{x}^{s}_n(k+m|k) - \mathbf{y}_{sp,k}(\boldsymbol{\Theta}_n) - \boldsymbol{\delta}_{y,k}(\boldsymbol{\Theta}_n) = \mathbf{0}, \quad (17)$$

$$V_k(\boldsymbol{\Theta}_n) \leq \tilde{V}_k(\boldsymbol{\Theta}_n), \quad (18)$$

where $\boldsymbol{\Theta}_N$ represents the nominal (or most likely) model of the set of L models, m is the control horizon, $\mathbf{y}(k+j|k)$ are the predictions of the controlled variables at time step $k+j$ given the information at time step k, $\Delta\mathbf{u}(k+j|k)$ are increments of the manipulated variables. $(\mathbf{u}_{\min}, \mathbf{u}_{\max})$, $\Delta\mathbf{u}_{\max}$ are the bounds of manipulated variables and increments on manipulated variables, respectively. $(\mathbf{y}_{\min}, \mathbf{y}_{\max})$ are the zone specification for the controlled variables, and $\mathbf{y}_{sp,k}$ are the output setpoints within the zone-control scheme imposed by (16). $\mathbf{Q}_\mathbf{y}$ and $\mathbf{R}$ are the weighting matrices of (ny) controlled variables and the increment of (nu) manipulated variables, respectively. $\bar{\mathbf{Q}}(\boldsymbol{\Theta}_N)$ is the terminal weight detailed. $\Delta\mathbf{u}_k = [\Delta\mathbf{u}^\top(k|k)\ldots\Delta\mathbf{u}^\top(k+m-1|k)]^T$ is the control actions vector; $\boldsymbol{\delta}_{y,k} \in \mathbb{R}^{ny}$ and $\boldsymbol{\delta}_{u,k} \in \mathbb{R}^{nu}$ are the slack variables used to guarantee the feasibility of the optimization *Problem 2*. $\mathbf{S}_\mathbf{y} \in \mathbb{R}^{ny \times ny}$ and $\mathbf{S}_\mathbf{u} \in \mathbb{R}^{nu \times nu}$ are the weighting matrices of the slack variables. $\mathbf{u}_{tg}$ are input targets and $\mathbf{Q}_\mathbf{u}$ is their weighting matrix. $V_k(\boldsymbol{\Theta}_n)$ is the actual cost function value for each model and $\tilde{V}_k(\boldsymbol{\Theta}_n)$ is the cost function obtained with a solution inherited from *Problem 2* at time step $k-1$ and translated to time k. More details in [23–25].

The cost-contraction constraint (18) is imposed for all models to guarantee robustly stabilizing properties of the controller. The predictive model must be on the canonical state-space based on the analytical form of the step response of the system outlined in [23], in which $\mathbf{x}^{s}_n(k|k)$ is artificial integrating states vector and $\mathbf{x}^{st}_n(k|k)$ is the stable states vector. The former is used to obtain an offset-free control law, originating from the incremental form of inputs and corresponding to the system outputs at the steady state. As a result, the terminal constraints (17) and (15) must be incorporated into the formulation to avoid an unbounded cost. Slack variables soften these constraints to guarantee the optimization problem's feasibility. It is worth noting that *Problem 2* is a nonlinear optimization problem. However, it is convex and less computationally intensive compared to non-convex optimization problems commonly found in nonlinear model predictive control formulations.

The artificial integrating states make it necessary to use a linear state estimator only to obtain these unmeasured states, and one Kalman filter was used for each of the L models—the state estimator block in Figure 2. The model states are considered measured.

3. The Implementation Method of Embedded MPC Strategies

This section presents the deployment steps on Teensy 4.1 (https://www.pjrc.com/store/teensy41.html, accessed on 6 March 2023) hardware for an embedded control strategy with hardware-in-the-loop (HiL) validation. It also explains how the aforementioned MPC strategies were C-code generated, tested in MATLAB environment, and deployed as a standalone static library in runtime on the microcontroller.

3.1. Code Generation for MPC Controllers

The first step was to write the controller code in MATLAB (R2020a) and solve them using an optimization algorithm. In this work, Sequential Quadratic Programming (SQP) was applied. Following the Model-Based Development approach [26], the next step was executing several simulations to tune the controllers and the optimization algorithm. In this sense, this work focused on defining the maximum limit of iterations of the optimization algorithm and its maximum number of objective function evaluations because they are required for the subsequent automatic generation of C++ code. Furthermore, it is essential to point out that these parameters directly influence the computational times of the embedded code on the hardware. A sensitivity analysis was carried out in several simulated scenarios to ensure the problems converge to the optimal solution.

The next step was generating and verifying the software for the applications referring to the projected control strategies. Then, the functions created for each controller were translated into MEX-type executables using the MATLAB Coder toolbox (https://www.mathworks.com/products/matlab-coder.html, accessed on 6 March 2023). Tests were performed with the control executables in the simulated environment, thus characterizing the software-in-the-loop (SiL) stage.

After performing all the previous steps, the C++ code was generated through the MATLAB Coder toolbox and integrated into the hardware that embedded the MPC strategies. Among the extensions available in Coder Toolbox, this work created static libraries (.lib) from the control functions to be integrated into the application embedded in the Teensy 4.1 microcontroller. Thus, it was not necessary to include external libraries in the project, e.g., matrix algebra package, since all mathematical operations related to the solution of *Problems 1* and 2 were encapsulated in the routines generated by the Coder Toolbox. With the embedded control routines, the hardware was tested in closed-loop through the HiL stage: the embedded controller was executed in real time, and only the process remained simulated in MATLAB.

The diagram in Figure 3 summarizes the necessary steps to embed MPC control strategies. This workflow facilitates application improvements and updates using the automatic code generation tool.

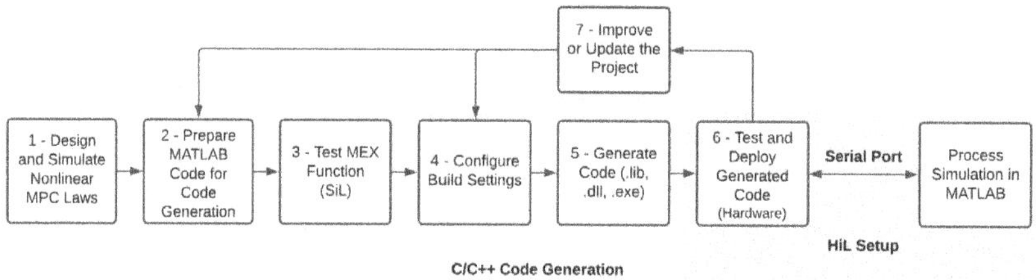

Figure 3. Diagram of the implementation process of embedded MPC strategies.

3.2. Hardware-in-the-Loop Setup

The hardware used in this work was the Teensy 4.1 microcontroller, which has an ARM cortex-M7 processor, 7936K Flash memory, 1024K RAM, 18 analog inputs, 35 PWM pins, 55 I/O pins, and eight serial ports. This hardware is suitable for advanced real-time control applications with computational resources superior to most microcontrollers. Preliminary tests were carried out on other hardware, which did not meet the requirements of this work. Therefore, the choice of Teensy 4.1 converged mainly due to its memory and processing resources.

In the HiL tests, the communication between the microcontroller and MATLAB was performed through the serial port, with a baud rate of 115,200. Therefore, the NMPC and RIHMPC problems were calculated in Teensy 4.1, and the optimal control actions were sent via serial to MATLAB, where the nonlinear model of the ESP-lifted oil-well process was simulated. After simulating the oil-production system, the process outputs were also sent to the microcontroller via the serial port, thus ending one communication cycle.

3.3. Solver and Controllers Tuning Parameters

Table 1 summarizes both controllers' tuning parameters. The zone configuration $(\mathbf{y}_{\min}(k) = [p_{in,sp}(k) \quad H_{min}(k)], \mathbf{y}_{\max}(k) = [p_{in,sp}(k) \quad H_{max}(k)])$ specifies setpoint tracking for the intake pressure and min/max head from the ESP operational envelope. Furthermore, a Runge–Kutta 4th order algorithm was used in the closed-loop system of both MPC controllers for the numerical integration of the nonlinear model of the ESP-equipped oil well, with a sampling time of $\Delta t = 4$ s.

Table 1. Tuning parameters of controllers and constraints on inputs.

	NMPC	RIHMPC
m	2	4
p	10	∞
$\mathbf{R}$	$\mathrm{diag}[10^{-3} \quad 10^{-3}]$	$\mathrm{diag}[10 \quad 1]$
$\mathbf{Q_y}$	$\mathrm{diag}[10^{3} \quad 10]$	$\mathrm{diag}[10^{3} \quad 10]$
$\mathbf{Q_u}$	$\mathrm{diag}[0 \quad 10]$	$\mathrm{diag}[0 \quad 1]$
$\mathbf{S_y}$	-	$\mathrm{diag}[10^{2} \quad 1]$
$\mathbf{S_u}$	-	$\mathrm{diag}[0 \quad 10]$
$\mathbf{u}_{\min}$	[35 Hz 0 %]	[35 Hz 0 %]
$\mathbf{u}_{\max}$	[65 Hz 100 %]	[65 Hz 100 %]
$\Delta\mathbf{u}_{\max}$	[2 Hz 2 %]	[2 Hz 2 %]

The multi-model set Θ that comprises the RIHMPC description of uncertainty was defined by linearization of the nonlinear model (1)–(3) for three different (steady-state) operating points: $f_{ss,1} = 35.29$ Hz, $z_{c_{ss,1}} = 14.53\%$; $f_{ss,2} = 55.44$ Hz, $z_{c_{ss,2}} = 31.58\%$; $f_{ss,3} = 65$ Hz, $z_{c_{ss,3}} = 98.97\%$. They were discretized with a sampling time of 4 s, making up $L = 3$ linear models in the canonical form. These models map a wide operating range of the system, representing probable operating conditions of an ESP-lifted oil-well system. Furthermore, the smallest number of models was chosen to avoid increasing the computational burden. The typical tuning parameters of the linear Kalman filters were defined as $\mathbf{Q}_{kf} = 0.5 \cdot \mathbf{I}_{14}$ and $\mathbf{R}_{kf} = \text{diag}[0.5 \quad 0.5]$ in the RIHMPC controller design.

The following main properties were used in the SQP solver compilation: maximum of 30 iterations; maximum of 470 objective function evaluations; primal and dual infeasibility threshold equal to 1×10^{-6}.

4. HiL Application of the Implementable Zone RIHMPC and NMPC Strategies in an ESP-Lifted Oil-Well System

This section investigates the capabilities of NMPC and RIHMPC strategies to perform the following tasks: (i) tracking the intake pressure to regulate the oil production, (ii) systematically treating the ESP head (down and up-thrust) operational envelope to ensure a safe operation, (iii) ensuring the feasibility of the control problem through an implementable control zone scheme, (iv) offset-free compensation of unmeasured disturbance, and (v) tracking targets for the production choke valve opening-like aiming to improve economic performance. For the sake of simplicity, only the results referring to the HiL stage were discussed in detail. Nevertheless, specific aspects of the other methodological steps were highlighted when necessary.

4.1. Control Performance under Setpoint Tracking and Disturbance Compensation

The first test scenario was characterized by three setpoint changes, mapping a wide operating region of the system, and severe variations in manifold pressure (p_m). The economic target ($\mathbf{u}_{tg,k}$) was turned off, just by setting $\mathbf{Q}_\mathbf{u} = \text{diag}([0 \quad 0])$, while the other tuning parameters were the same as in Table 1.

The manifold pressure (unmeasured disturbance) was systematically changed from 20 bar to 10 bar (at 200 s), then to 30 bar (at 700 s) and 35 bar (at 1200 s), while the intake pressure setpoints were defined at 38 bar (0 to 500 s), 60 bar (500 to 1000 s) and 85 bar (1000 to 1500 s). Furthermore, the head setpoint was defined by the optimization problems of the controllers in such a way that it remains within the range delimited by the operational envelope (zone control). The ESP-lifted oil-well system starts at a steady state corresponding to $\mathbf{u}(0) = [50 \text{ Hz}, \ 50 \ \%]$ and $\mathbf{y}(0) = [60 \text{ bar}, \ 592 \text{ m}]$. Please note that this scenario represents a case of plant–model mismatch due to the inclusion of unmeasured persistent disturbances.

Figure 4 shows the performance of the NMPC and RIHMPC controllers in the assigned task. Similar dynamic responses were observed during transient periods, and both controllers met the expected requirements related to intake pressure tracking and disturbance compensation.

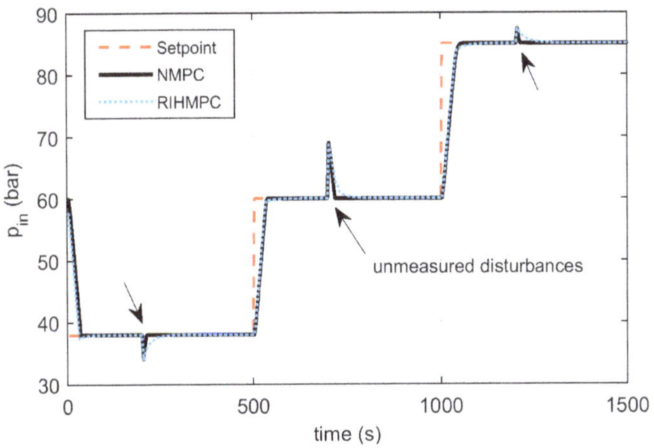

Figure 4. Dynamic of the intake pressure under the zone NMPC and RIHMPC controllers.

It is worth noting that considerable effort was devoted to the tuning stage of the controllers to obtain such similar control performance. Even so, the NMPC had a less damped response and drove the intake pressure back to the setpoint faster than the RIHMPC after disturbances, mainly due to the difference in its weights in the movement suppression matrices (**R**). Regarding the manipulated variables, Figures 5 and 6 show that the control actions did not violate the physical constraints of the actuators detailed in Table 1. The most aggressive actions of the NMPC due to its tuning are demonstrated in Figure 6, where variations of the MVs easily reached saturation during transients.

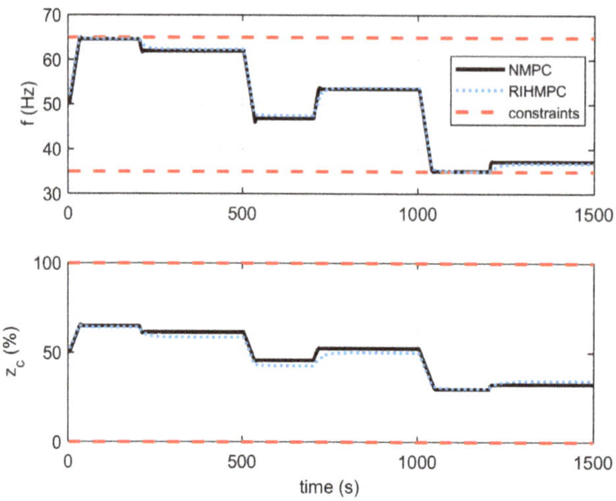

Figure 5. Manipulated variables and the input constraints for both controllers.

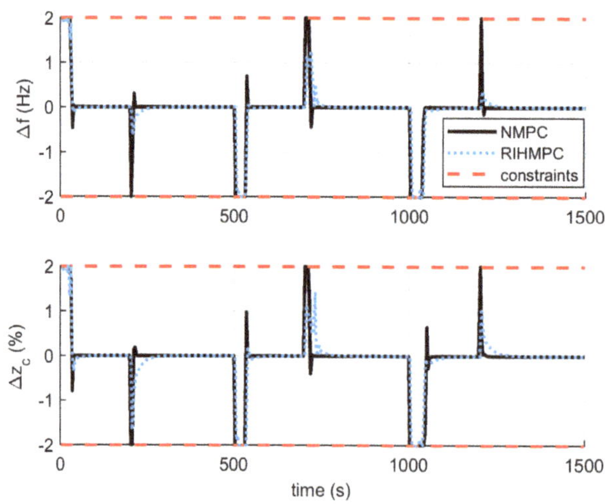

Figure 6. Manipulated variables increments and the respective constraints for both controllers.

The zone-control scheme is clearly shown in Figure 7, in which the setpoint is a decision variable constrained by the zone. This mechanism softens the optimization problems and contributes to its feasibility. This technique was strategically employed to accommodate the time-varying constraints of the ESP operational envelope since it gives additional degrees of freedom for the system to accommodate disturbances. Evidence of this property can be seen in the highlighted zoom of Figure 7, where after the disturbance in the manifold pressure at 200 s enters the plant, this was forced to operate momentarily outside the envelope. However, the setpoint remained within the zone in both controllers. After this short time, the control actions compensated for the disturbance and drove the process back to its desired zone.

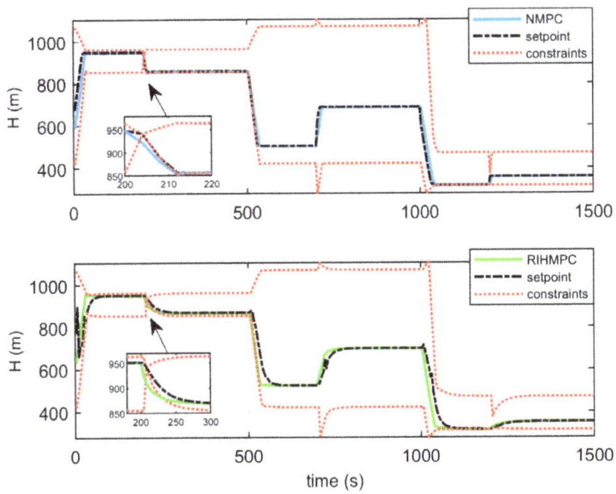

Figure 7. Head zone from the ESP operational envelope and artificial setpoints of the NMPC and RIHMPC controllers.

Figure 8 shows the ESP operational envelope, highlighting the constraint violation from another perspective. Despite this occurrence, the system was kept operating within the safe region most of the time, which is crucial to maintain the lifetime of the equipment. In this scenario, the residence time outside the envelope was approximately 8 s for both controllers investigated. In ESP-equipped oilfield installations, keeping the operation close to the up-thrust region is often desirable in order, for example, to maximize production. In such cases, constraining the limits of the operational envelope to a safety margin may be good practice. Nonetheless, the system was pushed to the limits to evaluate the feasibility of these embedded control techniques.

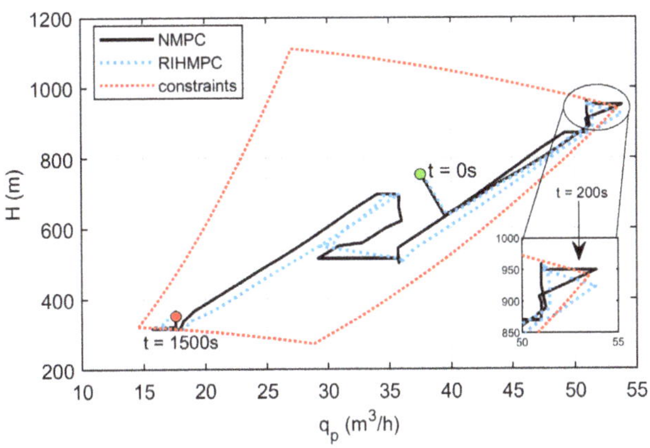

Figure 8. Trajectories over the ESP operational envelope for the NMPC and RIHMPC controllers.

Regarding the cost functions of the controllers depicted in Figure 9, their asymptotic behavior toward zero was observed after changing setpoints and the occurrence of disturbances. After a time, such functions reached their origin, which confirms that there was no offset error in the controllers. This property was achieved by the incremental model in the RIHMPC strategy and by the error correction included in the NMPC cost function.

Despite this asymptotic behavior, it is important to highlight that the NMPC formulation does not guarantee nominal or robust stability. On the other hand, the RIHMPC has a robustly stabilizing strategy only for the multi-linear models-based description of uncertainty. Although the proof cannot be taken in this plant–model mismatch scenario, the RIHMPC strategy provided adequate control actions, performing similarly to the NMPC strategy, which encompasses the phenomenological model.

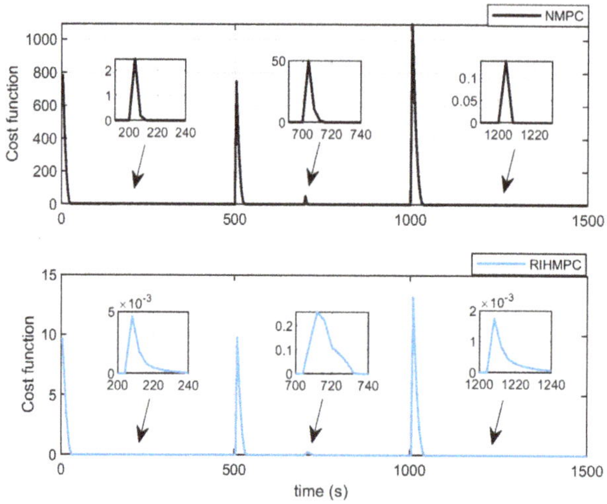

Figure 9. Cost function of both controllers.

The computational time performed throughout the HiL simulation by both control strategies on the host hardware Teensy 4.1 is shown in Figure 10. This is one validation to assess whether the controllers can operate in real time or if they need to be redesigned. The NMPC had an average time of 0.1680 s and a maximum time of 0.9505 s throughout this simulation, while the RIHMPC performed an average time of 0.2080 s and a maximum time of 1.3304 s. Naturally, the moments of more significant computational burden occurred during transient periods, in which the SQP solver performed more iterations to converge the NMPC and RIHMPC optimization problems, as shown in Figure 11. This indicates that both projected control laws met the real-time operation requirements with the ESP system, whose sampling time was 4 s. To demonstrate the repeatability of these results, a total of 10 simulations were performed with the same scenario, obtaining the mean (μ) and standard deviation (σ) of the average and maximum computational times of both controllers, which are shown in Table 2.

Figure 11 shows the number of iterations that the SQP solver performed to converge the NMPC and RIHMPC optimization problems. Only NMPC reached the maximum iteration limit during the transitory period. However, as shown in Figure 12, its number of objective function evaluations was smaller than the ones needed for the RIHMPC. From this result, it is clear that the RIHMPC control law required a computational effort slightly greater than the NMPC to be solved. This was also confirmed by the data in Table 2, in which the average and maximum times were greater for the RIHMPC.

However, it is important to point out that the RIHMPC has robust stability, with the introduction of the cost-contraction constraint (18), which naturally makes the optimization problem more complex. Still, warm starting the primal variables, the SQP solver turned out to be efficient for real-time operation, handling nonlinearities and converging to the optimal solution. In the aspect of optimality, another important property can be emphasized, the convex control law of the RIHMPC delivers a globally optimal solution, while in the NMPC strategy, there is no such guarantee.

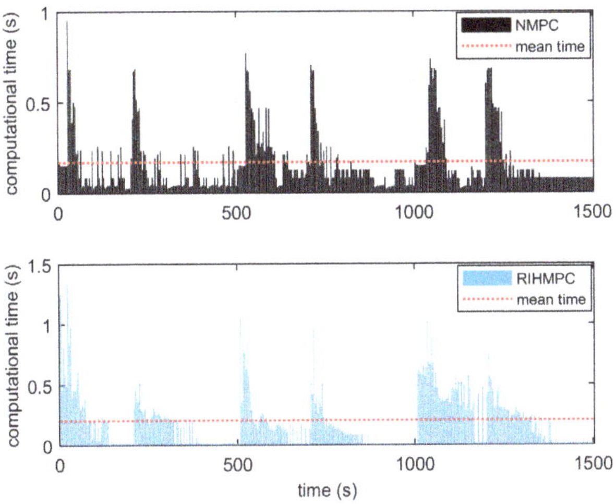

Figure 10. Computational performance between the NMPC and RIHMPC optimization problems.

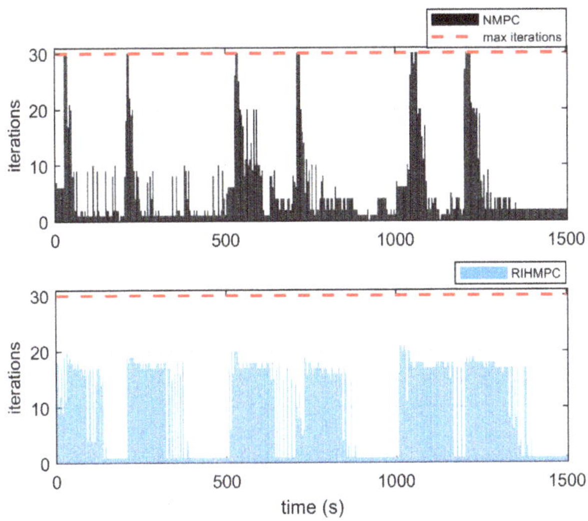

Figure 11. Number of iterations of the NMPC and RIHMPC optimization problems during the simulated scenario.

Table 2. Computational performance obtained for 10 simulations.

Index	NMPC	RIHMPC
avg. time (s)	$\mu = 0.1680$ and $\sigma = 0.0000$	$\mu = 0.2081$ and $\sigma = 0.0002$
max. time (s)	$\mu = 0.9505$ and $\sigma = 0.0000$	$\mu = 1.3303$ and $\sigma = 0.0001$

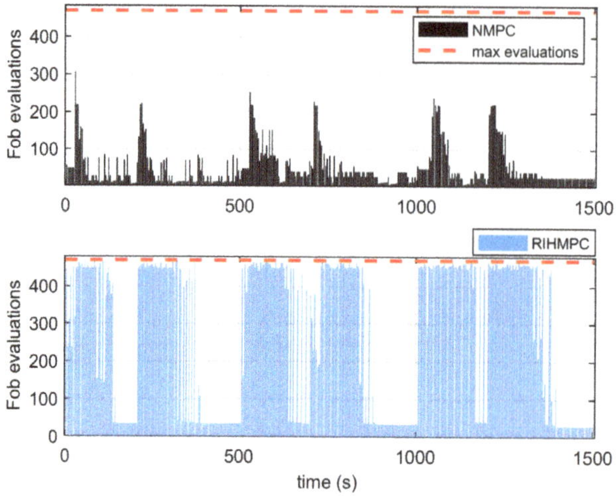

Figure 12. Objective function evaluations in the NMPC and RIHMPC optimization problems execution.

Although the NMPC and RIHMPC strategies presented similar control performance, e.g., computational time, they have very different demands in terms of hardware memory, which can be a determining factor for the feasibility or choice of applications according to available computational resources. The total flash and RAM used in each implementation is summarized in Table 3. The embedded RIHMPC application required about 80% more flash memory and 60% more RAM than the embedded NMPC application. This was mainly due to the need to store and process multidimensional matrices in the RIHMPC implementation, for example, the model matrices for each of the three controller models. In addition, code generation for the SQP solver via MATLAB Coder imposes the use of double-type variables, which also increased the storage memory requirement. As computational resources are not always compatible with such demand, it may be necessary to adopt reduced numerical precision strategies to deploy this controller on more limited hardware, adapting the methodology presented here. The NMPC, in turn, had a leaner coding and required few variables in storage, with the SQP solver and the numerical integration of the nonlinear model (Runge–Kutta 4th-order algorithm) being its main computational burden.

Table 3. Amount of memory used in the implementation of NMPC and RIHMPC controllers on Teensy 4.1.

Memory Type	NMPC	RIHMPC
Flash (bytes)	81,916	147,452
RAM (bytes)	149,888	246,432

4.2. Realistic Setpoint/Target Tracking Scenario under Measurement Noises

The following scenario imposed the controllers to a scenario more consistent with the operation of a real ESP-lifted oil well: the control aimed to track setpoints and economic targets in the manipulated variables of interest, in addition to the presence of unmeasured noises in the monitored variables. Therefore, the setpoint changes for the intake pressure were defined at 70 bar (0 to 500 s) and 38 bar (500 to 1000 s), in addition to an opening target of 90% for the choke valve active throughout the HiL simulation. Additionally, the unmeasured noises in the controlled variables had Gaussian distribution $\mathcal{N}(\mathbf{0}, \mathbf{W})$, with $\mathbf{W} = \text{diag}([1.90 \text{ bar}^2, 23.81 \text{ m}^2])$. Finally, to re-enforce the plant–model mismatch condition on both controllers, at instant 300 s, a systematic disturbance in the manifold pressure was introduced with an increase of 40% (from 20 to 28 bar) that lasted until the end of the test.

Figure 13 shows the intake pressure dynamics obtained by the controllers in the noisy scenario, where a considerable difference in performance was noted in the transitory period of the first setpoint change. Both NMPC and RIHMPC controllers stabilized the plant, but the NMPC was faster at driving the intake pressure to the desired setpoint with a transient of approximately 24 s, while the RIHMPC took around 140 s. This can be easily explained by the fact that the NMPC encompassed a nonlinear model that captured the plant dynamics very well. In contrast, the most probable RIHMPC model (Θ_1) is suitable for high-gain regions, and the activation of the economic target forced the controller to operate the plant with a low gain (large valve opening choke), as shown in Figure 14. However, it can be emphasized that this deliberate choice was intended to show that the controller remains stable within the multi-model-based uncertainty description.

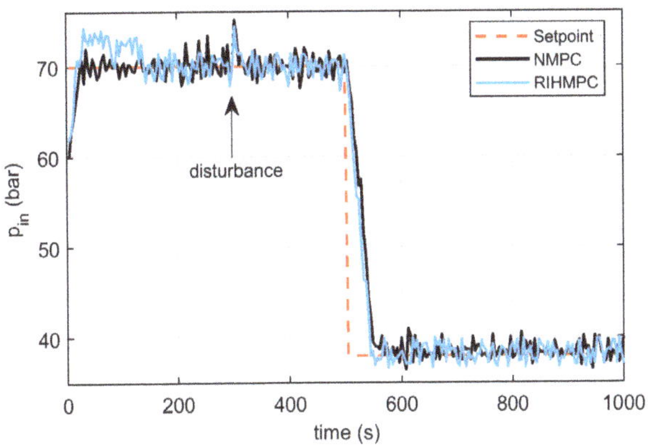

Figure 13. Intake pressure of both controllers under a realistic scenario.

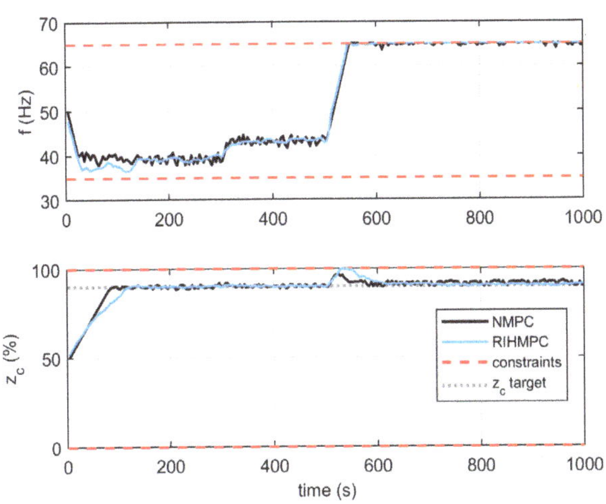

Figure 14. Manipulated variables, the input constraints, and economic target for both controllers under a realistic scenario.

With regard to manipulated variables, Figure 14 shows that the economic target in the choke valve was tracked whenever possible, with necessary deviations during the transitory period, since priority was given to tracking the controlled variables, according to

Table 1. Please note that at 300 s, the system had a degree of freedom to compensate for the disturbance solely by changing the frequency and maintaining the choke valve opening at the target of 90%. In the second setpoint change, a frequency saturation occurred, and the controllers moved the choke valve to stabilize the system but were able to drive it back to its target. In this way, both controllers sought to keep the valve as close as possible to the 90% opening and thus reduce energy losses, whose effectiveness has already been shown in [17].

Figure 15 shows the effectiveness of the zone-control strategy employed in both controllers, keeping the ESP head within its operational envelope constraints. After the last setpoint change, the system was operated in the down-thrust limit region with temporary envelope violations due to the noises that affect the process. This emphasizes that using a safety margin in the operational envelope limits can be a good practice to avoid safety violations, especially in noisy scenarios that portray real situations.

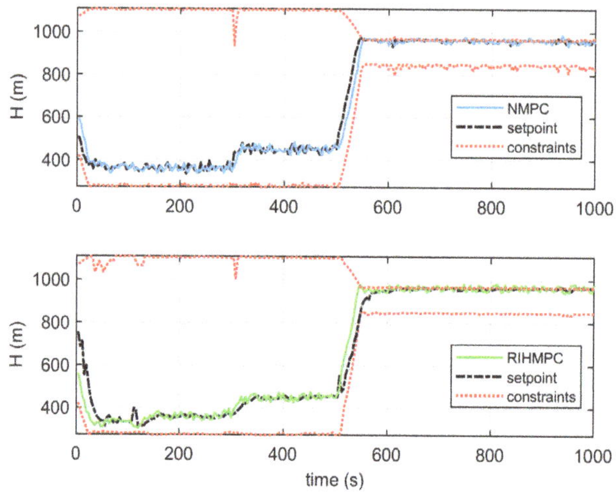

Figure 15. Head zone from the ESP operational envelope and artificial setpoints of the NMPC and RIHMPC controllers under a realistic scenario.

The cost functions of the controllers shown in Figure 16 indicated that the NMPC converged faster than the RIHMPC controller in all transient periods. Despite this, the latter delivered a performance very close to the former, which opens a question about the applicability of each one. In this regard, reflecting on obtaining the necessary models for each control law is essential. As is well known, obtaining linear models is generally less expensive and can follow several paths through identification techniques. On the other hand, nonlinear or phenomenological models are more laborious and require knowledge of the physical phenomena that describe the process, which may not be a viable choice for some applications.

Figure 17 shows the computational times obtained on the Teensy 4.1 hardware in this scenario. The NMPC had an average time of 0.4798 s and a maximum time of 0.6658 s throughout this simulation, while the RIHMPC performed an average time of 0.5243 s and a maximum time of 1.5848 s. In this respect, both met the real-time operation requirements with a sampling time of 4 s. Special attention must be given to the RIHMPC controller during transients because it reached the maximum registered value to comply with the predefined tolerances. This analysis was crucial to know if the computational performance of the hardware was compatible with the sampling intervals of the application. To verify the reliability of the results and that the embedded controllers presented repetitive behavior, a total of 10 simulations were performed with the same scenario, obtaining the mean (μ) and standard deviation (σ) of the average and maximum computational times of both controllers, which are shown in Table 4.

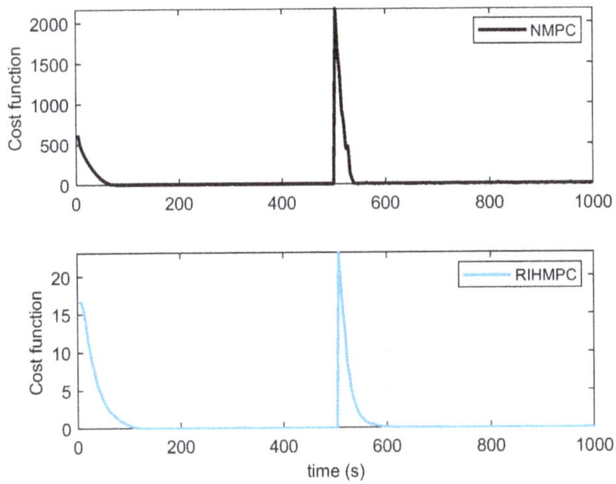

Figure 16. Cost function of both NMPC and RIHMPC controllers under a realistic scenario.

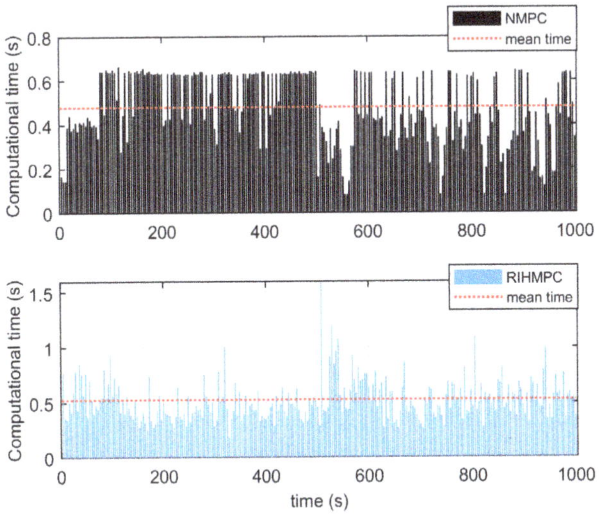

Figure 17. Solver computational time of both NMPC and RIHMPC controllers under a realistic scenario.

Table 4. Computational performance obtained for a total of 10 simulations in the noisy setpoint/target tracking scenario.

Index	NMPC	RIHMPC
avg. time (s)	$\mu = 0.4679$ and $\sigma = 0.0105$	$\mu = 0.5323$ and $\sigma = 0.0163$
max. time (s)	$\mu = 0.6720$ and $\sigma = 0.0190$	$\mu = 1.5815$ and $\sigma = 0.0097$

Finally, Figures 18 and 19 show the performance of the SQP solver to converge the optimization problems focusing on the number of iterations and objective function evaluations of the NMPC and RIHMPC, respectively. These were the main factors related to the computational times shown in Figure 17. To solve the NMPC control law, the limits of 30 iterations and 470 objective function evaluations were enough without infeasibility. Similarly, to solve the RIHMPC control law, the limit of 30 iterations was also sufficient. However, increasing the limit of objective function evaluations to 800 turned out to be necessary to avoid infeasibility. Thus, naturally, the computational times recorded in this scenario were greater than those recorded in the previous scenario, but they safely met the real-time requirements of the application.

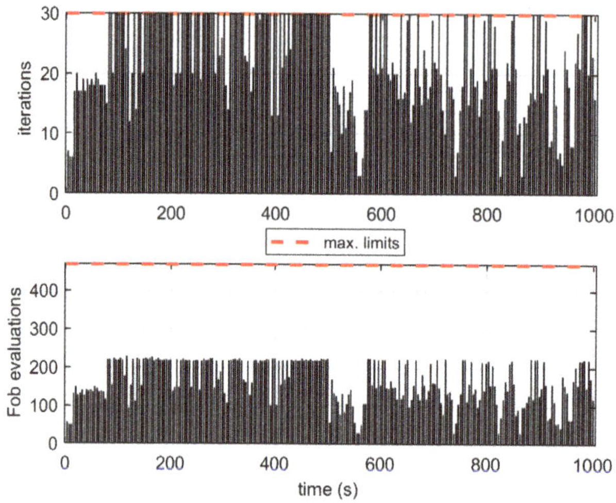

Figure 18. NMPC solver performance under the noisy scenario.

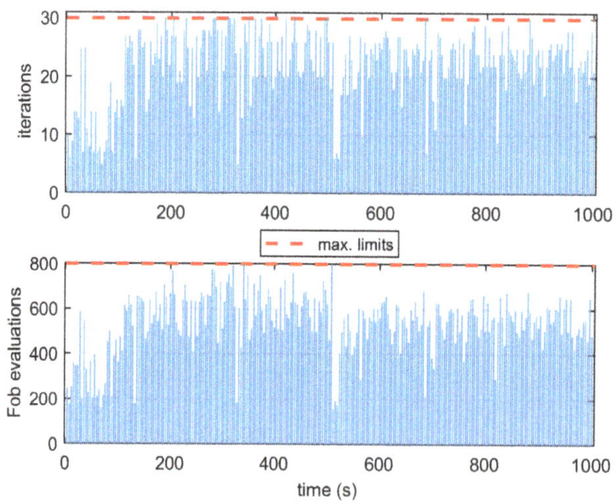

Figure 19. RIHMPC solver performance under the noisy scenario.

5. Conclusions

This paper dealt with the HiL application of robust and nonlinear MPC strategies embedded in a Teensy 4.1 microcontroller to control an ESP-lifted oil-well system. The novelty of this paper focused on the robustness and computational performance analysis of these MPC controllers with feasibility guarantee and their successful implementation in the limited hardware, which constitute a significant practical contribution to the field of control in E&P.

From the simulated scenarios, it was possible to show the effectiveness of the zone-control scheme employed in both controllers. It successfully accommodates the ESP operational envelope constraints and ensures the feasibility of the optimization problems. The embedded controllers were able to stabilize the intake pressure over a wide operating region, showing that the robust formulation, whose models are more easily obtained, can closely approximate the performance of the nonlinear formulation, even in a noisy and plant–model mismatch scenario.

The computational burden required to solve the control laws met real-time operation requirements. However, the robust formulation required a slightly higher computational time to converge, mainly due to the complexity imposed by the cost-contraction constraint. Regarding hardware memory, the RIHMPC controller used up to 80% more memory than the NMPC controller implementation due to its multi-model structure. Since these resources are not always abundant in industrial hardware, this is a factor that can limit its application. Even so, the RIHMPC formulation proved to be very promising for real-time operation in ESP-lifted oil-well systems, mainly when it is not feasible to obtain an accurate phenomenological model and when the necessary computational resources are available.

Author Contributions: B.A.S.: Conceptualization, Data curation, Formal analysis, Investigation, Methodology, Software, Validation, Visualization, Writing—original draft. V.S.M.: Conceptualization, Methodology, Visualization, Writing—original draft. D.D.S.: Conceptualization, Writing—original draft, Writing—review and editing. M.A.F.M.: Conceptualization, Funding acquisition, Project administration, Resources, Supervision, Writing—review and editing. All authors have read and agreed to the published version of the the manuscript.

Funding: This work was funded by ANP (Brazilian Agency of Oil, Natural Gas and Biofuels) in the scope of PRH 35.1. (Human Resources Training Program).

Institutional Review Board Statement: Not applicable.

Informed Consent Statement: Not applicable.

Data Availability Statement: Not applicable.

Acknowledgments: The authors thank the ANP (Brazilian Agency of Oil, Natural Gas and Biofuels) in the scope of PRH 35.1 and CAPES under Finance Code 001, for their financial support. Márcio Martins would also like to thank CNPq under grant 408339/2021-7.

Conflicts of Interest: The authors declare no conflict of interest.

Abbreviations

The following abbreviations are used in this manuscript:

ESP	Electric submersible pump
MPC	Model Predictive Controller
IHMPC	Infinite-Horizon Model Predictive Controller
RIHMPC	Robust Infinite-Horizon Model Predictive Controller
NMPC	Nonlinear Model Predictive Controller
HiL	Hardware-in-the-loop
PLC	Programmable logic controller
QP	Quadratic programming
SQP	Sequential quadratic programming
E&P	Exploration and production
EKF	Extended Kalman Filter
MV	Manipulated variables
I/O	Input output

References

1. Liang, X.; He, J.; Du, L. Electrical Submersible Pump System Grounding: Current Practice and Future Trend. *IEEE Trans. Ind. Appl.* **2015**, *51*, 5030–5037. [CrossRef]
2. Zhu, J.; Zhang, H.Q. A Review of Experiments and Modeling of Gas-Liquid Flow in Electrical Submersible Pumps. *Energies* **2018**, *11*, 180. [CrossRef]
3. Takacs, G. *Electrical Submersible Pumps Manual*, 1st ed.; Gulf Professional Publishing: Houston, TX, USA, 2009.
4. Krishnamoorthy, D.; Bergheim, E.M.; Pavlov, A.; Fredriksen, M.; Fjalestad, K. Modelling and Robustness Analysis of Model Predictive Control for Electrical Submersible Pump Lifted Heavy Oil Wells. *IFAC-PapersOnLine* **2016**, *49*, 544–549. [CrossRef]
5. Pavlov, A.; Krishnamoorthy, D.; Fjalestad, K.; Aske, E.; Fredriksen, M. Modelling and model predictive control of oil wells with Electric Submersible Pumps. In Proceedings of the 2014 IEEE Conference on Control Applications (CCA), Nice, France, 8–10 October 2014; pp. 586–592. [CrossRef]
6. Patel, K.; Bakhurji, A.; Salloum, H.; Kim, H.; Winarno, M.; Mubarak, S. Use of advanced process control for automating conventional oilfield operations. In Proceedings of the SPE Kingdom of Saudi Arabia Annual Technical Symposium and Exhibition 2018 (SATS 2018), Dammam, Saudi Arabia, 23–26 April 2018; p. 2020. [CrossRef]
7. Binder, B.J.; Kufoalor, D.K.; Pavlov, A.; Johansen, T.A. Embedded model predictive control for an electric submersible pump on a programmable logic controller. In Proceedings of the 2014 IEEE Conference on Control Applications (CCA 2014), Nice, France, 8–10 October 2014. [CrossRef]
8. Santana, B.A.; Fontes, R.M.; Martins, M.A. Controlador Preditivo de Horizonte Infinito Embarcado: Aplicação Hardware-in-the-loop a um Sistema de Bombeio Centrífugo Submerso. In Proceedings of the 15th Simpósio Brasileiro de Automação Inteligente, Online, 17–21 October 2021. [CrossRef]
9. Matos, V.S.; Santana, B.A.; Chagas, T.P.; Martins, M.A. Embedded predictive controller based on fuzzy linear parameter-varying model: A hardware-in-the-loop application to an ESP-lifted oil well system. *Digit. Chem. Eng.* **2022**, *5*, 100054. [CrossRef]
10. Binder, B.J.; Johansen, T.A.; Imsland, L. Improved predictions from measured disturbances in linear model predictive control. *J. Process. Control* **2019**, *75*, 86–106. [CrossRef]
11. Fontes, R.; Costa, E.; Abreu, O.L.; Martins, M.; Schnitman, L. On application of a zone IHMPC to an ESP-lifted oil well system. In Proceedings of the Anais do Congresso Brasileiro de Automática 2020, Online, 23–26 November 2020. [CrossRef]
12. Delou, P.d.A.; de Azevedo, J.P.; Krishnamoorthy, D.; de Souza, M.B.; Secchi, A.R. Model Predictive Control with Adaptive Strategy Applied to an Electric Submersible Pump in a Subsea Environment. *IFAC-PapersOnLine* **2019**, *52*, 784–789. [CrossRef]
13. Delou, P.; Souza, M.; Secchi, A.R. A Robust Adaptive MPC coupled with Kalman Filter for an Electric Submersible Pump System: A multi-model approach. In Proceedings of the I Brazilian Congress on Process Systems Engineering (PSE-BR 2019), Rio de Janeiro, Brazil, 20–22 May 2019; pp. 435–438.
14. Binder, B.J.; Pavlov, A.; Johansen, T.A. Estimation of flow rate and viscosity in a well with an electric submersible pump using moving horizon estimation. *IFAC-PapersOnLine* **2015**, *28*, 140–146. [CrossRef]
15. Santana, B.A.; Fontes, R.M.; Schnitman, L.; Martins, M.A. An Adaptive Infinite Horizon Model Predictive Control Strategy Applied to an ESP-lifted Oil Well System. *IFAC-PapersOnLine* **2021**, *54*, 176–181. [CrossRef]
16. Jordanou, J.P.; Osnes, I.; Hernes, S.B.; Camponogara, E.; Antonelo, E.A.; Imsland, L. Nonlinear Model Predictive Control of Electrical Submersible Pumps based on Echo State Networks. *Adv. Eng. Informat.* **2022**, *52*, 101553. [CrossRef]
17. Santana, B.A.; Fontes, R.M.; Martins, M.A. An implementable zone NMPC applied to an ESP-lifted oil well system: Handling the lack of measurements with nonlinear state estimator coupling. *J. Pet. Sci. Eng.* **2022**, *216*, 110816. [CrossRef]
18. Allamaa, J.P.; Listov, P.; Van Der Auweraer, H.; Jones, C.; Son, T.D. Real-time Nonlinear MPC Strategy with Full Vehicle Validation for Autonomous Driving. In Proceedings of the 2022 American Control Conference (ACC), Atlanta, GA, USA, 8–10 June 2022; pp. 1982–1987. [CrossRef]
19. Dini, P.; Saponara, S. Processor-in-the-Loop Validation of a Gradient Descent-Based Model Predictive Control for Assisted Driving and Obstacles Avoidance Applications. *IEEE Access* **2022**, *10*, 67958–67975. [CrossRef]
20. Adhau, S.; Patil, S.; Ingole, D.; Sonawane, D. Implementation and Analysis of Nonlinear Model Predictive Controller on Embedded Systems for Real-Time Applications. In Proceedings of the 2019 18th European Control Conference (ECC), Naples, Italy, 25–28 June 2019; pp. 3359–3364. [CrossRef]
21. Andersson, J.; Gillis, J.; Horn, G.; Rawlings, J.; Diehl, M. CasADi: A software framework for nonlinear optimization and optimal control. *Math. Program. Comput.* **2019**, *11*, 1–36. [CrossRef]
22. Wächter, A.; Biegler, L. On the implementation of an interior-point filter line-search algorithm for large-scale nonlinear programming. *Math. Program. Comput.* **2006**, *106*, 25–27. [CrossRef]
23. Odloak, D. Extended robust model predictive control. *AIChE J.* **2004**, *50*, 1824–1836. [CrossRef]
24. González, A.; Marchetti, J.; Odloak, D. Robust model predictive control with zone control. *IET Control Theory Appl.* **2009**, *3*, 121–135. [CrossRef]

25. Martins, M.A.; Odloak, D. A robustly stabilizing model predictive control strategy of stable and unstable processes. *Automatica* **2016**, *67*, 132–143. [CrossRef]
26. Forrai, A. *Embedded Control System Design: A Model Based Approach*; Springer: Berlin/Heidelberg, Germany, 2012.

processes

MDPI

Article

Unrestricted Horizon Predictive Controller Applied in a Biphasic Oil Separator under Periodic Slug Disturbances

Rodrigo Trentini [1,*], Alexandre Campos [2,*], Marcos Antonio Salvador [1], Yuri Matheus Scheuer [1] and Carlos Henrique Farias dos Santos [3]

1 Department of Electrical Engineering, Federal Institute of Santa Catarina—IFSC, R. dos Imigrantes 445, Jaraguá do Sul 89254-430, Brazil
2 Department of Petroleum Engineering, State University of Santa Catarina—UDESC, Av. Lourival Cesario Pereira s/n, Balneário Camboriú 88336-275, Brazil
3 Center for Engineering and Exact Sciences, State University of Western Paraná—UNIOESTE, Av. Tancredo Neves 6731, Foz do Iguaçu 85856-970, Brazil
* Correspondence: rodrigo.trentini@ifsc.edu.br (R.T.); alexandre.campos@udesc.br (A.C.)

Abstract: Multi-phase flow, characterised by the presence of both liquid and gas phases, often occurs in riser pipes during oil extraction. These flows can be problematic because they can cause oscillations due to the formation of bubbles within the pipes, which can negatively impact the safety and efficiency of offshore production operations. One solution to this problem is to use a gravitational oil separator, which is designed to dampen these oscillations. The separator is equipped with a control system that uses liquid level and gas pressure sensors to stabilise the flow by adjusting the positions of its valves. This paper presents the use of a specific type of model-based predictive controller to control the level and pressure of a biphasic oil separator, particularly in the presence of slug disturbances. The designs of the separator model and controller are discussed in detail, with a focus on the advantages of using an unrestricted horizon predictive controller, such as its ability to make predictions over a long horizon and its relatively low computational requirements. For the sake of comparison, a linear quadratic regulator is also evaluated. The simulation results demonstrate that the proposed control system is able to effectively regulate the separator's liquid level and gas pressure within a magnitude range of 10^{-4} m for the liquid level and 10^{-3} bar for the internal pressure. Aside from that, the dynamics of the closed-loop system is six times faster than the plant's for the liquid behaviour and 30 times faster for the pressure, while also presenting sharp attenuation characteristics for the input disturbances of nearly 50 dB for the pressure output and 68 dB for the liquid level.

Keywords: biphasic oil separator; slug disturbance; model-based predictive control; unrestricted horizon predictive controller

1. Introduction

In offshore oil production, a multi-phase mixture of oil, gas and water is transported from the seabed wells through submarine risers to the surface and onto a platform. The initial piece of equipment on the platform is the separator, which not only separates but also minimises oscillation in the incoming mixture flow to ensure optimal efficiency for the remaining platform equipment.

As an example, the oil production in Brazil is based mostly on offshore pre-salt exploration. For this, several challenges must be faced which are related to exploration in increasingly deep waters. In addition, the control and optimisation of processes is essential for the smooth functioning and financial return of projects with high investments, as is the pre-salt layer case. Deep water well existence further and further away from the shore generates piping lines (risers) of large extensions and of different geometries. Because of this, the occurrence of irregular flows with severe oscillations in pressure and flow rate, known as slug flow, becomes recurrent, causing operational problems during production.

Low efficiency in primary separation, overflow in the separators and reduced equipment life are among the possible problems caused by flows of this type.

Reducing the opening of the surface choke valve, through which oil comes from the wells, is a direct solution to this problem. However, this considerably decreases the production rate, especially in fields where pressure is lower, in addition to possible back pressure problems in wells. Thus, it is necessary to find alternatives that guarantee a stable flow with high productivity. Therefore, an alternative is to reduce part of this oscillation to the primary processing plants on the platform by using control strategies to stabilise the slugs. In this scenario, the gravitational separator is, in general, the first piece of equipment in production processing and should be responsible for stabilising these oscillations. According to Stewart and Arnold [1], in order to obtain a separator design suitable for surface production facilities, it is essential to maintain the quality and quantity of the produced hydrocarbon fluids and to avoid major operational problems in the downstream equipment.

In oil and gas separators, the gas and liquid at a specific temperature and pressure are separated from a hydrocarbon stream, and therefore their design, structure, and operation are critical [1]. Some minimum requirements exist for the design process for gas–liquid separators used in oil and gas processing plants [2]. Research on separators as a novel gas–liquid separator on offshore platforms [3] is evident nowadays. Additionally, the flow behaviour of crude oil in a battery (several separators operating at different pressures) of industrial crude oil and gas separators in the oil industry has been studied [4], as well as the evaluation of flow characteristics in an onshore horizontal separator using computational fluid dynamics [5].

Some studies addressed the basis of two-phase separator design selection and explored the applicability of artificial neural networks for the creation of intelligent systems to predict proper separator dimensions or an evolutionary computational approach in the design of horizontal gravity separators [6–8]. Other works analysed separator performance, such as for corrugated plate gas–liquid separators [9], horizontal multi-pipe separators (for subsea oil–water bulk separation) [10], and gas–liquid cylindrical cyclone separators [11]. Additionally, the design and capital cost optimisation of gravity separators were studied [12].

Oscillations in production are often called intermittent or slug flow and are characterised by a cyclic behaviour causing flow and pressure oscillations [13]. These conditions cause numerous operational problems in oil production, such as reductions in the lifespan of the equipment, reducing primary separation efficiency, separator overflow, burning gas to decrease the pressure inside separator, and in the worst case, emergency platform shutdowns [14].

Research on the cause and modelling of slugs is presented in the literature. For instance, Cozin et al. [15] showed that a two-phase slug flow might be characterised using artificial neural networks. A correlation between the phase velocity in the slug body and overall superficial flow velocity is built based on a slug closure model [16]. Mathematical models and numerical methods to study the inner flow field in oil–water disc separators aiming to optimise their design were presented in [17,18]. Aiming at dampening slug-induced oscillatory disturbances in the inflow to the gravity separator, slug handling with a virtual harp based on nonlinear predictive control for a gravity separator was presented in [19]. Aside from that, slugging frequency correlation was analysed for inclined gas–liquid flow in [20], where a novel correlation for the slugging frequency in inclined gas–liquid flow is proposed through experimental data, with the results suggesting that the liquid superficial velocity significantly influences the frequency for the range of flow conditions studied. Additionally, the vertical pipe properties, as characteristics of slug and churn turbulence, under varying flow parameters can be effectively understood using cross-nonlinear analysis, offering a new perspective for analysing the spatiotemporal evolution, instability, and complexity in oil–gas–water three-phase flow [21].

The conventional solution to reduce slug flow effects is to control the choke valve. However, this causes a back pressure increase in oil wells and decreases the production rate, and some solutions have been proposed to ensure stable flow and maximum oil production [22,23]. The feedback control loop is the most used method in the literature to manipulate the valves of separators. It is possible to implement a strategy to keep the level and pressure in the separator within a safe range of operation, increasing separation and reducing flow oscillations in the downstream equipment [24,25]. A control strategy letting the separator level oscillate within certain limits or bands was presented in [26]. As can be noticed, the most commonly used control technique is still the proportional–integral–derivative (PID) and its derivations.

The main aim of the feedback loop in a gravitational oil separator is to absorb disturbances from the input flow, keeping the pressure and level variables in acceptable value ranges. Nunes [27] proposed a control model based on bands, which allows the level to oscillate within certain limits. This control is based on direct level measurement instead of flow measurement, obtaining a simpler and better performance model. Matrix modelling of systems using state variables is the most efficient way to represent multiple-input multiple-output (MIMO) systems, in addition to being necessary for the modern control method's application. Minghui and Gensheng [28] described the temperature profile in a well through state spaces. Through this representation, it is possible to analyse the thermal behaviour during the circulation and closing phases, making it possible to propose control techniques for the temperature. The linear quadratic regulator (LQR) is an efficient modern control method which uses a cost function to optimise the controller. Liu et al. [29] proposed a control strategy to reduce the problem of riser recoil in deep waters, using the LQR, some results are obtained, reducing risks during operation. The LQR was also used in other publications for optimal damping stabilisation [30], singularly perturbed systems [31], linear system switching [32], and optimal reference tracking for three-phase separators [33].

Despite the amount of research that uses classical and modern control strategies, the implementation of predictive controllers in gravity separators is gaining prominence. Model-based predictive control (MPC) is a technique that uses optimisation (such as LQR) and still regards the so-called prediction horizon to determine the action plan from k (current time) until N_y (future fixed horizon). MPC's ability to anticipate future events and implement control actions accordingly, as well as its natural ability to deal with constraints, are the main keys to its success in several industry applications [34].

One of the first investigations on MPC in gravity separators was carried out by Stokke et al. [35]. The authors compared MPC with linear–quadratic–Gaussian (LQG) control and stated that the first was superior in performance. For instance, LQG control presented more sensitivity to noise. The prediction horizon was selected to be 30 samples, with a sampling time of 2 s. Notice that in this case, the prediction horizon was only 60 s, whilst the system's settling time was about 100 s. On the other hand, expanding the prediction horizon would increase further the computational cost, which was an issue for the authors. Mendes et al. [36] showed the implementation of nonlinear MPC (NMPC) in a three-phase oil separator. The performance of the proposed strategy was compared with a traditional PI controller used in the industry. The obtained comparative results showed that NMPC gave the better results, allowing good oscillation attenuation under slug perturbations. However, the prediction horizon was only set to 20 samples with a sampling time of 10 s, giving only 200 s of prediction within a model that needs several hours for settling. Likewise, Hansen et al. [37] used MPC for a three-phase gravity separator, with a comparison with H_∞ control. The authors cited that by applying the MPC solution, the gravity separator volume can be used as a buffer for the inflow and keep the level within the constraints under more severe conditions than the compared H_∞ control solution. The prediction horizon was set to 600 samples with a sampling time of 0.2 s (120 s of prediction). Jespersen and Yang [38] present a study where they compare a PID, an H_∞ control and a MPC for the control of a three phase oil separator. Experimental results based on a lab-scaled pilot plant illustrate the huge potential and benefit of the MPC to significantly improve the effluent water quality, particularly when the process is subject to slugging inflow. Here, the prediction horizon is also set to 600 samples with a sampling time of 0.2 s.

It is interesting to remark that the MPC prediction horizons of the cited works increased with the available computational power. Nevertheless, none of the cited papers used a prediction horizon near the plant settling time, which would be the optimal case according to the literature [39]. This happened due to the increased computational cost. The size of the MPC signal vector, which must be solved at every time step, is given by $z N_y$ [40], with N_y being the prediction horizon (in samples) and z being the order of the plant. We remark that the implementation of increased-size matrices and vectors into actual microprocessors is a challenge mainly due to the sampling time. Larger vectors or matrices require more processing time, and generally, the smaller the required sampling time for such calculations, the more expensive the microprocessor is. An interesting example of this drawback was cited by Backi et al. [19]. The authors stated that they recognised that real-time applicability is a crucial aspect of MPC implementation. To provide a comprehensive analysis, they utilised a detailed model that included droplet calculations for oil in water and water in oil. Nevertheless, the use of simpler linear models may be considered an option to reduce the computational time in actual applications. Additionally, Riberio et al. [41] presented a work where the existing offshore production system models, as described in the literature, underwent modifications to increase their usefulness and realism in the context of advanced control studies. The authors conducted a case study on a real plant to quantify the well-known benefits of MPC. The findings highlight MPC's ability to maintain quality despite perturbations, its intelligent and adaptive set point selection in response to economic considerations and perturbations, and its secure operation without constraint violations, even under perturbations. However, the computation cost of the algorithm is high. Therefore, to ensure computational efficiency, they modified the plant model to maintain cost-effectiveness.

It can be seen that despite the fact few works exist with MPC implementation for control of the level and pressure in oil separators, there are several papers which coped with MPC performing higher-level tasks on petroleum production platforms, such as the works of Neto and Secchi [42], Godhavn et al. [22], Campos et al. [43], and Willersrud et al. [44]. The convergence point of these four works, among others, is the use of MPC mainly for the supervision layer, where both higher sampling times and simplified models can easily be used. Differently, when using MPC for the primary control, the computation cost takes its toll, increasing the processing time and preventing high-frequency implementations of the algorithm.

Controversially, there are two light computing MPC models that reduce dramatically the computational cost in comparison with the most common ones (generalised predictive controller and dynamic matrix controller (GPC and DMC, respectively)). The predictive functional controller (PFC) was developed by Richalet et al. in 1978 [45]. It counts with a control algorithm which does not require any matrix inversion or numerical minimisation of a cost function, and it can be realised for transfer function models. For multivariable processes, one interesting light computing MPC model is the so-called unrestricted horizon predictive controller (UHPC) for SISO and MIMO cases [46,47]. Such a controller is applied for pitch control in a grid-connected wind turbine. In spite of the challenges of the plant model size and controller's sampling time, the UHPC reduces the computational cost, making it a preferable choice for power system applications [48]. Furthermore, the UHPC is used to ensure stability, robustness, reference tracking, and disturbance rejection even in the presence of modelling errors and noise in a continuous stirred tank reactor (CSTR), where greater margins of stability and noise attenuation are obtained in comparison with the GPC, being able to maintain the same loop response throughout the control system's operating range without degrading the performance of the controller, which is not the case with the GPC [49].

The UHPC might be a compelling option for the current oil industry. Its calculated matrices do not increase with N_y but solely with the order of the plant. In other words, it is possible to use small sampling times and long prediction horizons without overloading the computational cost, facilitating implementation in a microcontroller. Given that there are still no published works on UHPCs applied to gravity separators, the authors considered investigating, in this paper, the possibility of its implementation in a simulation environment, aiming to reduce level and pressure oscillations caused by slugs.

Section 2 presents both the process and slug modellings (Section 2.1) and the UHPC equating and design for the proposed plant (Section 2.2). Section 3 shows computer simulations of the proposed control strategy applied to a gravity two-phase separator, where four different simulation scenarios are studied: open loop, closed loop with input disturbances as slugs, closed loop for a reference change in the liquid level, and frequency response for the open- and closed-loop cases. Finally, the discussions and conclusions are presented in Section 4.

2. Materials and Methods

This section presents the modelling of the biphasic oil separator, detailing its nonlinear differential equations and the constant values used in this work. In addition, this section presents the UHPC and its equating and design for the closed-loop system, aimed at damping the slug disturbances.

2.1. Modelling of the Biphasic Oil Separator

During the extraction of oil from offshore locations, a phenomenon known as *multiphase flow* occurs in the riser pipes. This is a mixture of water, oil, and gas that exhibits an oscillatory behaviour. The flow of gas and liquid in these pipes often occurs in bursts, which can disrupt various processes in the production chain of these offshore platforms. This flow is caused by the formation of elongated gas bubbles within vertical lift ducts. These bubbles have a shape similar to projectiles, known as *Taylor bubbles*, which is depicted in Figure 1. The bubbles occupy almost the entire diameter of the tube, attached to a liquid thin sheet that slides through the tube. This blade has a slower speed than the bubbles and may even have a direction contrary to the flow in certain sections of the tube [50].

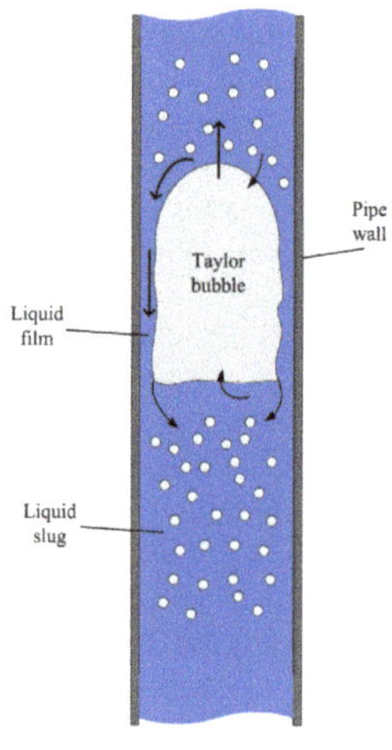

Figure 1. Schematic of a slug flow [50].

As the primary focus of this extraction is economic gain through the production of hydrocarbons (oil and gas), it is necessary to equip these fields (both maritime and terrestrial) with appropriate production facilities. These facilities are specifically designed to carry out the primary processing of these fluids [51].

At a primary processing facility, the first step is to separate the gas, which is less dense, from the liquids through gravitational forces. This process occurs in the initial stage, known as the *two-phase separator*, which is divided into two sections: the primary section and the secondary section. In the primary section, the fluid collides with deflectors that abruptly change its speed and direction, directing the liquid elements to the bottom of the reservoir. In the secondary section, the larger oil droplets present in the gaseous phase are allowed to settle out. However, the gaseous flow still holds some oil particles that are captured through porous means in the coalescence section, which is typically located near the gas outlet valve [1].

The separator typically consists of a cylindrical tank where a mixture of oil and dissolved gas enters and separates through the process inside the tank. The gas and oil then exit through the upper and lower valves, respectively. The pressure inside the tank and the oil level are the variables that are typically controlled in this process. This work assumes liquid incompressibility (constant density) and an isothermal process, and the gas is assumed to be inert or ideal. Figure 2 illustrates a schematic representation of the described system. In the figure, L_{in}, L_{out}, G_{in}, and G_{out} are the inflows and outflows of liquid and gas (m^3/s), respectively, C is the separator length (m), D is its diameter (m), h_L is the liquid level (m), P and T are the pressure (bar) and temperature (K) inside the separator, respectively, and x_L and x_G are the liquid and gas valve opening values, respectively, given from 0 (completely closed) to 1 (completely open).

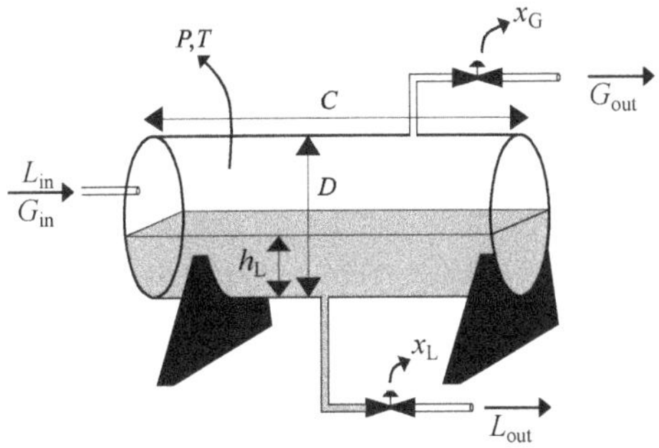

Figure 2. Biphasic gravitational separator.

The outlet liquid L_{out} and gas G_{out} flows adopt expressions for linear valves commonly used in the industry [52], such as

$$L_{\text{out}}(t) = 2.4 \cdot 10^{-4} x_L(t) C_{V_L} \sqrt{\frac{P(t) + (\rho_L g h_L(t) \cdot 10^{-5}) - P_1}{\rho_L / \rho_{H_2O}}}, \tag{1}$$

$$G_{\text{out}}(t) = 2.4 \cdot 10^{-4} x_g(t) C_{V_g} \sqrt{\frac{(P(t) - P_2) T (P(t) + P_2)(Mm)_0}{(Mm)_g / P(t)^2}}, \tag{2}$$

with g being the gravitational acceleration (m/s^2), C_{V_L} and C_{V_g} being the flow coefficients of the liquid and gas valves (gpm/psi), respectively, ρ_L and ρ_{H_2O} being the liquid and water densities (kg/m^3), respectively, P_1 and P_2 being the downstream pressures of the liquid and gas valves (bar), respectively, and $(Mm)_0$ and $(Mm)_g$ being the molar masses of air and gas (kg/mol), respectively.

Nunes et al. [51] showed that the equation for determining the volume of liquid within a two-phase cylindrical separator based on its geometric features, shown in Figure 2, is as follows:

$$V_L(t) = \frac{D^2 C}{4} \left(\arccos\left(\frac{D - 2h_L(t)}{D} \right) - \frac{2h_L(t)(D - 2h_L(t))\sqrt{D - h_L(t)}}{D^2} \right). \tag{3}$$

The same reference gives the equations for the liquid and gas balances inside the gravity separator as follows:

$$\frac{dh_L(t)}{dt} = \frac{L_{\text{in}}(t) - L_{\text{out}}(t)}{2C\sqrt{h_L(t)(D - h_L(t))}}, \tag{4}$$

$$\frac{dP(t)}{dt} = \frac{P(t)(G_{\text{in}}(t) - G_{\text{out}}(t) + L_{\text{in}}(t) - L_{\text{out}}(t))}{V - V_L(t)}, \tag{5}$$

with V being the total volume of the separator (m^3).

Finally, typical slug liquid and gas flow inputs that may be simplified to sine waves with a period of 2800 s are summed to their mean values ($\overline{L_{\text{in}}}$ and $\overline{G_{\text{in}}}$, respectively) as follows [51]:

$$L_{\text{in}}(t) = \overline{L_{\text{in}}} + 0.082 \sin\left(\frac{2\pi t}{2800}\right), \tag{6}$$

$$G_{\text{in}}(t) = \overline{G_{\text{in}}} + 0.075 \sin\left(\frac{2\pi t}{2800}\right). \tag{7}$$

Table 1 shows the constants of the separator model to be used in this work, whilst Table 2 presents its steady state values.

Table 1. Biphasic oil separator constants.

Symbol	Description	Value	Unity
Cv_L	Liquid valve flow coefficient	1025	gpm/psi
Cv_g	Gas valve flow coefficient	120	gpm/psi
ρ_L	Liquid density	850	kg/m^3
ρ_{H_2O}	Water density	999.19	kg/m^3
P_1, P_2	Valves downstream pressure	6	bar
T	Separator internal temperature	303.15	K
$(Mm)_0$	Air molar mass	0.039	kg/mol
$(Mm)_g$	Gas molar mass	0.021	kg/mol
D	Separator diameter	3	m
C	Separator length	8	m
V	Separator volume	56.55	m^3

Table 2. Biphasic oil separator steady state values.

Symbol	Description	Value	Unity
$\overline{x_L}$	Liquid valve opening	0.4375	1
$\overline{x_G}$	Gas valve opening	0.0536	1
$\overline{L_{\text{in}}}$	Liquid flow inlet	0.165	m^3/s
$\overline{L_{\text{out}}}$	Liquid flow outlet	0.165	m^3/s
$\overline{G_{\text{in}}}$	Gas flow inlet	0.1	m^3/s
$\overline{G_{\text{out}}}$	Gas flow outlet	0.1	m^3/s
$\overline{h_L}$	Internal liquid level	2	m
$\overline{P}$	Internal pressure	8	bar
$\overline{V_L}$	Volume of liquid	40.05	m^3

The two-phase oil separator nonlinear model presented in this section in Equations (1)–(7) is used for numeric simulations in Section 3, in addition to being used for the controller design after linearization in the following section.

2.2. The Unrestricted Horizon Predictive Controller

This section exploits the unrestricted horizon predictive controller (UHPC) and its equating and design for controlling the gravitational oil separator, which were presented previously.

The reader should note that the UHPC fits into the category of model-based predictive control (MPC). These type of controllers have been developed to address some of the limitations of traditional controllers, such as the input-output constraints, wind-up effect, and optimisation of control objectives [34]. The natural behaviour of MPC and its well-posed mathematical approach to dealing with constraints has led to it being widely accepted in both academia and industry [39].

Perhaps the most known and implemented MPC model is the generalised predictive controller (GPC), which was first presented by Clarke et al. [40]. Nevertheless, without its stochastic potential—which is rarely used—the GPC becomes similar to the predictive functional controller (PFC). The PFC is perhaps the most known member of the low computing cost MPC family [34]. The main difference between them is that the GPC is a multivariable controller, whilst the PFC is essentially a SISO one, therefore not being suitable for the application presented in this paper. Among these two MPC approaches lies the UHPC, which features stochasticity, multivariability, and a low computing cost. We also note that the UHPC is a general usage linear MPC model [46,47] such that its presented design does not aim at any specific application.

2.2.1. UHPC Equating

In contrast to traditional MPC models, the UHPC, similar to the PFC, utilises a quadratic cost function that considers only a single point in the future. For the general discrete state space system given by

$$x(k+1) = \mathbf{A}x(k) + \mathbf{B}u(k-d) + \mathbf{\Gamma}\xi(k), \tag{8}$$

$$y(k) = \mathbf{C}x(k) + \xi(k), \tag{9}$$

this cost is defined as

$$\mathcal{J} = x(k+N_y)^T\mathbf{Q}\,x(k+N_y) + u(k)^T\,\mathbf{R}\,u(k), \tag{10}$$

subjected to the diagonal weighting matrices $\mathbf{Q} = \mathbf{Q}^T \geq 0$ and $\mathbf{R} = \mathbf{R}^T \geq 0$. $\mathbf{A}$, $\mathbf{B}$, $\mathbf{C}$, and $\mathbf{\Gamma}$ are the state, input, output, and disturbance matrices in discrete form, respectively, while x, u, and ξ are the system state and input and noise vectors, respectively, and d is the transport delay.

Notice that regarding a single future point N_y means that any information between k and N_y is neglected by the UHPC during optimisation for the control law calculation. In practice, it implies that the UHPC calculates the control law, looking only to point N_y ahead. By doing so at every time step, the prediction window is entirely swept in N_y steps. Thus, the UHPC regards that if the system remains stable at sample time N_y, then it must be stable during all of the trajectory until that point. The main advantage of this approach is to reduce the computational cost. Differently, the GPC optimises the entire trajectory between k and N_y—its cost function concerns a sum (see [40])—which increases the computational cost. On the other hand, optimising the entire trajectory means that the GPC control law inherits more information from the system than the UHPC, which might make a difference for systems where control signal constraints are relevant.

By focusing on a single point N_y in the future in the system states, it is straightforward to obtain

$$\begin{aligned} x(k+N_y) = \mathbf{A}^{N_y}x(k) + \sum_{i=d}^{N_y}\mathbf{A}^{N_y-i}\mathbf{B}\,u(k-d+i) \\ + \sum_{i=1}^{d-1}\mathbf{A}^{N_y-i}\mathbf{B}\,u(k-d+i) + \sum_{i=1}^{N_y}\mathbf{A}^{N_y-i}\mathbf{\Gamma}\,\xi(k-1+i). \end{aligned} \tag{11}$$

From this point onward, the most common approach for MPC models is to split the parts related to the future (i.e., not known) from the current and past ones (i.e., known) [40].

For instance, the term related to $\xi(k)$ is uncertain since it represents the future of the noise. Therefore, it can be expressed as a combination of present and future components:

$$\sum_{i=1}^{N_y} \mathbf{A}^{N_y-i}\mathbf{\Gamma}\,\xi(k-1+i) = \underbrace{\mathbf{A}^{N_y-1}\mathbf{\Gamma}\,\xi(k)}_{\text{present}} + \underbrace{\sum_{i=2}^{N_y} \mathbf{A}^{N_y-i}\mathbf{\Gamma}\,\xi(k-1+i)}_{\text{future}}. \tag{12}$$

Similar to the stochastic term, the future of the control signal $\boldsymbol{u}(k)$ in Equation (11) is also uncertain if and only if $N_y \geq d$, which is always the case; otherwise, the prediction would be hampered by being smaller than the system delay. Now, the GPC and the UHPC differ. While the GPC also regards the future of the control signal (and thus its control signal is given by a vector), the UHPC separates this signal into present and future components, but it only utilises the known data (i.e., the terms that contain $\boldsymbol{u}(k)$). Specifically, we have

$$\sum_{i=d}^{N_y} \mathbf{A}^{N_y-i}\mathbf{B}\,\boldsymbol{u}(k-d+i) = \underbrace{\mathbf{A}^{N_y-d}\mathbf{B}\,\boldsymbol{u}(k)}_{\text{present}} + \underbrace{\sum_{i=d+1}^{N_y} \mathbf{A}^{N_y-i}\mathbf{B}\,\boldsymbol{u}(k-d+i)}_{\text{future}}. \tag{13}$$

By isolating $\xi(k)$ in Equation (9) to obtain $\xi(k) = \boldsymbol{y}(k) - \mathbf{C}\,\hat{\boldsymbol{x}}(k)$ and using only the available data for the predictor, the predicted state vector $\hat{\boldsymbol{x}}(k+N_y)$ given by Equation (11) is

$$\hat{\boldsymbol{x}}(k+N_y) = \mathbf{A}^{N_y}\boldsymbol{x}(k) + \mathbf{A}^{N_y-d}\mathbf{B}\,\boldsymbol{u}(k) + \sum_{i=1}^{d-1} \mathbf{A}^{N_y-i}\mathbf{B}\,\boldsymbol{u}(k-d+i) + \mathbf{A}^{N_y-1}\mathbf{\Gamma}\,\xi(k). \tag{14}$$

By combining some terms, it is possible to obtain

$$\hat{\boldsymbol{x}}(k+N_y) = \left(\mathbf{A}^{N_y} - \mathbf{FC}\right)\hat{\boldsymbol{x}}(k) + \mathbf{H}\,\boldsymbol{u}(k) + \mathbf{Y}_1\,\underset{\leftarrow}{\boldsymbol{u}}(k) + \mathbf{F}\,\boldsymbol{y}(k), \tag{15}$$

where $\mathbf{F} = \mathbf{A}^{N_y-1}\mathbf{\Gamma}$, $\mathbf{H} = \mathbf{A}^{N_y-d}\mathbf{B}$, $\underset{\leftarrow}{\boldsymbol{u}}(k)$ is a vector with past input data given by $\underset{\leftarrow}{\boldsymbol{u}}(k) = [\,\boldsymbol{u}(k-1) \quad \cdots \quad \boldsymbol{u}(k-d+1)\,]^T$, and $\mathbf{Y}_1 = \left[\,\mathbf{A}^{N_y-d+1}\mathbf{B} \quad \cdots \quad \mathbf{A}^{N_y-1}\mathbf{B}\,\right]$.

The just-obtained equation (Equation (15)) is called the *predictor* since it uses only past and current information to predict the states N_y steps ahead.

Moreover, some common inequalities that arise in the equating of most MPC models are the so-called Diophantine equations [34]. In some cases, the difficulty of having an algebraic solution to such equations prevents the further development of fully stochastic MPC models [53]. For the UHPC, the solutions to the Diophantine equations arise directly from the calculation of $\mathbf{F}$ (noise Diophantine equation) and $\mathbf{H}$ (input Diophantine equation), which is therefore another benefit of the chosen control strategy (refer to [47] for details).

Moreover, from this point forward, the following equations will assume $d = 1$, which causes the term $\mathbf{Y}_1$ to be eliminated. There are two main reasons for this assumption. First, the transport delay in separators is typically very small when the system is operating near its equilibrium points, and second, this assumption reduces the complexity of the controller without significantly degrading its performance.

It should also be noted that Equation (10) requires $\boldsymbol{x}(k+N_y)$. Since this information is not directly accessible due to the uncertainty of future noise and control signals, the employed solution is to utilise the available information from the cited predictor, transforming the cost function into

$$\begin{aligned} \hat{\mathcal{J}} = \mathbb{E}\{\mathcal{J}\} &= \mathbb{E}\left\{\boldsymbol{x}(k+N_y)^T\mathbf{Q}\,\boldsymbol{x}(k+N_y) + \boldsymbol{u}(k)^T\,\mathbf{R}\,\boldsymbol{u}(k)\right\}, \\ &= \hat{\boldsymbol{x}}(k+N_y)^T\mathbf{Q}\,\hat{\boldsymbol{x}}(k+N_y) + \boldsymbol{u}(k)^T\,\mathbf{R}\,\boldsymbol{u}(k), \end{aligned} \tag{16}$$

where $\mathbb{E}\{\cdot\}$ denotes the mathematical expectation operator. The expression in Equation (16) utilises the notion of expectation to signify that the control values are determined based on the available data up to and including the time t. As a result, the control design aspect of the UHPC entails solving a standard finite horizon optimal control problem [39].

The UHPC control law is obtained by minimising $\hat{\mathcal{J}}$ such that $\partial\hat{\mathcal{J}}/\partial u(k) = 0$. This, after some mathematical manipulation, results in

$$\boldsymbol{u}(k) = -\underbrace{\left(\mathbf{R} + \mathbf{H}^T\mathbf{Q}\mathbf{H}\right)^{-1}\mathbf{H}^T\mathbf{Q}}_{\mathbf{K}_0}\left(\mathbf{Y}_0\,\hat{\boldsymbol{x}}(k) + \mathbf{F}\,\boldsymbol{y}(k)\right), \tag{17}$$

with $\mathbf{Y}_0 = \mathbf{A}^{N_y} - \mathbf{FC}$. By combining the constant terms, one obtains

$$\boldsymbol{u}(k) = -\mathbf{K}_x\,\hat{\boldsymbol{x}}(k) - \mathbf{K}_y\,\boldsymbol{y}(k), \tag{18}$$

where $\mathbf{K}_x = \mathbf{K}_0\,\mathbf{Y}_0$ and $\mathbf{K}_y = \mathbf{K}_0\,\mathbf{F}$. The reader should note that the GPC signal for SISO systems is a vector where only the first position is used. For the multivariable case, its control signal becomes a matrix, and only the first position of each row is regarded. Controversially, the UHPC signal is only a scalar for the SISO case, whilst it is a vector for the multivariable one, in the same way as the LQR and LQG. Note that $\mathbf{K}_x$ is an $N_b \times N_a$ matrix, $\mathbf{K}_y$ is an $N_b \times N_c$ matrix, and $\hat{\boldsymbol{x}}(k)$ and $\boldsymbol{y}(k)$ are vectors of sizes N_a and N_c, respectively, with N_a being the system order, N_b being its number of inputs, and N_c being its number of outputs. Equation (18) shows that $\boldsymbol{u}(k)$ is a vector of a size N_b, with each row being the current control signal to be applied in each input of the plant. This is the main reason why the UHPC is considered a low-computing MPC model. The vectors and matrices used for its calculation have the size of the system order, which in this case is two. Note that the UHPC calculation burden is low in comparison with other multivariable MPC models, even with large prediction horizon, which do not impact the control signal calculation complexity.

2.2.2. UHPC Design for the Biphasic Oil Separator Model

Given that the expansion of the UHPC prediction horizon N_y does not burden the computational cost for its implementation, one might argue for an effective value for N_y. Trentini et al. [46] showed that for SISO plants, the increase in the prediction horizon makes the system's closed-loop poles move towards its open-loop ones. Complementary to that, Trentini [47] presented the results for multivariable systems using singular value plots. It was observed that the frequency response of the closed-loop system might easily be shaped to reach the regulation prerequisites. Both cited works have shown that it is not required to increase N_y further than the system's settling time.

Hence, it is necessary to determine the biphasic oil separator's settling time, which can be accomplished either experimentally or from its linear model. This work also regards the linearization of the separator model, given that the UHPC is a linear controller (i.e., it needs a linear model to be designed). Additionally, the reader should be aware that the separator model is approximately linear when working with small input or disturbance signals, which is exactly the case for the closed-loop system. This also allows the designer to linearise its model. This characteristic is exploited more deeply in Section 3.

Considering the nonlinear model given by Equations (1)–(7), and by applying first-order Taylor expansion, it is possible to obtain a continuous time second-order state space model with

$$\mathbf{A}_c = \begin{bmatrix} 0 & -0.0018 \\ 0 & -0.0307 \end{bmatrix}, \quad \mathbf{C}_c = \begin{bmatrix} 1 & 0 \\ 0 & 1 \end{bmatrix},$$

$$\mathbf{B}_c = \begin{bmatrix} -0.0167 & 0 \\ -0.1823 & -0.9019 \end{bmatrix}, \quad \boldsymbol{\Gamma}_c = \begin{bmatrix} 0.0442 & 0 \\ 0.4833 & 0.4833 \end{bmatrix},$$

where $\boldsymbol{x}_c(t) = \begin{bmatrix} h_L(t) & P(t) \end{bmatrix}^T$, $\boldsymbol{u}_c(t) = \begin{bmatrix} x_L(t) & x_G(t) \end{bmatrix}^T$ and $\boldsymbol{\xi}_c(t) = \begin{bmatrix} L_{\text{in}}(t) & G_{\text{in}}(t) \end{bmatrix}^T$. Both system controllability and observability are guaranteed, whilst its settling time—obtained from its eigenvalues—is 130 s. In addition, it is expected that the separator presents a natural integrator.

When considering a sampling time of $T = 0.1$ s to ensure all nonlinear dynamics, the UHPC prediction horizon is $N_y = 130/T = 1300$ samples. It should be noted that this value is significantly higher than any previously reported in the literature, and it is possible to obtain even higher values for N_y without overloading the computational capacity.

Similar to the LQR, the UHPC regards the weighting matrices $\mathbf{Q}$ and $\mathbf{R}$. The main difference is that $\mathbf{R}$ might be set to zero if one wants to use full control effort. Hence, for this work, these matrices are set to

$$\mathbf{Q} = 100 \cdot \mathbf{I}, \qquad \mathbf{R} = 0.$$

After system discretisation, matrices $\mathbf{F}$, $\mathbf{H}$, $\mathbf{Y}_0$, $\mathbf{K}_0$, $\mathbf{K}_x$, and $\mathbf{K}_y$ are calculated according to the description in Secion 2.2. Note that $\mathbf{F}$ and $\mathbf{H}$ are the solutions of the two arisen Diophantine equations for the noise and control signal, respectively. It is also worth mentioning that there is no need for the implementation of a Kalman filter, given that both states (level and pressure) are measured. In Figure 3, the control loop is depicted. Notice the disturbance inputs L_{in} and G_{in}, as well as the manipulated variables x_L and x_G and the measured signals h_L and P.

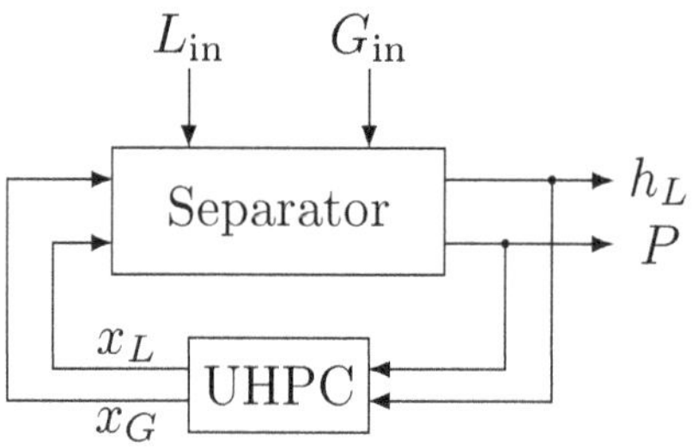

Figure 3. Control loop for the oil separator with UHPC.

3. Results

The biphasic oil separator nonlinear model presented in Equations (1)–(7) was implemented in MATLAB® and Simulink®. All constants and steady state values displayed in Tables 1 and 2 were considered for the simulation.

The separator's open-loop behaviour was first simulated using $\overline{L_{\text{in}}} = 0.165$ m^3/s and $\overline{G_{\text{in}}} = 0.1$ m^3/s, according to Table 2. Three different fixed valves openings were used:

Case 1: $x_L = 0.4375$ and $x_G = 0.0536$ (steady state values);

Case 2: $x_L = 0.5$ and $x_G = 0.0536$ (14% increase in liquid valve opening);

Case 3: $x_L = 0.4375$ and $x_G = 0.06$ (12% increase in gas valve opening).

Figure 4 presents the obtained results with the cited values, along with the simulation of the linear plant for comparison purposes. Graphs (a) and (b) present the results for the linear plant, whilst graphs (c) and (d) show the behaviour for the nonlinear one. Notice that both models presented a pressure drop for the cases which were not the steady state case, resulting in new magnitudes after the transient (around 130 s). On the other hand, the level of liquid presented a steady increase or decrease, depending upon the valve opening. Indeed, separator linearization showed that there was a natural integrator in the system, which was expected from the modelling.

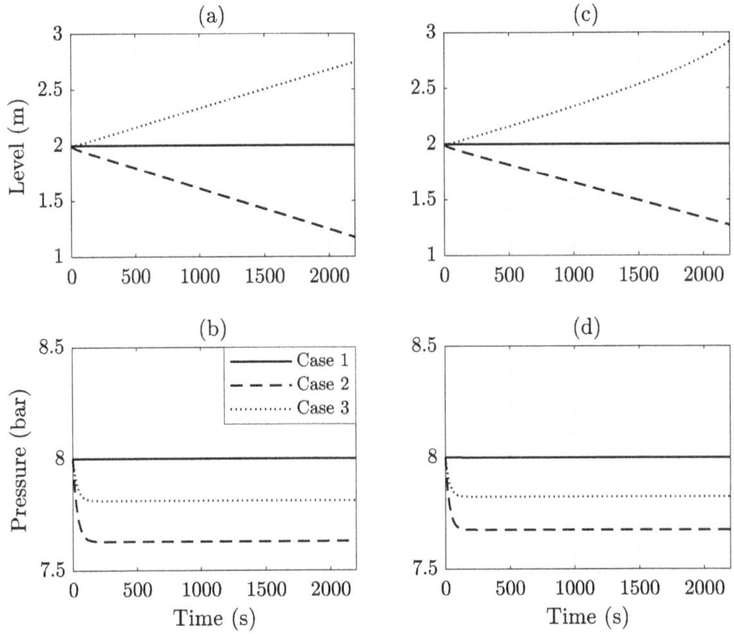

Figure 4. Simulation of the linear and nonlinear plants for different valve openings. (**a**,**b**) the results for the linear plant; (**c**,**d**) the behaviour for the nonlinear one.

The second simulation for the open-loop system was similar to the one we just presented, but now it adds the slugs to it according to Equations (6) and (7). Figure 5 presents the obtained results. Again, graphs (a) and (b) are the level and pressure for the linear model, whilst graphs (c) and (d) are the level and pressure for the nonlinear one.

Further, simulations were conducted to compare the effectiveness of the UHPC's control technique in regulating closed-loop behaviour for the system. The UHPC was implemented as discussed in Section 2.2, and for the sake of comparison, an LQR control scheme was also implemented. The choice for the LQR was made due to its high applicability in several multivariable processes with acknowledged success [54]. For the gravity separator shown in this paper, it was considered that $\mathbf{Q} = \mathbf{R} = 100 \cdot \mathbf{I}$. According to Figure 6, which displays the results obtained from the simulation using the same slugs as in the open-loop case, both the level and pressure exhibited acceptable behaviour, with oscillations occurring at a period determined by L_{in} and G_{in} and a magnitude to the order of 10^{-4} and 10^{-3}, respectively. In spite of these behaviours, the UHPC demonstrated more stable behaviour than LQR. Notice that, in Figure 6, graphs (a) and (b) are the level and pressure, respectively. Graphs (c) and (d) represent the control signals x_L and x_G, where a zoom is included to highlight the difference between the LQR and UHPC responses.

The next simulation experimented with a change in the level reference from 2 m to 1.5 m, which is presented in Figure 7. Again, graphs (a) and (b) are the system outupts (level and pressure), whilst graphs (c) and (d) are the respective control signals x_L and x_G. Here, the slugs were disabled to ease the analysis. The reference was changed in 10 s, where it was possible to observe abrupt changes in the valve openings. Notice that the liquid valve opened completely whilst the pressure valve closed entirely.

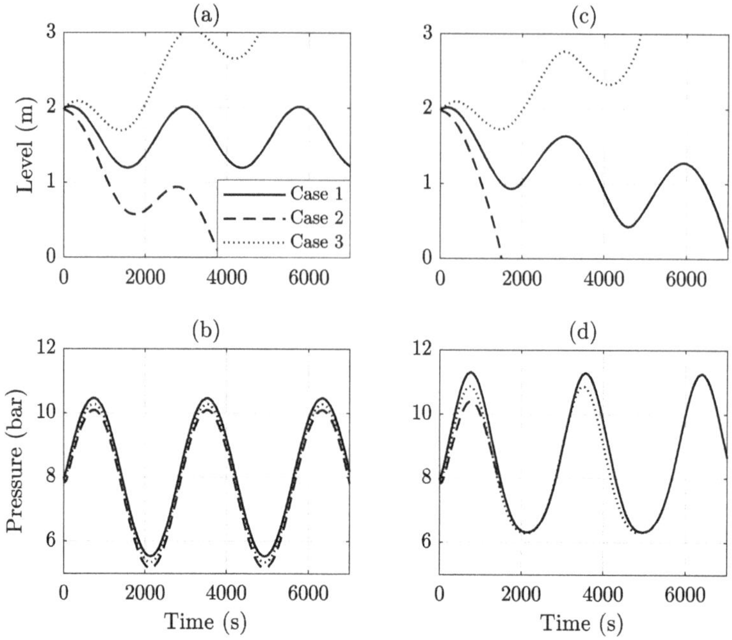

Figure 5. Simulation of the linear and nonlinear plants for different valve openings with slugs. (**a**,**b**) the level and pressure for the linear model; (**c**,**d**) the level and pressure for the nonlinear one.

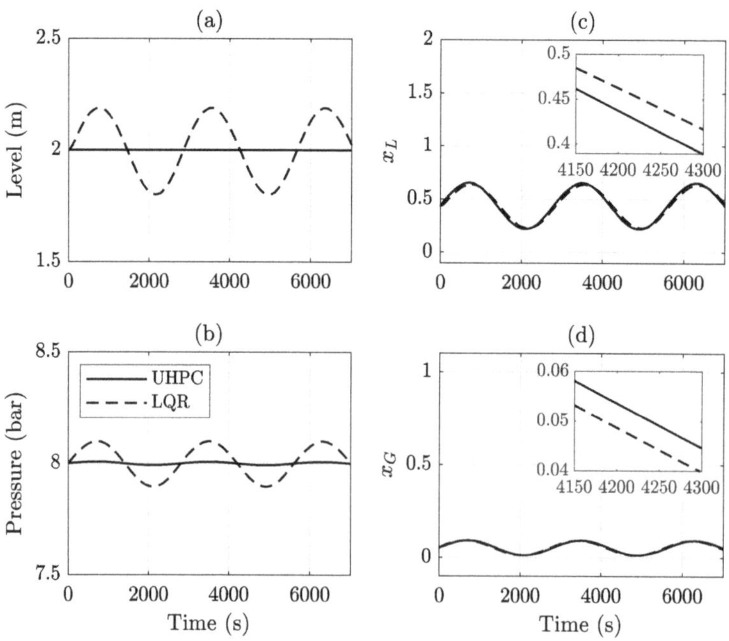

Figure 6. Simulation of the closed-loop system with slug disturbance. (**a**,**b**) the level and pressure, respectively; (**c**,**d**) the control signals x_L and x_G, where a zoom is included to highlight the difference between the LQR and UHPC responses.

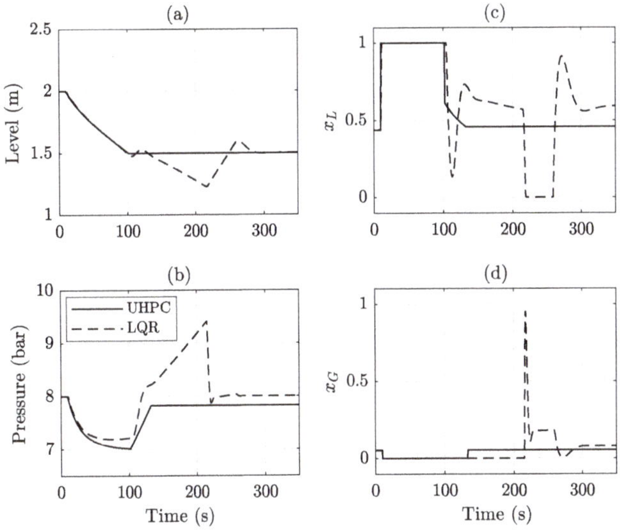

Figure 7. Simulation of the closed-loop system for step decrease in level reference. (**a**,**b**) the system outputs (level and pressure); (**c**,**d**) the respective control signals x_L and x_G.

Lastly, a frequency response simulation was obtained for the plant (open loop) and the closed loop. Considering a multivariable (two inputs and two outputs) and nonlinear plant, the usual procedure consists of the insertion of independent sinusoidal signals with increasing frequencies (chirp) at the disturbance inputs L_{in} and G_{in} separately. Three systems were simulated: an open-loop plant and two closed-loop systems, with one utilising the UHPC and the other utilising LQR.

Figure 8 presents the obtained results for the frequency response. The upper graph (a) shows the magnitude response when G_{in} was kept constant at its steady state value and L_{in} was a chirp signal with an amplitude of 0.082 m^3/s and frequency range from 0.1 mHz to 1 Hz within 28 hours of simulation. The bottom graph (b) depicts the evolution of the magnitude with an increase in frequency when G_{in} was regarded as a chirp signal with an amplitude of 0.075 m^3/s and the same frequency and time span as in the upper graph.

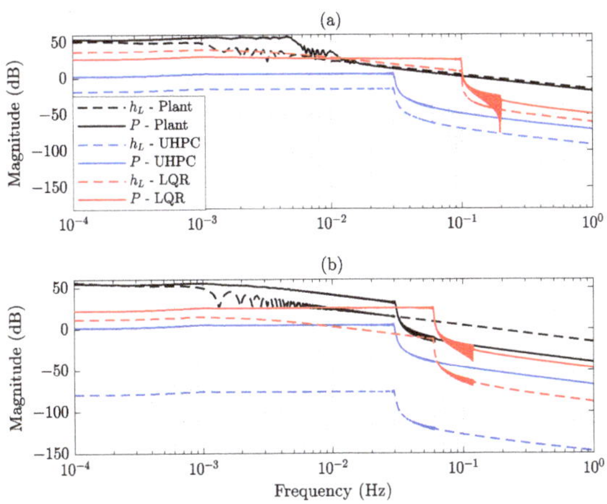

Figure 8. Frequency response of open- and closed-loop systems (**a**) with L_{in} as input and (**b**) with G_{in} as input.

4. Discussions and Conclusions

In this section, according to the obtained results, shown in Section 3, it is possible to address some discussion and highlight some paper contributions.

Considering the open-loop simulation, it can be observed in Figure 5 that both the liquid level and gas pressure presented oscillations due to the slugs. However, while the pressure just oscillated with the same period as the slugs, the level had a contrasting behaviour. Even when considering the valves' steady state openings (solid line), the level dropped until the separator was completely empty of liquid, which occurred shortly after 7000 s. Notice that this behaviour did not occur in the linear model. Instead, it oscillated around a fixed mean level (see Figure 4a). When keeping the gas valve at its original value and opening the liquid one approximately 14% more (dashed line), it can be seen that the level of liquid reached its minimum at nearly 1500 s, which is an expected result given that more liquid was leaving the separator. On the other hand, when keeping the liquid valve at its original opening and increasing by nearly 12% the value for the gas one, the level of liquid increased inside the separator until it reached its maximum value (3 m) shortly before 5000 s. This was also expected since more gas left the system, releasing more space for the liquid. For the pressure behaviour, notice that the dashed and dash-dotted lines are displayed only until the limits for the level were reached (1500 s and 4800 s, respectively). Note also that the mean value of the pressure is not its steady state.

Some improvements were observed when the loop was closed using the proposed controller. Figure 6 shows that their oscillations were within the limits of the valves, with the means being their steady state values. We call attention to the fact that despite the highly nonlinear behaviour presented in the previous section, the dynamical behaviour of the system was nearly linear within the cited magnitude of the oscillations, with control signals x_L and x_G working linearly without saturation. For this reason, the usage of input constraints for the optimisation problem was not required. We remark that not considering the constraints during the optimisation procedure means that it is not possible to comply with the input limits. In this work, this fact was verified through simulations, given that constraint handling was not implemented for the UHPC yet. The determination of the input and output constraints during the optimisation procedure might be explored in works where these limits are closer to being (or are) reached.

Moreover, the control signals for the valves (x_L and x_G) depicted in Figure 6c,d indicate that the UHPC anticipated the action over the liquid valve x_L while delaying the action over the gas valve. This phenomenon occurred because making decisions based on predictions among several potential future alternatives is generally more effective than relying solely on past and present data. For instance, the UHPC can produce superior results compared with a more accurate but non-predictive LQR response.

Additionally, aiming at obtaining more information from the proposed controller effect in the outputs, a step change is required in the level set point, and the system behaviour is shown in Figure 7. As cited, the liquid valve opened completely whilst the pressure valve closed entirely. This happened due to the drop in pressure inside the separator whenever there was a negative change in the level of liquid. The level reached its new reference nearly 100 s from when the liquid valve acted again. On the other hand, the pressure of the gas started increasing as soon as the level of liquid reached the reference. This increase regards the action of the liquid valve. Note that x_L reached its steady state value around 130 s, which was exactly the time when the pressure found its new stable value and the pressure valve opened again to keep it stable.

It is worth noting that the UHPC tended to produce smoother, less-oscillatory changes in both the process variables and valve control signals than LQR. This means that the UHPC can lead to more stable and predictable system behavior, while LQR may result in more abrupt and unpredictable changes.

Finally, in order to present the proposed control benefits relative to the frequency performance, the controlled plant closed-loop frequency response was obtained. Figure 8 shows that the plant had cutoff frequencies of approximately 1 mHz and 5 mHz for the liquid (h_L) and pressure (P) outputs, respectively. Additionally, the disturbance inputs were sharply amplified by the separator dynamics (around 50 dB in low frequencies). On the other hand, when the loop was closed using the UHPC, the system bandwidth increased to 30 mHz. This means that the closed-loop system responded 30 times faster for the liquid and 6 times faster for the pressure output, which is one of the main contributions of this paper. Moreover, it was observed that the UHPC had the potential to attenuate the impact of the low-frequency input disturbances: 3 dB for the pressure and −18 dB and −78 dB for the liquid level for input disturbances L_{in} and G_{in}, respectively.

It is important to notice that LQR presented a wider bandwidth than the UHPC. On the other hand, its attenuation of external disturbances was not as effective as the UHPC.

In summary, this paper presented a novel methodology for the implementation of a model-based predictive controller in a biphasic oil separator. The unrestricted horizon predictive controller (UHPC) is shown to be an interesting option for actual implementations in the oil industry due to its simplicity, state feedback structure, and low computational effort in comparison with other members of the MPC family.

This work introduced the detailed modelling of the separator, as well as the modelling of common slug disturbances. These oscillations, which cause the accumulation of bubbles during oil extraction, are dangerous, and their impacts on the remaining platform equipment must be diminished in the gravitational separator through the action of the liquid and gas valves. Aside from that, the results for three different valve openings using the nonlinear model were presented.

It is shown that the UHPC was able to perform a fairly stable operation for the separator without overloading the prospective processor, even with the application of the modelled slug disturbances. The UHPC's difference with other MPC models is mainly due to its intrinsic stochastic characteristic and also the implementation of its cost function. Both features together turned the UHPC into a predictive controller that uses less computer processing than the most common ones, even with the expansion of the controller's predictive horizon until (or more than) the system's settling time. In addition, a simulation for a 25% step-shaped decrease in the liquid level reference was explored, where it was shown that the UHPC was able to lead the level to its new operating point within less than 100 s.

Future works on this theme should include constraint calculations for the UHPC, mainly concerning its control signals. A comparison with other MPC models might also be considered. Moreover, even though the actual implementations in platforms are complex, such an implementation could also be performed in future research.

Author Contributions: Conceptualisation, R.T. and A.C.; methodology, R.T. and C.H.F.d.S.; software, R.T.; validation, R.T. and A.C.; formal analysis, A.C., M.A.S. and C.H.F.d.S.; investigation, R.T., A.C. and Y.M.S.; resources, R.T., A.C. and Y.M.S.; data curation, R.T.; writing—original draft preparation, R.T., A.C., Y.M.S. and C.H.F.d.S.; writing—review and editing, R.T., A.C., M.A.S. and Y.M.S.; visualisation, R.T., A.C., M.A.S. and Y.M.S.; supervision, A.C. and M.A.S.; project administration, R.T. All authors have read and agreed to the published version of the manuscript.

Funding: This research received no external funding.

Institutional Review Board Statement: Not applicable.

Informed Consent Statement: Not applicable.

Data Availability Statement: The data presented in this study are available on request from the corresponding author. The data are not publicly available due to the lack of an institutional repository.

Conflicts of Interest: The authors declare no conflict of interest.

References

1. Stewart, M.; Arnold, K. *Gas-Liquid and Liquid-Liquid Separators*; Gulf Professional Publishing: Houston, TX, USA, 2008.
2. Bahadori, A. *Natural Gas Processing*; Gulf Professional Publishing: Boston, MA, USA, 2014; pp. 151–222. [CrossRef]
3. Liu, Y.; Wang, C.; Cai, J.; Lu, H.; Huang, L.; Yang, Q. Pilot application of a novel Gas–Liquid separator on offshore platforms. *J. Pet. Sci. Eng.* **2019**, *180*, 240–245. [CrossRef]

4. Constant-Machado, H.; Leclerc, J.P.; Avilan, E.; Landaeta, G.; Añorga, N.; Capote, O. Flow modeling of a battery of industrial crude oil/gas separators using 113mIn tracer experiments. *Chem. Eng. Process. Process Intensif.* **2005**, *44*, 760–765. [CrossRef]
5. Acharya, T.; Casimiro, L. Evaluation of flow characteristics in an onshore horizontal separator using computational fluid dynamics. *J. Ocean. Eng. Sci.* **2020**, *5*, 261–268. [CrossRef]
6. Ayala, H.L.F.; Alp, D.; Al-Timimy, M. Intelligent design and selection of natural gas two-phase separators. *J. Nat. Gas Sci. Eng.* **2009**, *1*, 84–94. [CrossRef]
7. Mostafaiyan, M.; Saeb, M.R.; Alorizi, A.E.; Farahani, M. Application of evolutionary computational approach in design of horizontal three-phase gravity separators. *J. Pet. Sci. Eng.* **2014**, *119*, 28–35. [CrossRef]
8. Grødal, E.O.; Realff, M.J. Optimal design of two-and three-phase separators: A mathematical programming formulation. In Proceedings of the SPE Annual Technical Conference and Exhibition, Society of Petroleum Engineers, Huston, TX, USA, 3–6 October 1999.
9. Song, J.; Hu, X.; Zhang, J.; Chen, J. Experimental study on performance of two types of corrugated plate gas-liquid separators. *Energy Procedia* **2017**, *142*, 3282–3286. [CrossRef]
10. Skjefstad, H.S.; Stanko, M. Experimental performance evaluation and design optimization of a horizontal multi-pipe separator for subsea oil-water bulk separation. *J. Pet. Sci. Eng.* **2019**, *176*, 203–219. [CrossRef]
11. Hreiz, R. On the effect of the nozzle design on the performances of gas–liquid cylindrical cyclone separators. *Int. J. Multiph. Flow* **2014**, *58*, 15–26. [CrossRef]
12. Ahmed, T.; Russell, P.A.; Makwashi, N.; Hamad, F.; Gooneratne, S. Design and capital cost optimisation of three-phase gravity separators. *Heliyon* **2020**, *6*, e04065. [CrossRef]
13. Pedersen, S.; Durdevic, P.; Yang, Z. Challenges in slug modeling and control for offshore oil and gas productions: A review study. *Int. J. Multiph. Flow* **2017**, *88*, 270–284. [CrossRef]
14. Sausen, A.; Sausen, P.; Campos, M. Chapter 5—The Slug Flow Problem in Oil Industry and Pi Level Control. In *New Technologies in the Oil and Gas Industry*; Gomes, J.S., Ed.; IntechOpen: London, UK, 2012; pp. 103–116. [CrossRef]
15. Cozin, C.; Vicencio, F.E.C.; de Almeida Barbuto, F.A.; Morales, R.E.M.; da Silva, M.J.; Arruda, L.V.R. Two-Phase Slug Flow Characterization Using Artificial Neural Networks. *IEEE Trans. Instrum. Meas.* **2016**, *65*, 494–501. [CrossRef]
16. Dong, X.; Tan, C.; Dong, F. Gas–Liquid Two-Phase Flow Velocity Measurement With Continuous Wave Ultrasonic Doppler and Conductance Sensor. *IEEE Trans. Instrum. Meas.* **2017**, *66*, 3064–3076. [CrossRef]
17. Zhao, Z.; Shi, B. Numerical simulation of oil-water separation process in disc separator. In Proceedings of the 2011 International Conference on Remote Sensing, Environment and Transportation Engineering, Nanjing, China, 24–26 June 2011; pp. 8369–8372. [CrossRef]
18. Nemoto, R.H.; Baliño, J.L. Modeling and simulation of severe slugging with mass transfer effects. *Int. J. Multiph. Flow* **2012**, *40*, 144–157. [CrossRef]
19. Backi, C.J.; Krishnamoorthy, D.; Skogestad, S. Slug handling with a virtual harp based on nonlinear predictive control for a gravity separator. *IFAC-PapersOnLine* **2018**, *51*, 120–125. [CrossRef]
20. Hernandez-Perez, V.; Abdulkadir, M.; Azzopardi, B.J. Slugging Frequency Correlation for Inclined Gas-liquid Flow. *Int. J. Chem. Mol. Eng.* **2010**, *4*, 10–17.
21. Liu, W.; Han, Y.; Wang, D.; Zhao, A.; Jin, N. The Slug and Churn Turbulence Characteristics of Oil–Gas–Water Flows in a Vertical Small Pipe. *Z. FüR Naturforschung* **2017**, *72*, 817–831. [CrossRef]
22. Godhavn, J.M.; Fard, M.P.; Fuchs, P.H. New slug control strategies, tuning rules and experimental results. *J. Process Control* **2005**, *15*, 547–557. [CrossRef]
23. Mokhatab, S.; Poe, W.A.; Mak, J.Y. *Handbook of Natural Gas Transmission and Processing*, 4th ed.; Gulf Professional Publishing: Huston, TX, USA, 2019; pp. 103–176. [CrossRef]
24. Havre, K.; Dalsmo, M. Active Feedback Control as a Solution to Severe Slugging. *Spe Prod. Facil.* **2002**, *17*, 138–148. [CrossRef]
25. Pinto, D.D.D.; Araújo, O.Q.; de Medeiros, J.L.; Nunes, G.C. Slug Control Structures for Mitigation of Disturbances to Offshore Units. In *10th International Symposium on Process Systems Engineering: Part A*; Elsevier: Amsterdam, The Netherlands, 2009; Volume 27, pp. 1305–1310. [CrossRef]
26. Nunes, G.C.; Coelho, A.A.R.; Sumar, R.R.; Mejía, R.I.G. Band Control: Concepts and Application in Dampening oscillations of Feed of Petroleum Production Units. *IFAC Proc. Vol.* **2005**, *38*, 123–128. [CrossRef]
27. Nunes, G.C. *Design and Analysis of Multivariable Predictive Control Applied to an Oil-Water-Gas Separator: A Polynomial Approach*; University of Florida: Gainesville, FL, USA, 2001.
28. Minghui, W.; Gensheng, L. Wellbore Temperature Prediction and Control Through State Space Model. In Proceedings of the SPE Kingdom of Saudi Arabia Annual Technical Symposium and Exhibition, OnePetro, Dammam, Saudi Arabia, 25–28 April 2016.
29. Liu, X.; Liu, Z.; Wang, X.; Zhang, N.; Qiu, N.; Chang, Y.; Chen, G. Recoil control of deepwater drilling riser system based on optimal control theory. *Ocean. Eng.* **2021**, *220*, 108473. [CrossRef]
30. Veremey, E.I. Optimal damping stabilisation based on LQR synthesis. *Int. J. Syst. Sci.* **2020**, *52*, 1359–1372. [CrossRef]
31. Shim, K.H.; Sawan, M.E. Linear quadratic regulator design for singularly perturbed systems by unified approach using d operators. *Int. J. Syst. Sci.* **2001**, *32*, 1119–1125. [CrossRef]
32. Lincoln, B.; Bernhardsson, B. LQR optimization of linear system switching. *IEEE Trans. Autom. Control* **2002**, *47*, 1701–1705. [CrossRef]

33. Patarroyo-Montenegro, J.F.; Andrade, F.; Guerrero, J.M.; Vasquez, J.C. A Linear Quadratic Regulator With Optimal Reference Tracking for Three-Phase Inverter-Based Islanded Microgrids. *IEEE Trans. Power Electron.* **2021**, *36*, 7112. [CrossRef]
34. Rossiter, J.A. *Model-Based Predictive Control: A Practical Approach*; CRC Press: Boca Raton, FL, USA, 2017.
35. Stokke, S.; Strand, S.; Sjong, D. Model predictive control (MPC) of a gas-oil-water separator train. In *Methods of Model Based Process Control*; Springer: Berlin/Heidelberg, Germany, 1995; pp. 701–713.
36. Mendes, P.; Normey-Rico, J.E.; Plucenio, A.; Carvalho, R. Disturbance estimator based nonlinear MPC of a three phase separator. *IFAC Proc. Vol.* **2012**, *45*, 101–106. [CrossRef]
37. Hansen, L.; Durdevic, P.; Jepsen, K.L.; Yang, Z. Plant-wide optimal control of an offshore de-oiling process using mpc technique. *IFAC-PapersOnLine* **2018**, *51*, 144–150. [CrossRef]
38. Jespersen, S.; Yang, Z. Performance Evaluation of a De-oiling Process Controlled by PID, H_∞ and MPC. In Proceedings of the IECON 2021—47th Annual Conference of the IEEE Industrial Electronics Society, Toronto, ON, Canada, 13–16 October 2021.
39. Bitmead, R.R.; Gevers, M.; Wertz, V. *Adaptive Optimal Control—The Thinking Man's GPC*; Prentice Hall: Hoboken, NJ, USA, 1990.
40. Clarke, D.; Mohtadi, C.; Tuffs, P. Generalized predictive controller—Part I: The basic algorithm. *Automatica* **1987**, *23*, 137–148. [CrossRef]
41. Ribeiro, C.; Miyoshi, S.; Secchi, A.; Bhaya, A. Model Predictive Control with quality requirements on petroleum production platforms. *J. Pet. Sci. Eng.* **2016**, *137*, 10–21. [CrossRef]
42. Neto, S.S.; Secchi, A. Nonlinear Model Predictive Controller Applied to an Offshore Oil and Gas Production Facility. In Proceedings of the OTC Brasil, OnePetro, Rio de Janeiro, Brasil, 24–26 October 2017.
43. Campos, M.C.M.M.; Ribeiro, L.D.; Diehl, F.C.; Moreira, C.A.; Bombardelli, D.; Carelli, A.C.; Junior, G.M.J.; Pinto, S.F.; Quaresma, B. Intelligent System for Start-Up and Anti-Slug Control of a Petroleum Offshore Platform. In Proceedings of the Offshore Technology Conference, Houston, TX, USA, 1–4 May 2017. [CrossRef]
44. Willersrud, A.; Imsland, L.; Hauger, S.O.; Kittilsen, P. Short-term production optimization of offshore oil and gas production using nonlinear model predictive control. *J. Process Control* **2013**, *23*, 215–223. [CrossRef]
45. Richalet, J.; Rault, A.; Testud, J.; Papon, J. Model predictive heuristic control: Applications to industrial processes. *Automatica* **1978**, *14*, 413–428. [CrossRef]
46. Trentini, R.; Silveira, A.; Kutzner, R.; Hofmann, L. On the unrestricted horizon predictive control—A fully stochastic model-based predictive approach. In Proceedings of the 2016 European Control Conference (ECC) IEEE, Aalborg, Denmark, 29 June–1 July 2016; pp. 1322–1327.
47. Trentini, R. Contributions to the Damping of Interarea Modes in Extended Power Systems: A Turbine Governor Approach with the Help of the Unrestricted Horizon Predictive Controller. Ph.D. Thesis, Gottfried Wilhelm Leibniz Universität, Hannover, Germany, 2017.
48. De Lima, J.C., Jr.; Trentini, R.; Prioste, F. Comparative study of the pitch control of a wind turbine using Linear Quadratic Gaussian and the Unrestricted Horizon Predictive Controller. *Int. Trans. Electr. Energy Syst.* **2021**, *31*, e12720.
49. De Castro, L.A.; Silveira, A.d.S.; Araújo, R.d.B. Unrestricted horizon predictive control applied to a nonlinear SISO system. *Int. J. Dyn. Control* **2022**, *11*, 286–300. [CrossRef]
50. Yan, K.; Che, D. Hydrodynamic and mass transfer characteristics of slug flow in a vertical pipe with and without dispersed small bubbles. *Int. J. Multiph. Flow* **2011**, *37*, 299–325. [CrossRef]
51. Nunes, G.C.; de Medeiros, J.L.; Araújo, O.; de Queiroz Fernandes, O. *Modelagem e Controle da Produçao de Petróleo: Aplicaçoes em Matlab*; Editora Blucher: São Paulo, Brazil, 2010.
52. Perry, R.H.; Green, D.W. *Perry's Chemical Engineer's Handbook*, 6th ed.; McGraw-Hill Book Co.: New York, NY, USA, 1985.
53. Silveira, A.S.; Coelho, A.A. Generalised minimum variance control state-space design. *IET Control Theory Appl.* **2011**, *5*, 1709–1715. [CrossRef]
54. Berkovitz, L.D. *Optimal Control Theory*; Springer Science & Business Media: Berlin/Heidelberg, Germany, 2013; Volume 12.

processes

MDPI

Article

Experimental Study on Constant Speed Control Technology of Hydraulic Drive Pavers

Xiujie Yin [1,*], Shujun Yin [2], Hong Zhu [1] and Zhihao Zhang [1]

[1] School of Construction Machinery, SHAN DONG JIAO TONG University, Jinan 250357, China
[2] Highway School, Chang'an University, Xi'an 710075, China
* Correspondence: 206047@sdjtu.edu.cn; Tel.: +86-13065091268

Abstract: The paver needs superior constant speed performance when paving the pavement. In order to effectively reduce the paver speed fluctuation of the paver, and the wandering deviation from the predetermined track during the paving operation, a control scheme of paver travelling system based on GNSS, Global Navigation Satellite System, is proposed; the scheme can realize open-loop control, closed-loop control, and deviation correction control according to the driver's choice. During closed-loop control, the setting value and the PID controller output of the left wheel are combined to control the speed of the left wheel, as is the closed-loop control of the right wheel. During the deviation correction control, the coordinate provided by the RTK GNSS receiver and the predetermined trajectory line are used to calculate the lateral deviation of the paver. The lateral deviation is input to the right wheel navigation correction PID algorithm. After the calculation, the correction value of the right wheel speed is obtained, which is input to the right wheel PID controller for the deviation correction control. In this paper, the low constant speed performance of the paver, such as during straight driving, turning driving, and driving when resistance changing, was studied by means of experiments. The test results show that when the test paver was running at a speed of more than 2 m/min, the average speed was almost the same. The higher the average speed was, the more stable the speed was. When the paver was less than 1 m/min, its speed fluctuation tended to increase, and its constant speed performance could not be guaranteed. When the test paver hit a movable obstacle at a speed of 5 m/min, which changed the driving resistance, the average speed of the left and right wheels decreased significantly, with a change of about 2.8%, and there was no significant change in the speed fluctuation of the left and right wheels. At the same time, the wandering deviation test proves that the strait-line travelling wandering deviation was basically controlled within 2.5 cm. Without driver intervention, the wandering deviation of the test paver travelling 50 m decreased by about 97.4%, and the constant speed control fluctuation was within 0.2% when the paver travelled at the speed of 5 m/min.

Keywords: hydraulic drive; overshoot; constant speed; PID

1. Introduction

At present, mobile machines in some industries need to have good low constant speed performance [1] to meet their work requirements, such as the paving of pavers. It is difficult for the mechanical transmission chassis to meet the requirements. Generally, the full hydraulic transmission system with two pumps and two motors is selected. As shown in Figure 1, the chassis drive scheme is composed of two closed hydraulic circuits composed of variable pumps and variable motors. Each hydraulic circuit controls a drive wheel. The hydraulic circuits on both sides can be independently controlled to realize the forward, backward, steering, stepless speed changes and other actions of the paver. Compared with mechanical transmission chassis, pavers with full hydraulic transmission chassis have the characteristics of stepless speed change. Further, constant speed control is very important, especially since it is still required to achieve stable speed control in low-speed operation.

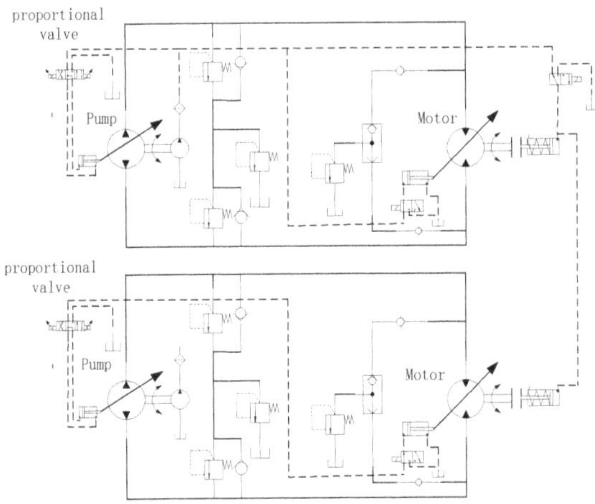

Figure 1. Hydraulic transmission schematic diagram of paver.

In 2020, Shi Chendi et al. [2] studied a hydraulic drive spindle speed control based on a fuzzy adaptive PID controller. The fuzzy algorithm was used to detect and adjust the parameters of a traditional PID controller to form a fuzzy adaptive PID controller. It was concluded that the maximum deviation of the designed fuzzy adaptive PID controller was 3.83% less than that of the traditional PID controller, and it could control the speed of the hydraulic drive spindle more accurately. In 2018, Liu Ting et al. [3] designed a PID controller based on the RBF neural network to control the output speed of the pump control motor system. Using Matlab software, the simulation analysis of hydraulic pump control motor system proved that RBF neural network controller had better dynamic characteristics and load disturbance adaptability in controlling motor output speed. In 2021, Huang Yajun et al. [4] proposed a four-pump and four-motor hydraulic drive system and determined the circuit diagram of the hydraulic drive system. Using AMESim software, the simulation model of the unilateral hydraulic drive circuit was established, and the dynamic response of the 565 kW full hydraulic bulldozer was studied. During the starting process of the bulldozer, when the paver speed was 2.6 km/h, the bulldozer still started smoothly. In 2021, Jiang Zhenhong et al. [5] studied the hydrostatic transmission system of a tractor. By using the power bond graph theory, they established a bond graph model of the tractor hydrostatic transmission system. By using SIMULINK simulation software, they simulated and analyzed the dynamic characteristics of the tractor hydrostatic transmission system. They believe that with a certain depth of tillage, the smaller the throttle opening is during normal operation of the tractor, the smaller the fluctuation of fuel consumption rate will be when the soil-specific resistance changes

To sum up, the research on the constant speed control system of hydraulic transmission mostly adopts the simulation method [6–9]. For pavers, the constant speed travelling performance is more important. If the speed was not constant, it would mean that there was an acceleration of the paver, which would affect the force balance of the paver's screed, the screed would be easy to vibrate, and then the flatness of the paved road would be affected. It would also mean that the number of vibrations within the same travel distance of the paver was different, and that the compactness of the paved road would be uneven. A new control scheme of the paver travelling system is adopted in this paper. The scheme can realize open-loop control, closed-loop control, and deviation correction control according to the driver's choice. When the closed-loop control is used, the setting value obtained from control panel and the PID controller output of the left wheel are combined to control the speed of the left wheel, as is the closed-loop control of the right wheel. The control function of the PID controller accounts for 10%, and the direct control of the speed setting value accounts for 90%. Of course, other combinations can also be used. This closed-loop control scheme can effectively avoid the speed fluctuation caused by PID regulation, while retaining the closed-loop control function of PID. During the deviation correction control, the coordinate provided by the RTK GNSS receiver and the predetermined trajectory line are used to calculate the lateral deviation of the paver. The lateral deviation is input to the right wheel navigation correction PID algorithm and then used to correct the right wheel speed. In addition, this paper will use the experimental method to study the low constant speed performance of a hydraulic transmission paver.

2. Design and Discussion of Paver Travelling Speed Control Scheme

2.1. Control Principle

The paver travelling control system includes a PLC controller, two speed sensors installed on the output shafts of the hydraulic motors that drive the left and right wheels of the paver, and two electromagnetic proportional valves installed on the variable displacement pumps. The solenoid valves can control the hydraulic oil flow of the variable displacement pumps. The common closed-loop constant speed control is shown in Figure 2a. The speed feedback value of the left wheel is subtracted from the speed set value of the left wheel, and the deviation is input to the PID algorithm module. The PID algorithm is used to calculate and output PWM, which controls the variable pump electric proportional valve, and further controls the speed of the hydraulic motor. The same is true for right wheel control.

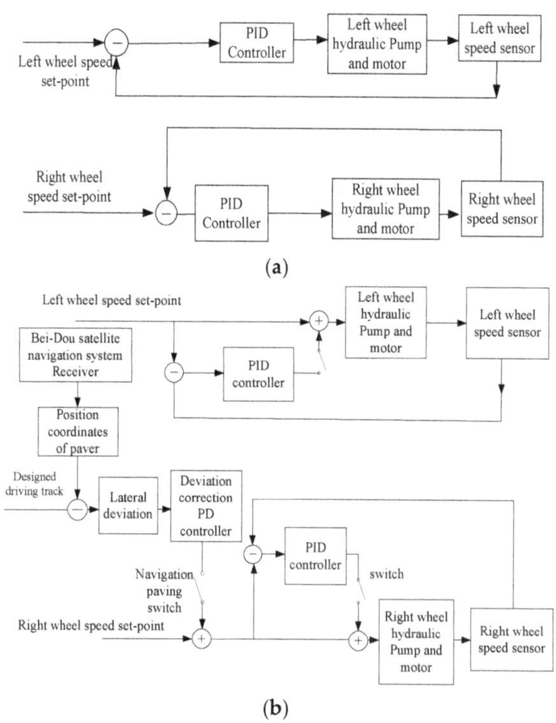

Figure 2. Travelling control scheme of paver. (**a**) Current travelling control scheme. (**b**) New control scheme.

In order to effectively reduce the driving wandering deviation of the paver and keep the speed of the paver constant, a new control scheme of the paver travelling system based on Global Navigation Satellite System is needed. The scheme has speed open-loop control, speed closed-loop control, and deviation correction control. As shown in Figure 2b, a RTK GNSS mobile station is installed on the paver, and a base station is set near the work site. The mobile station is connected with the paver PLC through RS232 to output the actual position coordinates of the paver.

When both the navigation paving switch and the closed-loop switch are in the off position, the closed-loop control and the deviation correction control are not involved in the paver speed control. In this case, the control system calculates the running speed setting values of the left and right wheels according to the speed knob of the paver, and the setting values of the left and right wheels directly control the electromagnetic proportional valves of the variable pumps (Figure 1). When the closed-loop switch is closed, the control system is in the closed-loop state. The output of the PID controller and the speed set value are added together to jointly control the electromagnetic proportional valve of the variable displacement pump (Figure 1), and further control the speed of the hydraulic motor. When the navigation paving switch is closed, the coordinates provided by the RTK mobile station and the predetermined trajectory line are input to the lateral deviation calculation module to obtain the lateral deviation of the paver, which is input to the right wheel navigation correction PID algorithm. Further, the correction value of the right wheel is obtained after the navigation correction PID algorithm operation. The correction value and the right wheel speed setting value are combined to form a new right wheel speed setting value of the paver. The output of the right wheel motor speed sensor is fed back to the input port of the right wheel PID algorithm module, and compared with the new set value of the running speed of the right wheel to form a closed-loop control. When the navigation paving switch is disconnected, the deviation correction control will not participate in the right wheel travelling control. Therefore, the paver travelling control has the following three states:

(1) Open-loop control: when the paver is in the open-loop state, the left and right wheel speed setting values are calculated according to the speed knob on the panel, and then the left and right wheel speed setting values directly control the displacement pump (Figure 1);
(2) Closed-loop constant speed control: when the paving operation requires closed-loop control, the set values of the left and right wheels are added to their PID controller outputs to control the variable displacement pump;
(3) Navigation correction control: when the paver is in the correction control state, the right wheel navigation correction algorithm module performs the calculation according to the principle that the lateral deviation tends to zero, and the algorithm module outputs the correction value of the right wheel, which is combined with the right wheel speed setting value to form a new setting value, so as to realize the deviation correction control of the paver.

To sum up, in this scheme, when travelling at high speed, the paver does not need constant speed control and deviation correction control. In this case, open-loop control is generally selected. When the paver is in usual paving operation, the constant speed closed-loop control is selected. When the paver is in a high-quality operation situation that requires constant speed and deviation correction control, the deviation correction control can be selected. The scheme can easily realize the conversion of open-loop control, closed-loop control, and navigation correction control, which is convenient to improve the operating efficiency of the controller. When the PID controller participates in the speed control of the paver, it means that the PID controller is limited to a small range. It effectively prevents the influence on the travelling speed of the paver when the overshoot of the control system is too large, and effectively ensures the speed control accuracy of the entire control system. When the paver uses navigation correction paving, the accuracy of constant speed control can be further improved according to the actual driving speed of the left and right wheels. In addition, the RTK GNSS mobile station is used to rectify the wandering deviation of the paver in real time, thus effectively ensuring that the paver travels along the preset travelling track.

2.2. Calculation Method of Paver Speed Setting Value

In this scheme, the controller calculates the speed setting values of the left and right wheels according to the speed setting knob A and steering setting knob B on the control panel, as shown in Figure 3. When the paver is paving along a straight line, the steering setting knob B is in the middle position, and the speed settings of the left and right wheels are equal, that is,

$$V_L = V_R = V \tag{1}$$

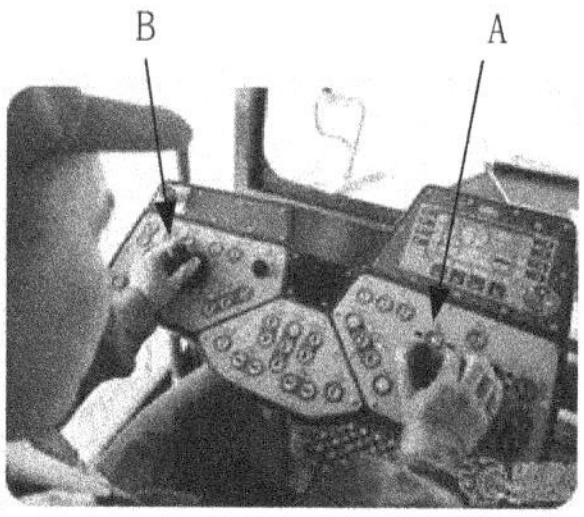

Figure 3. Control panel of paver. A—Speed setting knob, and B—Steering setting knob.

According to the forward/backward switch on the control panel, the paver can move forward or backward along a straight line.

When the steering setting knob B deviates to the left or right from the middle position, the paver will be in a turn paving. It is assumed that the running speed of the paver center set by the current speed setting knob A is V.

When the paver turns left,

$$V_L < V_R \tag{2}$$

where

$$V_L = V - \delta V, \tag{3}$$

$$V_R = V + \delta V \tag{4}$$

If the paver turns to the right,

$$V_L = V + \delta V, \tag{5}$$

$$V_R = V - \delta V \tag{6}$$

$$\delta V = \frac{B_X V}{a B_{MAX}} \tag{7}$$

where

V—Paver center speed;
V_L—Paver left wheel speed;
V_R—Paver right wheel speed;
δV—Speed difference between wheel and paver center;
B_X—Actual value by steering setting knob B;
B_{MAX}—Maximum value of steering setting knob B;
a—Paver turning coefficient, a = 2 in this paper.

3. Design of New PID Controller

3.1. Speed Closed-Loop PID Design

The PID controller can performs proportional, integral and differential operations on deviation e(t).

The expression is:

$$u(t) = K_P\left[e(t) + \frac{1}{T_i}\int_0^t e(t)dt + T_D\frac{de(t)}{dt}\right] \tag{8}$$

where

K_P—Proportional coefficient;

T_i—Integral action time;

T_D—Differential action time;

Discretize Equation (8),

$$u(k) = K_P\left[e(k) + \frac{T}{T_i}\sum_{i=0}^{k} e(i) + \frac{T_D}{T}(e(k) - e(k-1))\right] \tag{9}$$

$$u(k-1) = K_P\left[e(k-1) + \frac{T}{T_i}\sum_{i=0}^{k-1} e(i) + \frac{T_D}{T}(e(k-1) - e(k-2))\right] \tag{10}$$

By subtracting Equation (10) from Equation (9), the discrete PID control increment expression can be obtained as:

$$\Delta u(k) = K_P[e(k) - e(k-1)] + K_i e(k) + K_d[e(k) - 2e(k-1) + e(k-2)] \tag{11}$$

where

K_P—Proportional coefficient;

K_i—Integral coefficient;

K_d—Differential coefficient.

According to this paper, the actual assignment range of control board PWM is 0~65,535, and the output of speed sensor is pulse signal of 0~1000 pulse/s. After A/D conversion, the output signal of lift sensor is 0~1024.

The preliminary assignment method of PID controller parameters in this paper is as follows:

$$K_P = \frac{65535}{1024} = 64$$

Accaccording to experience,

$$K_i = \frac{K_d}{4} = 1.6$$

$$K_d = \frac{K_P}{10} = 6.4$$

3.2. Deviation Correction PD Controller Design

The differential control is adopted for the deviation correction controller, and the expression is:

$$V(t) = K_{PC}\left[T_{DC}\frac{del(t)}{dt}\right] \tag{12}$$

where

K_{PC}—Proportional coefficient of deviation correction control;

T_{DC}—Differential time constant of deviation correction control;

$V(t)$—Right wheel speed correction value of paver;

$el(t)$—Distance difference between left and right wheels of paver.

The expression of $el(t)$ is

$$el(t) = K_{bC}b(t) \tag{13}$$

where

$b(t)$—lateral deviation, the lateral distance between the paver center position and the preset travelling line.

Substitute Equation (13) into Equation (12),

$$V(t) = K_{PC}K_{bC}\left[T_{DC}\frac{db(t)}{dt}\right] \tag{14}$$

Discdiscretize Equation (14),

$$V(k) = K_{bC}K_{PC}\left[\frac{T_{DC}}{T}(b(k) - b(k-1))\right],$$

$$V(k-1) = K_{bC}K_{PC}\left[\frac{T_{DC}}{T}(b(k-1) - b(k-2))\right],$$

The inincremental expression of the deviation correction controller is obtained as follows:

$$\Delta V(k) = K_{DC}[b(k) - 2b(k-1) + b(k-2)] \tag{15}$$

where

K_{DC}—differential coefficient, $K_{DC} = K_{bC} \times K_{PC}$.

According to the Chinese National Standard GB/T16277-2021 for pavers, the maximum deviation of the paver 50 m is less than 50 cm without the driver's control. Therefore, the actual value range of the lateral deviation in this paper is 0–50 cm. In order to prevent large fluctuations in the speed of the right wheel from affecting the stability of the paver's speed, the correction value of the paver's right wheel speed is limited to 10% of the actual travelling speed of the right wheel in this paper. According to the maximum paving speed of 24 m/min, the output of the speed sensor is within 0~1024 pulses/s. Therefore, the maximum correction value of the right wheel speed is not more than 100 pulses/s. The preliminary assignment method of the parameters of the deviation correction controller in this paper is as follows:

$$K_{DC} = \frac{100}{50} = 2$$

4. Test

4.1. Materials and Methods

4.1.1. Test Site and Equipment

The test site and equipment were as follows:

The runway was dry and hard gravel pavement. The straight distance of the runway was more than 250 m. The curved part of the runway was more than 100 m. The width of the runway was more than twice the width of the test paver. The longitudinal gradient of the runway was not more than 1%, and the transverse gradient of the runway was not more than 1.5%.

The paver RP951 manufactured by Xuzhou Construction Machinery Group in Xuzhou, China was selected as the test object. This paver was driven by a closed hydraulic system with two pumps and two motors. The two electric proportional variable pumps A4VG40EP manufactured by Rexroth in Germany. and the two electric variable displacement motors A6VE80EZ manufactured by Rexroth in Germany. were configured to form two independent closed hydraulic systems.

4.1.2. Test Data Collection Method

In order not to affect the operation of the paver, the left and right wheel motor speed sensor pulse signals are also connected to another EPEC2023 controller manufactured by

Epec Oyin Seinajoki, Finland, which is responsible for collecting the speed sensor data of the hydraulic motor of the paver and uploading the collected speed pulse signals to the notebook through the CAN bus.

4.1.3. Constant Speed Control Test

(1) Test of constant speed control when paver travelling straight

The first step was to connect the left wheel and right wheel motor speed sensors of the paver RP951 to the data acquisition system according to the method described in 4.1.2. Then, the second step was to download the software to the original paver controller and power on the data acquisition system to collect data. Further, the travelling speed of the test paver was set at about 5 m/min. Then, the paver was started and driven along a straight line for several minutes. The third step was to replace the controller with the new control scheme and repeat the above steps.

(2) Test of constant speed control when paver turning

The first step was to start the paver. The speed of the paver was set at about 5 m/min. The second step was to quickly turn the steering knob to the maximum position and power on the data acquisition system to collect data. The paver turned automatically and drove for about 50 m. The third step was to replace the controller with a new scheme and repeat the above steps.

Further, the speed was set at about 5 m/min, and then the steering knob was quickly turned to 1/2 of the maximum position, and the above turning test of paver was repeated.

(3) Test of the influence on the low speed stability of the paver when the driving resistance changes

A forklift was placed about 3 m in front of the test paver. The forklift weighed 3 tons and carried 3 tons of iron blocks. The iron blocks were placed on the ground; at the same time, the forklift was braked to increase the friction. The driving resistance of the paver would change when the paver hit the forklift. The second step was to start the test paver and set its speed at about 5 m/min and power on the data acquisition system to collect data. Further, the third step was to replace the controller with a new scheme and repeat the above steps.

(4) Test of low speed stability of paver at different speeds

We kept the paver running straight and powered on the data acquisition system to collect data. In order to test the constant speed performance of the paver with the new control scheme during the driving of the paver, the driving speed of the paver was changed in the order of 2-4-6-5-1 m/min. At the same time, the data of the speed sensors were recorded.

4.1.4. Test of Deviation Correction Control

The test distance of paver's straight-line travelling wandering deviation is 50 m according to the Chinese National Standard GB/T16277-2021 for pavers. A starting line, a terminal line, and a lane-marking line were drawn on the test road. The test paver entered from the starting line. The direction of the paver was adjusted to make the longitudinal centerline of the paver coincide with the lane-marking line as much as possible. The paver ran at normal paving speed and passed the test area without driver intervention. A water dripping device was fixed on the paver, and the track formed by water droplets on the road was used as the travelling track line of the paver. The lateral distance between the track line and the lane marking line was measured every 3 m.

Figure 2b and the controller with the original control scheme shown in Figure 2a were tested on XCMG RP951 paver. The lateral distance between the actual travelling line of the paver and the lane marking line was measured and recorded every 3 m; the measured data are shown in Figure 4, where the horizontal ordinate is the distance between the actual position of the paver and the starting line, and the ordinate is the lateral distance between the actual position of the paver and the lane marking line. In order to reduce the error, the initial track extension line was used as the preset travelling line for the paver travelling straight. The deviation value of the paver is equal to the difference between its actual travelling track line and the preset travelling track line in the same coordinate system.

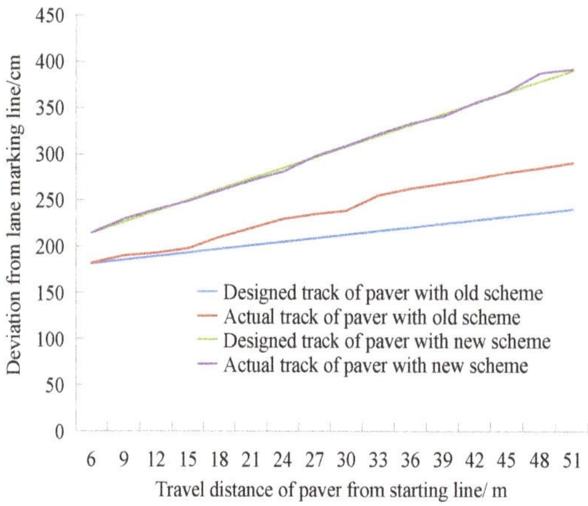

Figure 4. Wandering deviation correction test of paver with new scheme and old scheme.

4.2. Test Results and Discussion

As shown in Figure 4. The paver straight driving wandering with a new deviation correction control scheme was about 2.4 cm when it travelled 50 m. On the contrary, that of the paver with the original control scheme was about 89 cm. In the original control scheme, the straight driving wandering of the paver occurred in two stages. Firstly, at the beginning, the main factors were the difference in the minimum current of the left and right wheel solenoid valves, the difference in the pump and motor of the left and right wheels under small hydraulic oil flow, the difference in the driving resistance of the left and right wheels, and the difference in the wheel slip rate of the left and right wheels. Secondly, during constant speed driving, the main influencing factors were the different resistance and the different wheel slip rates of the left and right wheels. Especially in the beginning, once deflection is formed, the error will be amplified after travelling a distance. On the contrary, in the new navigation correction control scheme, the influencing factors are included in the deviation correction control where the lateral deviation tends to zero, and are finally eliminated by modifying the speed setting value of the right wheel.

In addition, from the test results, the wandering deviation of the paver with the new navigation correction control scheme was about 2.4 cm, on the contrary, that of the paver with the original control scheme reached about 89 cm when driving straight. The main reason was that when the paver with the new scheme used the navigation correction control, on the one hand, the constant speed control was based on the speed of the left wheel; on the other hand, according to the centimeter level positioning accuracy of the RTK GNSS receiver manufactured by Huace Navigation Co., Ltd. in Shanghai, China, the right wheel speed setting value could be slightly adjusted. This not only ensures the realization of constant speed control and navigation correction control, but also avoids their mutual interference.

The constant speed test results of the paver are shown in Figures 5–8. The ordinate is the speed of the left and right hydraulic motors of the paver. The speed unit is the number of pulses output by the speed sensor per second, and the horizontal coordinate is the paver travelling time.

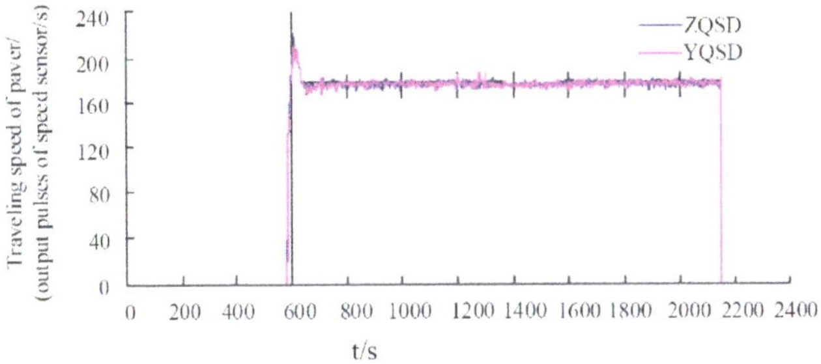

(a)

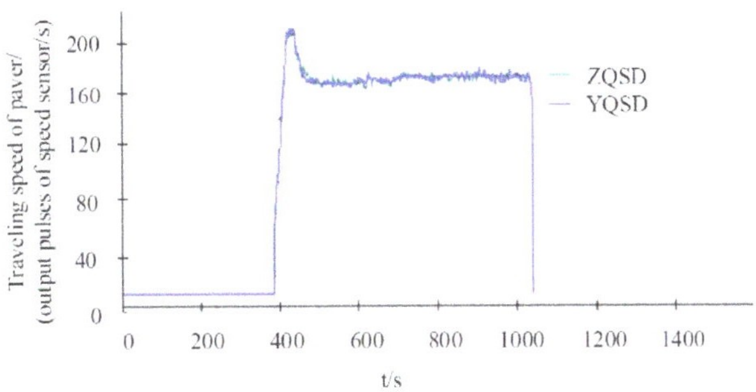

(b)

Figure 5. Constant speed test of paver when driving straight. (**a**) Constant speed test of the paver with the new control scheme when driving straight. (**b**) Constant speed test of the paver with the old control scheme when driving straight. ZQSD—Speed of the left wheel, and YQSD—speed of the right wheel.

The test results of the paver travelling straight are shown in Figure 5. The test speed of the paver was about 5 m/min, which is the usual paving speed (at this speed, the speed sensor outputs about 180 pulses/s). As shown in Figure 5a, under the control system with the new scheme, the average travelling speed was 180 pulses/s, the whole process was stable without change, and the left and right wheel speeds were almost equal. The control accuracy can be expressed by the speed fluctuation, which was basically kept within $\pm$10 pulses/s. When the step signal is input, the peak speed of the left wheel can reach 238 pulses/s, and the overshoot can reach 32.2%, but the peak speed of the right wheel can reach 210 pulses/s, and the overshoot can reach 16.7%. As shown in Figure 5b, under the control system with the original scheme, the average driving speed was 170 pulses/s, the average value of the whole process changed slowly, the speed of left and right wheels was almost equal, and the speed fluctuation was basically kept within $\pm$ 10 pulses/s. When the step signal is input, the peak speed can reach 200 pulses/s, and the overshoot can reach 18%; by comparing the data of the new and old schemes, the new scheme adopted the control method of open loop and PID closed loop in parallel. The PID control function becomes smaller, which can reduce the overshoot of the control system and keep the average speed stable.

The test results of the paver turning are shown in Figure 6. The running speed of the paver center was about 5 m/min. As shown in Figure 6a, under the control of the new scheme, the speed of the left and right wheels changed rapidly after the paver steering knob on the control panel was quickly set to the maximum position, the speed of the left wheel decreased, and the speed of the right wheel increased. The speed curves of the left wheel and the right wheel are almost symmetrical, which is consistent with the differential

Formula (4) and (5). In this process, the right wheel speed had an overshoot and the peak value reached 420 pulses/s. The overshoot was about 16.7%. The average value was stable at 360 pulses/s, which meant that the right wheel of the paver was equal to 9 m/min, and the speed fluctuation was basically within ±10 pulses/s. There was an overshoot at the beginning of the speed change of the left wheel. The minimum speed was close to 0, and the average value was stable at 37 pulses/s (this speed is equivalent to the left wheel speed of 1 m/min). The speed fluctuation could be kept within ±10 pulses/s. However, it also indicates that when the speed was less than 1 m/min, the speed fluctuation had a large proportion in the average speed.

As shown in Figure 6b, under the control of the original scheme, after the paver steering knob on the control panel was also quickly set to the maximum position, the right wheel speed had an overshoot, which was significantly greater than the overshoot shown in Figure 5a; the average speed was about 360 pulses/s, and the speed fluctuation was also within ± 10 pulses/s. The speed curve of the left wheel was basically similar to that of the left wheel in Figure 5a.

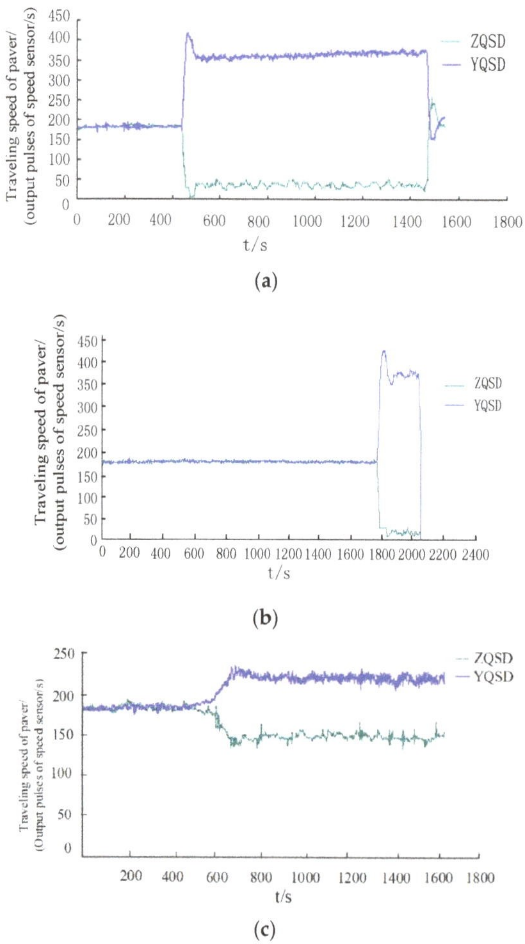

Figure 6. *Cont.*

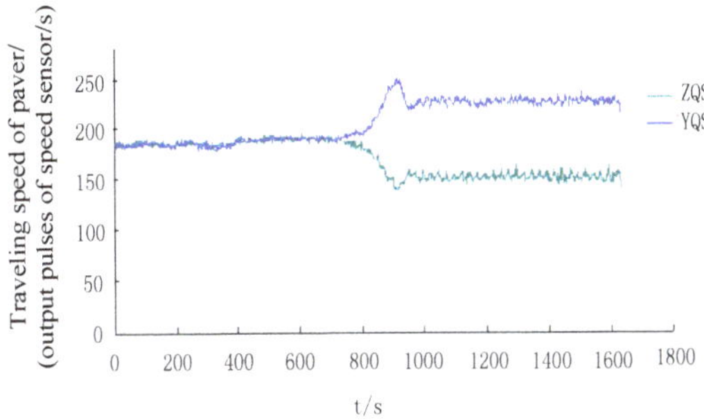

(d)

Figure 6. Constant speed test of paver when turning. (**a**) Constant speed test of the paver with a new control scheme when turning (maximum position of steering knob). (**b**) Constant speed test of the paver with the old control scheme when turning (maximum position of steering knob). (**c**) Constant speed test of the paver with the new control scheme when turning, (the steering knob is turned to 1/2 of the maximum position). (**d**) Constant speed test of the paver with the old control scheme when turning (the steering knob is turned to 1/2 of the maximum position). ZQSD—Speed of the left wheel, YQSD—speed of the right wheel.

As shown in Figure 6c, under the control of the new scheme, the speed of the left and right wheels changed rapidly after the steering knob was quickly set to 1/2 of the maximum position. Similarly, the speed of the left wheel decreased and the speed of the right wheel increased. Moreover, the speed changing curves of the left and right wheels are also symmetrical, which also conforms to the differential Formulas (4) and (5). At the same time, in Figure 5c, it can be observed that there was no overshoot of the speed of the left and right wheels, and the speed fluctuation was within $\pm$ 10 pulses/s. The average speed of the right wheel was stable at 215 pulses/s, which was equivalent to about 6 m/min. The average speed of the left wheel was stable at 145 pulses/s, which was equivalent to about 4 m/min.

As shown in Figure 6d, under the control of the original scheme, the steering knob was quickly set to 1/2 of the maximum position, and the speed of the right wheel had an overshoot. The peak speed reached 250 pulses/s, and the overshoot was about 4.3%. The average speed of the right wheel was stable at 230 pulses/s, which was about 6 m/min, and the speed fluctuation was within $\pm$10 pulses/s. At the same time, after an overshoot of the left wheel speed, the peak value reached 140 pulses/s downward, the overshoot was about 3.7%, and the average value was stable at 150 pulses/s, which was equivalent to about 4 m/min. By comparing the new and old control schemes, under the control of the new scheme, the overshoot was smaller and the speed was more stable when the paver was turning.

Because the dumping truck in front of the paver is often replaced, the driving resistance of the paver will change. The simulation test results, from when the driving resistance of the paver changes, are shown in the Figure 7. The travelling speed was set at about 5 m/min. As shown in Figure 7a, when the paver encountered obstacles, the speed decreased instantaneously by about 10 pulses/s, and then quickly returned to the set value.

As shown in Figure 7b, the paver was in a slight turning state, the left wheel speed was slightly higher than the right wheel speed, and the left and right wheel speed fluctuations were within $\pm$10 pulses/s. It can be seen from the data in Figure 6b that the paver hit an obstacle after driving for about 700 s. At this time, the average speed of the left and right wheels decreased significantly by about 5 pulses/s, and the speed fluctuation of the left and right wheels did not change significantly. The reasons are as follows: Firstly, the new scheme combined the PID closed loop and knob direct control in parallel. Secondly, in the new scheme, the reason why the PID closed loop did not change the output is that the average speed decreased by 5 pulses/S and was still within the dead zone, so that the control system did not change the output. The original scheme adopted full PID closed-loop control, and the speed was reduced by 10 pulses/s, which exceeded the dead zone of closed-loop control. The PID closed loop worked so that the paver speed could quickly return to the set value.

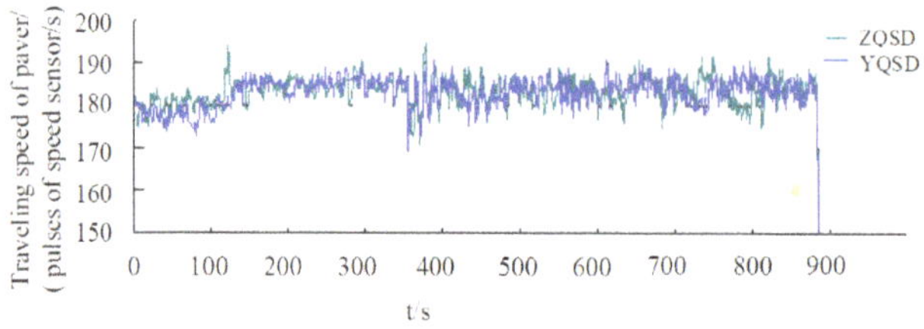

(a)

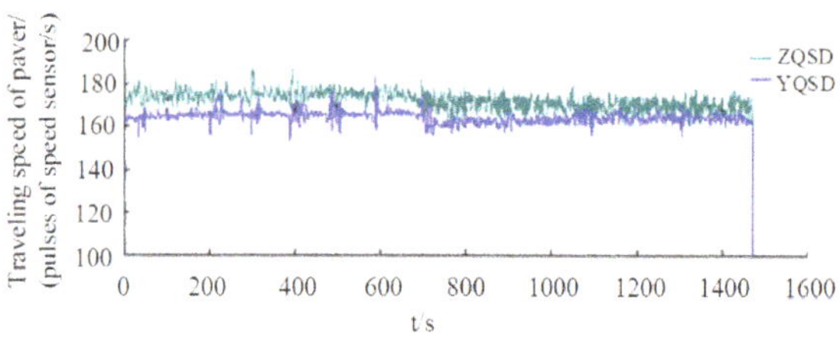

(b)

Figure 7. Constant speed test of paver when driving resistance changes. (**a**) Constant speed test of the paver with a new control scheme when driving resistance changes. (**b**) Constant speed test of the paver with the old control scheme when driving resistance changes. ZQSD—Speed of the left wheel, and YQSD—speed of the right wheel.

The test results, from when the paver was driving straight at different speeds, are shown in Figure 8. When the test paver ran at a speed of more than 2 m/min, which was equivalent to the average of 75 pulses/s, the test results show that the average speeds of the left and right wheel were almost the same, and the speed fluctuation was less than ±10 pulses/s. When the test paver ran at a speed of less than 1 m/min (which was equivalent to the average of 37 pulses/s), the speed fluctuation of the right wheel was significantly greater than that of the left wheel, and the speed fluctuation of the right wheel could reach ±10 pulses/s. The main reason was that there were some manufacturing errors in the hydraulic components of the test paver, which made it impossible for them to be completely consistent. Especially in the case of low speed and small flow, the proportion of hydraulic system leakage gradually increases.

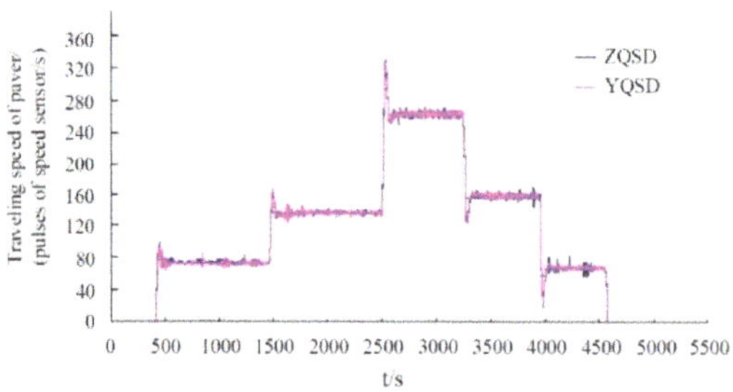

Figure 8. Constant speed test of the paver with the new control scheme when speed changes. ZQSD—Speed of the left wheel, and YQSD—speed of the right wheel.

5. Conclusions

Aiming at the advanced applications of full hydraulic pavers, such as constant speed control, deviation correction control, automatic driving, etc., a set of paver travelling control schemes based on the Global Navigation Satellite System was provided, which effectively solved the problem that the paver easily deviates from the pre-arranged route. In this scheme, when the paver deviated from the pre-arranged route, the right wheel was slightly adjusted to realize the real-time deviation correction control. At the same time, a travelling control method using open-loop control and closed-loop PID control in parallel was proposed, and the constant speed performance and deviation correction performance of the hydraulic drive paver under the new control scheme were studied by means of experiments. The conclusions are as follows:

(1) When the test paver with the new control scheme ran at a speed of more than 2 m/min, the test results show that the average speed was almost the same, and the speed fluctuation was less than ± 10 pulses/s. The higher the average paver speed was, the smaller the proportion of speed fluctuation was, and the more stable the speed was.
(2) When the test paver was less than 1 m/min, its speed fluctuation tended to increase, and the speed fluctuation reached ± 15 pulses/s. The constant speed of the paver can no longer be guaranteed.
(3) When the paver encountered a movable obstacle, the average speed of the left and right wheels decreased significantly, and the fluctuation of the speed of the left and right wheels did not change significantly. When the running resistance changes, it has a certain impact on the constant speed of the paver, and when the running resistance remains unchanged, it has no impact on the constant speed performance of the test paver.
(4) The travelling system of the paver with satellite navigation could effectively reduce the paver straight-line driving wandering. The wandering deviation of the paver could be controlled within 2.5 cm, and the fluctuation of low constant speed control was less than ± 10 pulses/s. The next work will further study the PID parameter matching of the paver travelling control.

Author Contributions: X.Y.: Conceptualization, supervision, funding acquisition, writing—original draft, writing—review and editing. S.Y.: investigation, validation, data curation, writing—original draft, writing—review and editing. H.Z.: investigation, data curation, writing—review and editing. Z.Z.: investigation, data curation, and writing—review and editing. All authors have read and agreed to the published version of the manuscript.

Funding: This research received no external funding.

Data Availability Statement: The data are not publicly available due to the privacy restrictions.

Conflicts of Interest: The authors declare no conflict of interest.

References

1. Tang, X.X.; Zhao, D.X.; Huang, H.D. Three-stage with geometric ratios hydrostatic-mechanical compound transmission for construction vehicle. *J. Jilin Univ.* **2006**, *36*, 56–61.
2. Shi, C.D.; Zhang, Y.C. Research on speed control of hydraulic drive spindle based on fuzzy adaptive PID controller. *Chin. J. Constr. Mach.* **2020**, *18*, 516–520.
3. Liu, T.; Zhang, Y. Research on Neural Network PID Control in Speed Control System of Hydraulic Pump Motor. *Commun. Power Technol.* **2018**, *35*, 4–7.
4. Huang, Y.J.; Wang, X.; Wang, Y.Y.; Zhao, Y. Dynamic Response Analysis of Driving System of 565 kW Hydraulic Bulldozer. *Chin. Hydraul. Pneum.* **2021**, *45*, 116–121.
5. Jiang, Z.H.; Xia, C.G. Study on Characteristic of Hydrostatic Transmission System of Tractor. *J. Agric. Mech. Res.* **2021**, *43*, 249–254.
6. Jiao, S.J.; Hao, P.; Long, S.G. Study of the velocity of asphalt paver. *China J. Highw. Transp.* **2003**, *16*, 124–126.
7. Bai, D.P.; Ding, J.; He, P.Y. Research on Intelligent Control Technique of the Paver's travelling System. *Mach. Tool Hydraul.* **2014**, *42*, 55–57.

8. Han, S.J.; Zhao, D.X. Energy saving strategy for automatic transmission of engineering pavers based on four-parameters. *J. Jiangsu Univ.* **2006**, *27*, 505–508.
9. Zhao, D.X.; Cui, G.J.; Li, D.B. Shift quality of transmission system for construction paver. *J. Jiangsu Univ.* **2008**, *29*, 386–389.

processes

MDPI

Article

Deep Transfer Learning for Approximate Model Predictive Control

Samuel Arce Munoz, Junho Park, Cristina M. Stewart, Adam M. Martin and John D. Hedengren *

Department of Chemical Engineering, Brigham Young University, Provo, UT 84602, USA
* Correspondence: john_hedengren@byu.edu

Abstract: Transfer learning is a machine learning technique that takes a pre-trained model that has already been trained on a related task, and adapts it for use on a new, related task. This is particularly useful in the context of model predictive control (MPC), where deep transfer learning is used to improve the training of the MPC by leveraging the knowledge gained from related controllers. One way in which transfer learning is applied in the context of MPC is by using a pre-trained deep learning model of the MPC, and then fine-tuning the controller training for a new process automation task. This is similar to how an equipment operator quickly learns to manually control a new processing unit because of related skills learned from controlling the prior unit. This reduces the amount of data required to train the approximate MPC controller, and also improves the performance on the target system. Additionally, learning the MPC actions alleviates the computational burden of online optimization calculations, although this approach is limited to learning from systems where an MPC has already been developed. The paper reviews approximate MPC formulations with a case study that illustrates the use of neural networks and transfer learning to create a multiple-input multiple-output (MIMO) approximate MPC. The performance of the resulting controller is similar to that of a controller trained on an existing MPC, but it requires less than a quarter of the target system data for training. The main contributions of this paper are a summary survey of approximate MPC formulations and a motivating case study that includes a discussion of future development work in this area. The case study presents an example of using neural networks and transfer learning to create a MIMO approximate MPC and discusses the potential for further research and development in this area. Overall, the goal of this paper is to provide an overview of the current state of research in approximate MPC, as well as to inspire and guide future work in transfer learning.

Keywords: deep learning; transfer learning; model predictive control; approximate model predictive control

1. Introduction

Model predictive control (MPC) requires the development of a dynamical mathematical model and the solution of optimization subroutines in real time. These requirements represent a challenge because industrial processes are often too complex to be modeled from first principles, and first-principles models, when available, can be computationally too expensive to be solved in real time. System identification through statistical methods and historical data can expedite the development of a dynamic model when a first-principles formulation is not available [1]. While a linear model can often be solved in real time, many industrial processes exhibit nonlinear dynamics, and linear MPC becomes insufficient. This has motivated the use of deep neural networks as a non-linear method for process control.

Deep-learning architectures designed to handle sequential data, such as recurrent neural networks (RNNs), long short-term memory (LSTM) networks, and transformers, are responsible for remarkable results in the tasks of natural language processing [2], speech recognition [3], and, more recently, protein folding [4]. These networks are also being used for system identification and model predictive control of evaporation processes [5], pharmaceutical manufacturing [6] and reactor control [7]. Although deep neural networks are capable of capturing complex dynamics, they represent a challenge when they are used to calculate optimal control actions. As noted by Schweidtmann et al. [8], the consideration of complex data-driven models has long been limited to local or stochastic solution approaches [9–11], or fit-for-purpose solutions limited to low-dimensional problems and state-of-the art solvers [12,13]. While these approaches focus on using the deep neural network model as part of the optimization routine, other researchers have focused on using the deep learning models to generate explicit, or approximate MPC formulations [14–17]. These architectures, however, depend on the development of an MPC in the first place so that enough training data are generated.

This study contributes to the development of improved approximate MPC formulations by using deep learning models and transfer learning, a research area in the field of machine learning that aims to transfer the knowledge learned from one problem to a similar problem.

The contributions of this paper are as follows:

- A non-comprehensive survey of the work done on approximate MPC formulations.
- A case study on a multiple-input multiple-output (MIMO) system to show how deep learning and transfer learning can be used to develop an approximate MPC controller.

1.1. Previous Work on Approximate MPC Formulations

Table 1 summarizes a non-comprehensive survey on approximate MPC formulations. A large portion of the research done in this area has focused on decreasing the computational cost of MPC formulations, applied to equivalent systems. Alessio and Bemporad [18], for example, summarize different approaches to solving the MPC optimization problem offline for a given range of operating conditions applied on a system of interest, focusing mostly on piecewise affine functions. Alessio and Bemporad [18] also provide a novel algorithm that reduces the complexity of explicit MPC. In their approach, extensive off-line simulations of closed-loop MPC collect the most recurrent combinations of active constraints and generate critical regions. These critical regions are used to calculate or extrapolate a suboptimal input move in real time.

Table 1. A non-comprehensive survey on explicit MPC, MPC emulators and approximate MPC.

Articles	MPC Informed for Equivalent Systems	Equivalent System with Neural Network	Transfer Learning for Similar Systems	Transfer Learning for Unique Systems
Explicit MPC: 329 Articles, 22 with Neural Networks or Deep Learning				
Tøndel et al. [19]	x			
Alessio and Bemporad [18]	x			
Csekő et al. [20]	x	x		
Grosso et al. [21]	x	x		
Katz et al. [22]	x	x		
Chen et al. [23]	x	x		
MPC Emulation: 357 Articles, 10 with Neural Networks or Deep Learning				
Zheng et al. [24]	x			
Moness and Moustafa [25]	x			
Wang et al. [26]	x	x		
Yan and Wang [27]	x	x		
Novak and Dragicevic [28]	x	x		

Table 1. *Cont.*

Articles	MPC Informed for Equivalent Systems	Equivalent System with Neural Network	Transfer Learning for Similar Systems	Transfer Learning for Unique Systems
Approximate MPC: 24 Articles, 12 with Transfer Learning				
Hofer et al. [29]	x			
Pin et al. [30]	x	x		
Yang et al. [16]	x	x		
Gan et al. [31]	x	x		
Wang et al. [32]		x	x	
Wang and Hong [33]	x	x	x	
Chen et al. [34]	x	x	x	
Han et al. [35]	x	x	x	
Chen et al. [36]	x	x	x	
This study	x	x	x	x

Grosso et al. [21] explore the use of neural networks to analyze complex drinking water networks (DWNs). In their approach, the authors provide a novel quasi-explicit MPC formulation, where heavy computational tasks are replaced by non-linear modules using neural networks, and the prediction horizon and weight matrices of the MPC are continuously adjusted based on a trade-off between risk and economic cost. This allows for the effective scaling of MPC to large, multi-objective applications. In the work of Tøndel et al. [19], polyhedral partitions of the state space are used as a novel approach to improve the efficiency of explicit solutions, avoiding unnecessary partitioning and QP problem solving. Similarly, Csekő et al. [20] propose the use of radial basis function (RBF)-based neural networks as an approximation of the MPC controller and compare that to a Kalman filter applied to an explicit MPC, showing that the neural network approximation can provide fast control but at the expense of deteriorated performance. Additionally, Katz et al. [22] show a novel approach where deep neural networks approximate the non-linear parts of a multiparametric mixed integer linear programming (mpMILP), making the formulation tractable while maintaining the linear aspects components of the optimization problem, such as box constraints.

Other examples include the work of Moness and Moustafa [25], where a switched model predictive controller (SMPC) is used in a physical wind turbine emulator, and the work of Zhu et al. [37], where an improved finite-set MPC is proposed to achieve simplified control structure and faster dynamic response in an islanded alternating current microgrid. In both cases, the novel use of simplified MPC architectures that rely on switching between control strategies from a pool of prediction models result in faster controller response. Similarly, Novak and Dragicevic [28] use a finite-set MPC formulation for power electronics, but with the introduction of a supervised learning approach to imitate the controller. This contribution further alleviates the computational burden of finite-set MPC formulations by introducing an approximate MPC that is independent of the non-linearity of the system.

Examples of deep learning-based approximations of MPC include the work of Wang et al. [26], where a machine learning (ML)-based emulation of an MPC is used for modular multilevel converters, and the work of Yan and Wang [27], where recurrent neural networks (RNNs) and a collective neurodynamic optimization approach are used to create non-linear model predictive control (NMPC) without linearization. The novel neurodynamic approach presented by [27] further augments the use of deep learning-based MPC formulations by improving the starting points of each RNN using the information of global and local best known solutions and a particle swarm optimization approach. This approach effectively combines global search and precise local search capabilities, and converges to global optimal solutions in simulation studies.Other examples include the work of Pin et al. [30], where a NN is used to capture the non-linearities of a discrete-time NMPC model with hard constraints for similar performance in offline control. Similarly,

Hofer et al. [29] manipulate an approximate MPC to mirror performance of a full MPC while lowering the computational cost of the onboard control of unmanned aerial vehicles. Their novel method uses a continuous time parametrization of the state and input trajectory to derive a compact description of the optimal control problem, which outperforms a linear quadratic regulation approach in simulation studies.

Regarding the use of transfer learning for approximate control formulations, an offline approximate MPC from Gan et al. [31] controls a hybrid energy ship using a random forest model, performing nearly as well as the MPC policy used to train the model. Han et al. [35] build a knowledge-data-driven MPC using fuzzy NN for waste-water treatment with transfer learning mechanisms to adapt to similar systems. Their novel transfer learning mechanism improves the control performance by integrating state information of the current model with historical model data to offset the data shortage in the learning process. This transfer of knowledge of model parameters has the potential to benefit other water-related applications where deep learning has shown promising results but collecting data is difficult, such as intermittent streamflow predictions in arid regions [38–40].

Similarly, Wang and Hong [33] offer a comprehensive review of building control using reinforcement learning and transfer learning to imitate the performance of MPC. This contributions are valuable in applications where MPC has proven very useful but the expertise required for each application limits its scalability [41]. In a successful example, Chen et al. [34] inform control of a room in Shanghai with limited data using layers from a NN model trained intensively on a room in Beijing. In their approach, a deep neural network with more than 40,000 trainable parameters was pre-trained with multi-year data from the room in Beijing and re-trained with only 15 days of data from the room in Shanghai, resulting in high-quality predictions of the air temperature and relative humidity of the room in Shangai. In other publications, Yang et al. [16] train a RNN with a nonlinear autoregressive exogenous model structure on data from test rooms controlled with MPC, reporting significant improvement from original control, Chen et al. [36] captures MPC behavior for building control in a RL model which then informs models for other buildings. The performance of their learning application improves during operation, reinforced with additional data.

As described in this survey, the research performed on approximate MPC formulations has focused primarily on reducing the computational time of solving the optimization problem. An important aspect of this study is that it presents an approximate MPC that not only reduces the computational time of a traditional MPC formulation, but that it is also designed to learn from different representations of knowledge. This is similar to how an operator can quickly learn to manually control a new processing unit due to their prior experience and skills gained from controlling a similar unit. In other words, the operator's previous experience with a related task enables them to more easily adapt to and perform the new task. This concept is similar to how transfer learning in machine learning allows a model that has been trained on a related task to be adapted and fine-tuned for use on a new, related task, thereby reducing the amount of training data and improving performance. For example, deep learning facilitates learning from complex systems, such as the decision criteria followed by an operator that is manually controlling a process. Additionally, transfer learning facilitates learning from systems with similar dynamics, similar to how an operator can learn to control a new process by using prior knowledge.

1.2. Transfer Learning

Transfer learning is a research area in the field of machine learning that aims to transfer the knowledge learned from one problem to a similar problem. As noted by Yang et al. [42], transfer learning is rooted in AI, psychology, educational theory and cognitive science, and it has been addressed through frameworks, such as analogical problem solving [43,44] and high-level structural similarity in analogical reasoning [45]. Foundational work by Thrun and Pratt [46] introduces the term *learning to learn*, and proposes methods that can change the way machine learning algorithms generalize. These methods are compared to how humans learn to generalize from exposure to a continual stream of learning tasks. According to Thrun and Pratt [46], a machine learning algorithm is capable of *learning to learn* (i.e., generalize) if its performance at each task improves with experience and with the number of tasks, given a family of tasks, training experience for each of these tasks and a family of performance measures.

Transfer learning can be formally defined with the notations introduced by Pan and Yang [47]. A domain $\mathbb{D}$, consists of a feature space $\mathcal{X}$ and marginal probability distribution $\mathbb{P}^X$, where each input instance $\mathbf{x} \in \mathcal{X}$. Given a specific domain, $\mathbb{D} = \{\mathcal{X}, \mathbb{P}^X\}$, a task $\mathbb{T}$ has two components: a label space $\mathcal{Y}$ and a function $f(\cdot)$ (denoted by $\mathbb{T} = \{\mathcal{Y}, f(\cdot)\}$). The function $f(\cdot)$ is a function that can make predictions on unseen instances $\mathbf{x}^*$s.

In a two-domain scenario, there is one source domain $\mathbb{D}_s$ and one target domain $\mathbb{D}_t$, where the *source domain labeled data* are denoted by $\mathcal{D}s = \{(\mathbf{x}_{s_i}, y_{s_i}\}_{i=1}^{n_s}$, where $\mathbf{x}_{s_i} \in \mathcal{X}_s$ is the data instance and $y_{s_i} \in \mathcal{Y}_s$ is the corresponding class label. Similarly, $\mathcal{D}_t = \{(\mathbf{x}_{s_i}, y_{s_i}\}_{i=1}^{n_s}$ is denoted as the *target domain labeled data*, where the input $\mathbf{x}_{t_i} \in \mathcal{X}_t$ is the input, and $y_{t_i} \in \mathcal{Y}_t$.

Transfer learning has shown success in various fields of study. In the work of [48] a pre-trained network was found to best evaluate body posture as an indicator of sleep quality. Utilizing a transfer learning approach, researchers successfully identified four predefined sleep poses during 12 participants' overnight sleep episodes. When assessed against manually evaluated sleep positions, the pre-trained network using transfer learning outperformed other pre-trained networks. While the standard position sensor was estimated to have 88.2% accuracy, their deep learning video scoring stood at 95.1% accuracy.

Transfer learning has also been used to improve swimming pool safety by using computer vision to produce automated surveillance systems that might identify drowning objects, as applied to a motion detection sensor. Alotaibi [49] utilizes a specialized neural network based on a ResNet-50 model. Researchers successfully enhanced the performance of the proposed system, overcoming the limitations of a small available data size. While a DCNN learning model achieved 95% accuracy upon evaluation, the specialized model was found to have 99% accuracy.

In the field of process control, transfer learning has been used as an approximation of a system with complex dynamics where data are scarce. In the work of Chen et al. [34] for example, a neural network model with 42,902 parameters is used to learn the dynamics of natural ventilation in a building, based on multi-year data from a source system. The model is then retrained with only 200 trainable parameters and only 15 days of data from the target system, which has different floor area, building material, and window size. The resulting model shows a reduced mean squared error, almost one order of magnitude smaller than the models that are only trained on source data or target data. A similar approach is presented by Wang et al. [32], where a growing deep belief network (GDBN) network is used to identify the dynamics of a continuous stirred-tank reactor (CSTR). In their work, transfer learning is used to dynamically optimize the structure of a GDP, by using the pre-trained layers of an initialized deep belief network (DBN) to help the GDBN automatically determine its optimal size and achieve high performance in system identification.

Although both examples show promising results on the use of transfer learning for system identification, they still depend on an efficient MPC to control the system. This article shows a case study in which a deep neural network is used to identify the control policy of a source system, similar to the work discussed on deep learning-based explicit MPC, and deep learning-based MPC emulators, and deploy on a target system. Transfer learning, in this case, is used to enhance the performance of the approximate MPC on the target system, without the need for additional steps to identify the dynamics of the new system or solving an optimization subroutine to calculate the controller outputs.

As noted by Tan et al. [50], deep learning models are *data-dependent*, and they require large amounts of data for training and testing. Ideally, the training and test data must be independent and identically distributed (i.i.d) to accurately assess the predictive capabilities of the model. However, in many situations, it is difficult to collect large amounts of high quality data for deep learning. In these situations, transfer learning is attractive because it relaxes the assumption of i.i.d. training and data sets, suggesting that a model can be pre-trained on a training set and fine tuned on a target test set belonging to a different domain.

Tan et al. [50] further categorize transfer learning into the following:

- Instances-based;
- Mapping-based;
- Adversarial-based;
- Network-based.

An *instances-based* transfer learning approach takes instances from a source domain that can be used in a target domain by adjusting the associated weights. This approach has been successful in text classification applications [51].

In a *mapping-based* approach, the two domains are combined in a new data set and fed to a union network. Matasci et al. [52] use this approach to improve image classification in a remote sensing application.

An *adversarial-based* approach finds patterns that are applicable to the source and target domain, while discriminating between the source and target tasks. Li et al. [53] use this approach in a fault diagnosis application, where plant data are scarce but simulation data are easy to generate.

Finally, a *network-based* approach reuses parts of a network that has been pre-trained with source domain data and learns from the target domain data. As an example, Chen et al. [34] use this approach to train an MPC for smart-building control where training data are scarce but a model pre-trained on a similar building is available. This study presents a *network-based* approach, pre-training a neural network with data from a source system, and fine-tuning with data from a target system, as shown in Figure 1.

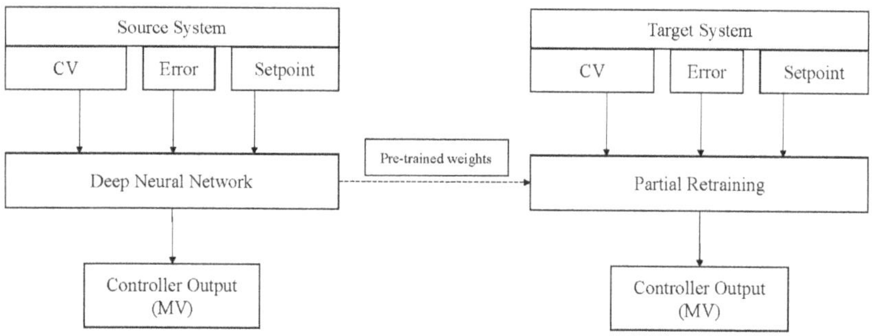

Figure 1. A network-based transfer-learning approach to develop an approximate MPC for a target system using data from a source system.

2. Materials and Methods

The source system in this study is a dual tank arrangement [54] (See Figure 2). The gravity-drained dual tank system uses pump speed and valve position to regulate the liquid level in an upper tank and a lower tank. The upper tank drains to the lower tank, and the lower tank drains to a reservoir. The control system adjusts the pump speed to return the liquid to the tanks and uses a valve to split the flow of liquid between the top and bottom tanks. The valve is adjusted by an MPC that regulates the liquid level in the tanks, ensuring that the tanks remain at a predetermined level. The liquid that flows into the upper tank cascades down to the lower tank. This creates a second-order response between the pump action and lower liquid tank level because of the dynamics of the first tank. The pump speed and valve position are adjusted to maintain the desired liquid levels in both the upper and lower tanks. This system uses the force of gravity and a combination of valve and pump to regulate the liquid level in the upper and lower tanks, ensuring that the tanks remain at their desired levels and that any excess liquid is properly drained to the reservoir. The system benefits from MPC regulation, similar to an application of model predictive control for improved water recovery and throughput stability for tailings reprocessing [55] that adjusts the inlet flow to maintain density and surge tank level.

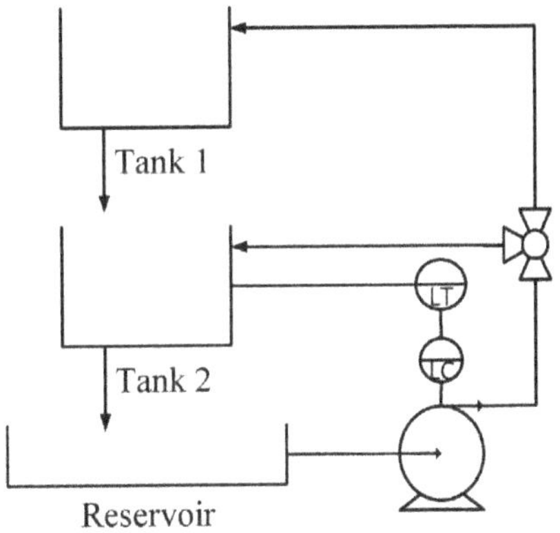

Figure 2. MIMO level control on a dual tank system. Adapted from [54].

The liquid level system is simulated using a first-principles model with Equations (1) and (2), where H_1 and H_2 are the levels (0–100%) of tank 1 and 2 respectively, c_1 and c_2 are inlet valve coefficient and tank outlet coefficient respectively, v corresponds to the valve output, and p corresponds to the pump output, both scaled from 0% to 100%. The same equations are used to generate the predictions used for MPC. Note that this MPC is intentionally formulated as an idealization. While this is not representative of an MPC in practice, it highlights the main goal of this transfer learning exercise, which is to obtain accurate control in the target system. Therefore, the design of the source system is considered part of a feature engineering exercise to gather fit-for-purpose data:

$$\frac{dH_1}{dt} = c_1(1-v)p - c_2\sqrt{H_1} \tag{1}$$

$$\frac{dH_2}{dt} = c_1 vp + c2\sqrt{H_1} - c2\sqrt{H_2} \tag{2}$$

The simulation and MPC were developed using GEKKO [56,57], a dynamic simulation and optimization suite for Python. The MPC developed used a squared error objective function and the level on both tanks are controlled by adjusting the speed of the pump (%) and the valve opening (%).

The target system in this study consists of a small scale temperature control unit [58], where the power input is regulated to heat up a small transistor to a desired temperature, as shown in Figure 3. The temperature control lab implements MPC with an Arduino, two heaters, and two temperature sensors to maintain desired temperature setpoints. The heaters transfer thermal energy to the temperature sensors through conduction and convection while also losing heat to the surroundings with radiative and convective heat transfer. This pocket-sized lab is a hardware benchmark [59] for model identification and controller development for testing new methods [60–62] and reinforcing existing methods in control theory through educational studies [58,63,64].

The dynamics of this system are described by Equations (3) and (4), where m, C_p and T are the mass, heat capacity and temperature of the transistor, U is an overall heat transfer coefficient, A is the surface area of the transistor, and T_∞ is the ambient temperature, with Q_1 being the power input to the transistor:

$$mC_p\frac{dT_1}{dt} = UA(T_\infty - T_1) + Q_1 \tag{3}$$

$$mC_p\frac{dT_2}{dt} = UA(T_\infty - T_2) + Q_2 \quad (4)$$

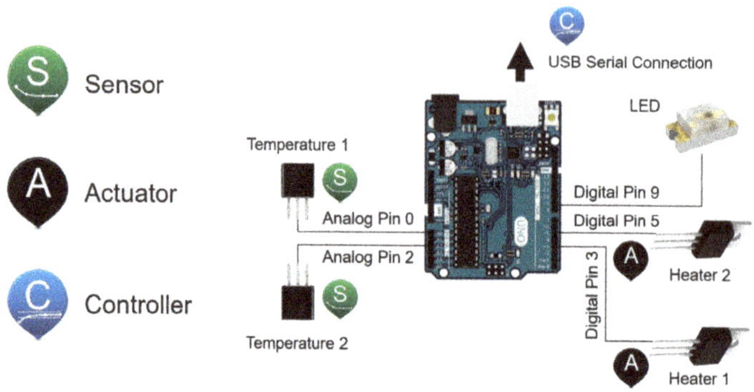

Figure 3. Small-scale temperature control unit [65].

An LSTM network developed in TensorFlow [66] is used to learn the control actions of the source and target systems by taking as inputs the measured quantities ($Y1$ and $Y2$), the setpoints ($Y1_{sp}$ and $Y2_{sp}$) and the controller error ($Y1 - Y1_{sp}$ and $Y2 - Y2_{sp}$). The outputs of the network ($\hat{U}1$ and $\hat{U}2$) are approximate control actions for the actuators. The architecture of the LSTM network is shown in Table 2.

Table 2. LSTM model summary.

Layer	Output Dimension	Number of Parameters
LSTM	$Batch \times 5 \times 100$	42,800
Dropout	$Batch \times 5 \times 100$	0
LSTM	$Batch \times 5 \times 100$	80,400
Dropout	$Batch \times 5 \times 100$	0
LSTM	$Batch \times 100$	80,400
Dropout	$Batch \times 100$	0
Dense	$Batch \times 2$	202
Total trainable parameters		203,802

Three scenarios are analyzed in this study: In the first scenario, the LSTM network learns the control policy from a target system with a linear MPC (2500 points). This serves as a reference to evaluate the performance of an approximate MPC model derived from source system data. In this case, the inputs to the network are the measured temperature of each sensor ($T1$ and $T2$), the setpoints ($T1_{sp}$ and $T2_{sp}$) and the controller error ($T1 - T1_{sp}$ and $T2 - T2_{sp}$). The outputs of the network correspond to approximate power inputs to each transistor ($U1$ and $U2$). The linear MPC uses an autoregressive model with exogenous inputs (ARX), and both the development of the model and the solution of the optimization subroutines are handled with GEKKO [56]. The temperature setpoints (°C) and time between setpoint changes (seconds) are selected at random from $U(40, 60)$ and $U(120, 360)$, respectively.

In the second scenario, 50,000 points of data are gathered from the source system, where the setpoints (%) are selected at random from $U(20, 80)$ for each tank level, and the time between set point changes (seconds) are selected at random from $U(120, 360)$. An LSTM network is then used to learn the control policy from 50,000 points of data from the source system, taking as inputs the measured level from each tank ($H1$ and $H2$), the setpoints ($H1_{sp}$ and $H2_{sp}$) and the controller error ($H1 - H1_{sp}$ and $H2 - H2_{sp}$). The resulting model is then deployed as an approximate MPC to the target system.

In the third scenario, the same model developed in the second scenario is used as a starting point to be re-trained using 600 points of data from the target system. This is achieved by saving the weights of the model developed in the second scenario and retraining all weights for 300 epochs. The resulting model is then deployed as approximate controller to the target system in an online test.

In all scenarios, the deployed models are tested on the same set of setpoint changes. The temperature setpoints (°C) and time between setpoint changes (seconds) are initially selected at random from $U(30,70)$ and $U(120,360)$ respectively. The mean squared error (MSE) between the measured temperature and the temperature setpoint is used as a metric to compare the performance of each controller. An overview of the transfer learning process is shown in Figure 4.

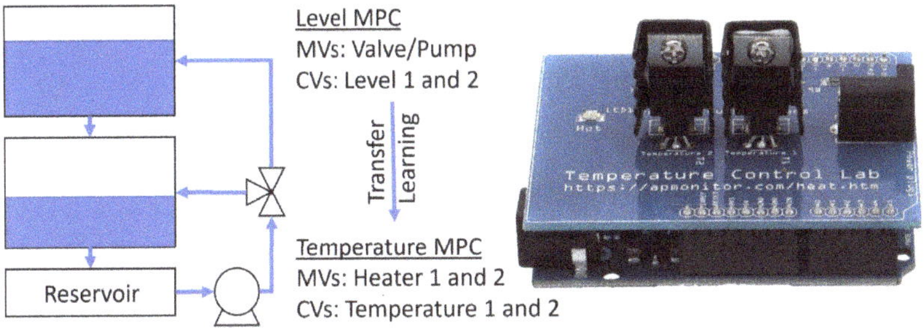

Figure 4. Overview of the transfer learning between liquid level MPC and temperature MPC.

3. Results and Discussion

Table 3 shows the MSE for the three scenarios described in Section 2. The MSE for a linear MPC is also included as reference. All training data are generated with simulated temperature response, while final tests are performed with the Arduino hardware in closed-loop MPC mode.

Table 3. Mean squared error (measured temperature vs. temperature setpoint).

Model	Linear MPC	LSTM Emulator	LSTM Transfer	LSTM Transfer Re-Train
MSE	27.75	102.52	131.82	102.43

In the first scenario, an LSTM network learns the control policy from the target system, which is controlled by a linear MPC. Figure 5 shows the fit of the model in a training and a validation set after 300 epochs. The model is then deployed as an approximate MPC to the target system in an online test. Figure 6 shows the performance of the approximate MPC, which follows the general behavior of the original linear MPC but has an MSE that is more than three times higher. This example of deep learning-based approximate MPC is the least complex of the three scenarios presented; it requires large amounts of training data, and fine tuning of the parameters of the neural network (e.g., learning rate, batch size, number of layers and nodes), but these can be adjusted by monitoring the learning metrics during training.

In the second scenario, an LSTM is trained with 50,000 points of data from the source system and deployed to the target system. Figure 7 shows the resulting performance of the approximate controller. Although the MSE is larger than that of the first scenario, the controller is able to maintain the general behavior from the MPC that was controlling the source system.

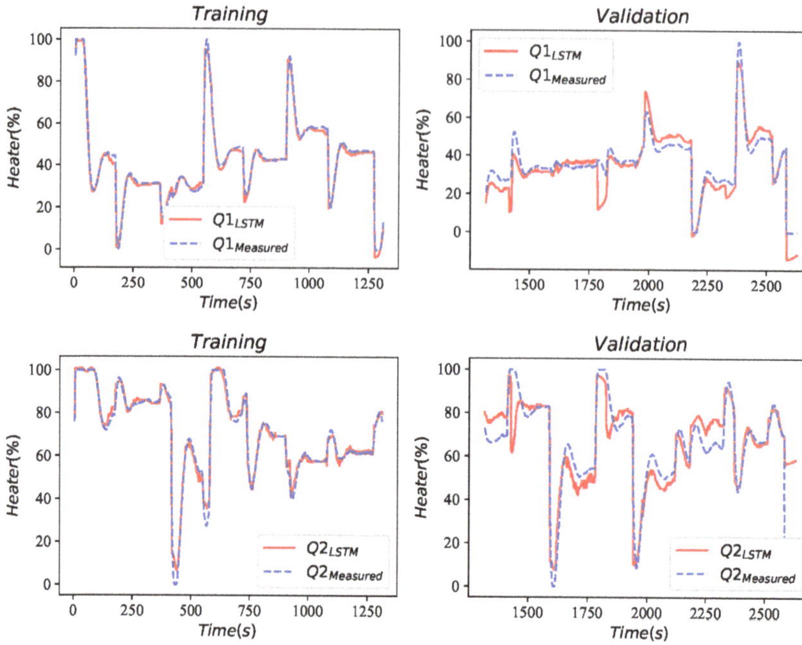

Figure 5. Training and validation fit from training an LSTM to learn the control actions of the target system. The model is trained on 2600 points of data for 300 epochs.

This is the first example of transfer learning in this study, and it shows that it is possible to develop an approximate MPC for a target system by using data from a source system, even if the dynamics and gains are not the same. In this test, the LSTM network is capable of extracting high-level features that are transferable from one system to another, but it is important to note that the selection of inputs to the approximate MPC (Y, Sp, $Error$) is also a key abstraction that helps the LSTM network generalize to other scenarios. The complexity of this technique increases due to the introduction of source data. In this case, the training metrics are affected not only by the parameters of the neural network, but also by the amount and quality of the source data.

In the third scenario, an LSTM is trained with 50,000 points of data from the source system, and partially retrained with 600 points of data from the target system under linear MPC. Figure 8 shows the model fit after training for 300 epochs. The model is then deployed to the target system in an online test as shown in Figure 9. This approximate MPC has an MSE of 99.45, which is similar to the performance in scenario 1, except this is using only a quarter of the data for training.

This is the second example of transfer learning in this study. It presents a framework for developing an approximate MPC by pre-training an LSTM network with source data and fine tuning with target data as shown in Algorithm 1.

Note that for transfer learning to offer improved control of the target system, the controller on the source system must exhibit improved performance over the initial control of the target system, as measured by the MSE. Additionally, this framework takes into account the number of points from the target data as the only tuning parameter to control the learning. In this case, 600 points was found by trial and error to be a satisfactory quantity. The effect of additional parameters such as learning rate, or batch size, along with optimization algorithms to find the optimal tuning parameters that minimize the MSE on the deployed controller should be explored in future publications.

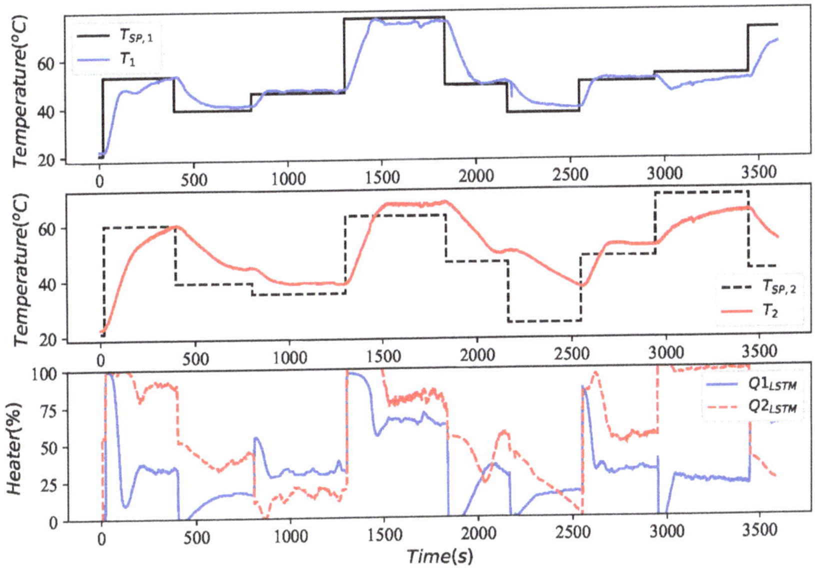

Figure 6. Scenario 1. Temperature and heater output of an approximate MPC control deployed online to a small scale temperature control (target system). The approximate MPC control is trained using only target data.

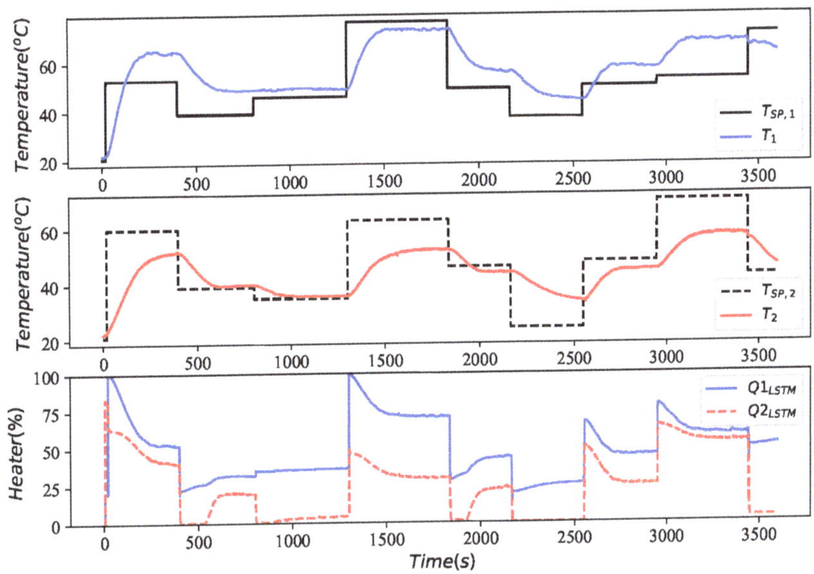

Figure 7. Scenario 2. Temperature and heater output of an approximate MPC control deployed online to a small-scale temperature control (target system). The approximate MPC control is trained using only source data.

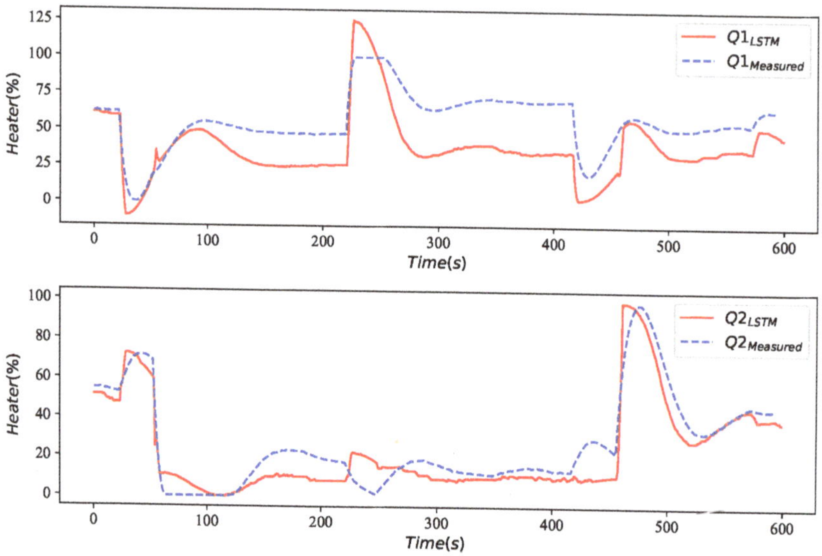

Figure 8. Validation fit after an LSTM network trained on source domain data is re-trained with 600 points of target domain data for 300 epochs.

Algorithm 1 A framework for approximate MPC using LSTM networks and transfer learning.

CV, SP and $Error$ data are collected from a target system.
The control performance is measured by $MSE_{Initial}$
CV, SP and $Error$ data are collected from a source system.
An LSTM network is pre-trained with data points from the source system.
The LSTM network is fine tuned with data points from the target system.
The performance of the fine tuned LSTM network is measured by the $MSE_{Approx.MPC}$ after being deployed.
while $MSE_{Approx.MPC} > MSE_{Initial}$ **do**
    Fine tune the controller with additional points from the target system.
end while

Scenario 3 shows an improvement of 28% compared to deploying to the target system without retraining (scenario 2). It is important to note that the training data used to develop the approximate MPC from source data used random changes in the setpoint, with $U(20, 80)$, while the training data from the target system that were used to partially retrain the approximate MPC used random changes in the setpoint, with $U(40, 60)$. Figure 9 shows that at around 1500 seconds, the approximate MPC extrapolates to meet a setpoint past the range of the target domain data. This is also observed in Figure 7, except the MSE is larger in that case. The complexity of this technique increases even further, and the performance of the learning metrics is now affected by the amount of quality of both the source data and the target data.

Additional future work is needed to remove the persistent offset through the use of state estimation or control variable bias in the approximate MPC and transfer learning framework [67]. The state estimation removes the persistent offset between the setpoint and process variable and would further improve the MSE results. It is purposefully left out of this study to isolate the benefit of transfer learning of only the controller.

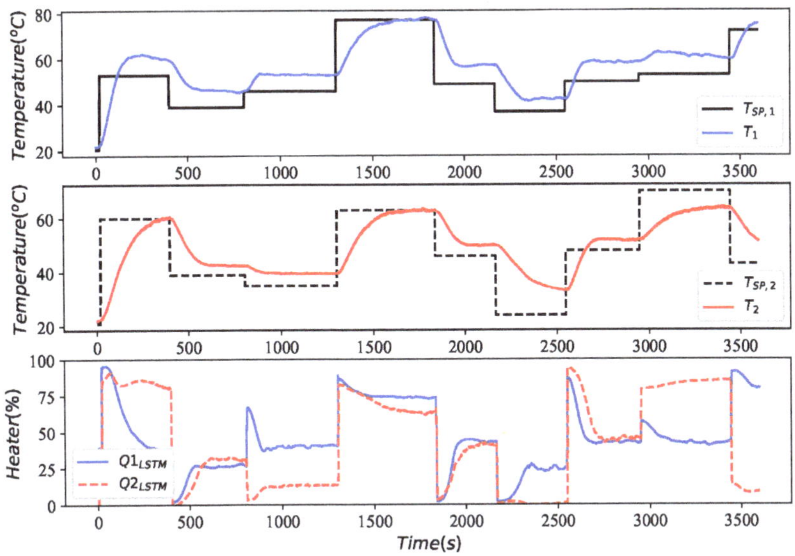

Figure 9. Scenario 3. Temperature and heater output of an approximate MPC control deployed online to a small scale temperature control (target system). The approximate MPC control is trained using 50,000 points of source data and fine tuned with 600 points of target data.

4. Conclusions

This study explored the use of LSTM networks to create an approximate MPC that can control a MIMO system. A deep learning-based control trained with data from a linear MPC maintains the behavior of the source MPC, but the MSE is 102.52 compared to 22.75 from linear MPC. Additionally, a framework for approximate MPC using LSTM networks and transfer learning was demonstrated by training an approximate MPC control on closed-loop data from a MIMO level control system and deploying to a small-scale MIMO temperature control unit. An approximate MPC trained on source domain data alone can control the temperature unit within 23% error (MSE is 131.82) compared to the approximate MPC trained on target domain data (MSE is 102.52). However, when the approximate MPC trained on source domain data is partially retrained on a small portion of the target domain data (600 points), the performance improves, and it is within 1% error (MSE is 102.43) compared to the approximate MPC trained on target domain data (MSE is 102.52). Furthermore, the transfer-learning based controller developed in this study can help extrapolate in some instances to setpoint changes outside the domain of the target data. While this study shows promising results using an LSTM network, future work should explore the use of other network architectures, such as Transformers, which have been successfully used for similar transfer learning tasks in the field of natural language processing [68] and process automation [69]. More work is needed on transfer learning for state estimation and methods to remove the potential offset from approximate MPC and other persistent errors from incomplete learning. Additionally, methods to quantify, generalize, and optimize the transfer of learning for approximate MPC formulations should be developed.

Author Contributions: Conceptualization, methodology, validation and formal analysis: S.A.M.; software, J.P. and S.A.M.; writing—review and editing, S.A.M., A.M.M., C.M.S., J.P. and J.D.H.; visualization, S.A.M. and A.M.M.; supervision, J.D.H. All authors have read and agreed to the published version of the manuscript.

Funding: This research was funded by Reliable Controls Corporation.

Institutional Review Board Statement: Not applicable.

Informed Consent Statement: Not applicable

Acknowledgments: The authors acknowledge the funding and technical support provided by Reliable Controls Corporation.

Conflicts of Interest: The authors declare no conflict of interest.

References

1. Ljung, L. Perspectives on system identification. *Annu. Rev. Control* **2010**, *34*, 1–12.
2. Otter, D.W.; Medina, J.R.; Kalita, J.K. A Survey of the Usages of Deep Learning for Natural Language Processing. *IEEE Trans. Neural Netw. Learn. Syst.* **2021**, *32*, 604–624. [CrossRef]
3. Zhang, Z.; Geiger, J.; Pohjalainen, J.; Mousa, A.E.D.; Jin, W.; Schuller, B. Deep Learning for Environmentally Robust Speech Recognition: An Overview of Recent Developments. *ACM Trans. Intell. Syst. Technol.* **2018**, *9*, 49:1–49:28. [CrossRef]
4. Jumper, J.; Evans, R.; Pritzel, A.; Green, T.; Figurnov, M.; Ronneberger, O.; Tunyasuvunakool, K.; Bates, R.; Žídek, A.; Potapenko, A.; et al. Highly accurate protein structure prediction with AlphaFold. *Nature* **2021**, *596*, 583–589. [CrossRef]
5. Al Seyab, R.; Cao, Y. Nonlinear system identification for predictive control using continuous time recurrent neural networks and automatic differentiation. *J. Process Control* **2008**, *18*, 568–581. [CrossRef]
6. Wong, W.C.; Chee, E.; Li, J.; Wang, X. Recurrent Neural Network-Based Model Predictive Control for Continuous Pharmaceutical Manufacturing. *Mathematics* **2018**, *6*, 242. [CrossRef]
7. Schwedersky, B.B.; Flesch, R.C.C.; Dangui, H.A.S. Nonlinear MIMO system identification with echo-state networks. *J. Control Autom. Electr. Syst.* **2022**, *33*, 743–754. [CrossRef]
8. Schweidtmann, A.M.; Esche, E.; Fischer, A.; Kloft, M.; Repke, J.U.; Sager, S.; Mitsos, A. Machine Learning in Chemical Engineering: A Perspective. *Chem. Ing. Tech.* **2021**, *93*, 2029–2039. [CrossRef]
9. Henao, C.A.; Maravelias, C.T. Surrogate-based superstructure optimization framework. *AICHE J.* **2011**, *57*, 1216–1232. [CrossRef]
10. Fahmi, I.; Cremaschi, S. Process synthesis of biodiesel production plant using artificial neural networks as the surrogate models. *Comput. Chem. Eng.* **2012**, *46*, 105–123. [CrossRef]
11. Boukouvala, F.; Hasan, M.M.F.; Floudas, C.A. Global optimization of general constrained grey-box models: New method and its application to constrained PDEs for pressure swing adsorption. *J. Glob. Optim.* **2017**, *67*, 3–42. [CrossRef]
12. Wilson, Z.T.; Sahinidis, N.V. The ALAMO approach to machine learning. *Comput. Chem. Eng.* **2017**, *106*, 785–795. [CrossRef]
13. Zhang, Q.; Grossmann, I.E.; Sundaramoorthy, A.; Pinto, J.M. Data-driven construction of convex region surrogate models. *Optim. Eng.* **2016**, *17*, 289–332. [CrossRef]
14. Zomorodi, H.; Landers, R.G. Extrusion based additive manufacturing using Explicit Model Predictive Control. In Proceedings of the 2016 American Control Conference (ACC), Boston, MA, USA, 6–8 July 2016; pp. 1747–1752. [CrossRef]
15. Mu, J.; Rees, D. Approximate model predictive control for gas turbine engines. In Proceedings of the 2004 American Control Conference, Boston, MA, USA, 30 June–2 July 2004; Volume 6, pp. 5704–5709. [CrossRef]
16. Yang, S.; Wan, M.P.; Chen, W.; Ng, B.F.; Dubey, S. Experiment study of machine-learning-based approximate model predictive control for energy-efficient building control. *Appl. Energy* **2021**, *288*, 116648. [CrossRef]
17. Pon Kumar, S.S.; Tulsyan, A.; Gopaluni, B.; Loewen, P. A Deep Learning Architecture for Predictive Control. *IFAC-PapersOnLine* **2018**, *51*, 512–517. [CrossRef]
18. Alessio, A.; Bemporad, A. A survey on explicit model predictive control. In *Nonlinear Model Predictive Control*; Springer: Berlin/Heidelberg, Germany, 2009; pp. 345–369.
19. Tøndel, P.; Johansen, T.A.; Bemporad, A. An algorithm for multi-parametric quadratic programming and explicit MPC solutions. *Automatica* **2003**, *39*, 489–497. [CrossRef]
20. Csekő, L.H.; Kvasnica, M.; Lantos, B. Explicit MPC-based RBF neural network controller design with discrete-time actual Kalman filter for semiactive suspension. *IEEE Trans. Control Syst. Technol.* **2015**, *23*, 1736–1753. [CrossRef]
21. Grosso, J.M.; Ocampo-Martínez, C.; Puig, V. Learning-based tuning of supervisory model predictive control for drinking water networks. *Eng. Appl. Artif. Intell.* **2013**, *26*, 1741–1750. [CrossRef]
22. Katz, J.; Pappas, I.; Avraamidou, S.; Pistikopoulos, E.N. Integrating deep learning models and multiparametric programming. *Comput. Chem. Eng.* **2020**, *136*, 106801. [CrossRef]
23. Chen, J.; Chen, Y.; Tong, L.; Peng, L.; Kang, Y. A backpropagation neural network-based explicit model predictive control for DC–DC converters with high switching frequency. *IEEE J. Emerg. Sel. Top. Power Electron.* **2020**, *8*, 2124–2142. [CrossRef]
24. Zheng, C.; Dragičević, T.; Blaabjerg, F. Model predictive control-based virtual inertia emulator for an islanded alternating current microgrid. *IEEE Trans. Ind. Electron.* **2020**, *68*, 7167–7177. [CrossRef]
25. Moness, M.; Moustafa, A.M. Real-time switched model predictive control for a cyber-physical wind turbine emulator. *IEEE Trans. Ind. Inform.* **2019**, *16*, 3807–3817. [CrossRef]
26. Wang, S.; Dragicevic, T.; Gontijo, G.F.; Chaudhary, S.K.; Teodorescu, R. Machine learning emulation of model predictive control for modular multilevel converters. *IEEE Trans. Ind. Electron.* **2020**, *68*, 11628–11634. [CrossRef]
27. Yan, Z.; Wang, J. Nonlinear model predictive control based on collective neurodynamic optimization. *IEEE Trans. Neural Netw. Learn. Syst.* **2015**, *26*, 840–850. [CrossRef] [PubMed]

28. Novak, M.; Dragicevic, T. Supervised imitation learning of finite-set model predictive control systems for power electronics. *IEEE Trans. Ind. Electron.* **2020**, *68*, 1717–1723. [CrossRef]
29. Hofer, M.; Muehlebach, M.; D'Andrea, R. Application of an approximate model predictive control scheme on an unmanned aerial vehicle. In Proceedings of the 2016 IEEE International Conference on Robotics and Automation (ICRA), Stockholm, Sweden, 16–21 May 2016; pp. 2952–2957.
30. Pin, G.; Filippo, M.; Pellegrino, F.; Fenu, G.; Parisini, T. Approximate model predictive control laws for constrained nonlinear discrete-time systems: Analysis and offline design. *Int. J. Control* **2013**, *86*, 804–820. [CrossRef]
31. Gan, M.; Hou, H.; Wu, X.; Liu, B.; Yang, Y.; Xie, C. Machine learning algorithm selection for real-time energy management of hybrid energy ship. *Energy Rep.* **2022**, *8*, 1096–1102. [CrossRef]
32. Wang, G.; Jia, Q.S.; Qiao, J.; Bi, J.; Zhou, M. Deep learning-based model predictive control for continuous stirred-tank reactor system. *IEEE Trans. Neural Netw. Learn. Syst.* **2020**, *32*, 3643–3652. [CrossRef]
33. Wang, Z.; Hong, T. Reinforcement learning for building controls: The opportunities and challenges. *Appl. Energy* **2020**, *269*, 115036. [CrossRef]
34. Chen, Y.; Tong, Z.; Zheng, Y.; Samuelson, H.; Norford, L. Transfer learning with deep neural networks for model predictive control of HVAC and natural ventilation in smart buildings. *J. Clean. Prod.* **2020**, *254*, 119866. [CrossRef]
35. Han, H.; Liu, Z.; Liu, H.; Qiao, J. Knowledge-data-driven model predictive control for a class of nonlinear systems. *IEEE Trans. Syst. Man, Cybern. Syst.* **2019**, *51*, 4492–4504. [CrossRef]
36. Chen, B.; Cai, Z.; Bergés, M. Gnu-rl: A precocial reinforcement learning solution for building hvac control using a differentiable mpc policy. In Proceedings of the 6th ACM International Conference on Systems for Energy-Efficient Buildings, Cities, and Transportation, New York, NY, USA, 13–14 November 2019; pp. 316–325.
37. Zhu, L.; Takami, G.; Kawahara, M.; Kanokogi, H.; Matsubara, T. Alleviating parameter-tuning burden in reinforcement learning for large-scale process control. *Comput. Chem. Eng.* **2022**, *158*, 107658. [CrossRef]
38. Koycegiz, C.; Buyukyildiz, M. Calibration of SWAT and Two Data-Driven Models for a Data-Scarce Mountainous Headwater in Semi-Arid Konya Closed Basin. *Water* **2019**, *11*, 147. [CrossRef]
39. Liu, Q.; Liu, Y.; Niu, J.; Gui, D.; Hu, B.X. Prediction of the Irrigation Area Carrying Capacity in the Tarim River Basin under Climate Change. *Agriculture* **2022**, *12*, 657. [CrossRef]
40. Forghanparast, F.; Mohammadi, G. Using Deep Learning Algorithms for Intermittent Streamflow Prediction in the Headwaters of the Colorado River, Texas. *Water* **2022**, *14*, 2972. [CrossRef]
41. Privara, S.; Široky, J.; Ferkl, L.; Cigler, J. Model predictive control of a building heating system: The first experience. *Energy Build.* **2011**, *43*, 564–572. [CrossRef]
42. Yang, Q.; Zhang, Y.; Dai, W.; Pan, S.J. *Transfer Learning*; Cambridge University Press: Cambridge, UK, 2020. [CrossRef]
43. Carbonell, J.G. A computational model of analogical problem solving. In Proceedings of the IJCAI—7th International Joint Conference on Artificial Intelligence, Vancouver, BC, Canada, 24–28 August 1981; Volume 81, pp. 147–152.
44. Winston, P.H. Learning and reasoning by analogy. *Commun. ACM* **1980**, *23*, 689–703. [CrossRef]
45. Forbus, K.D.; Gentner, D.; Markman, A.B.; Ferguson, R.W. Analogy just looks like high level perception: Why a domain-general approach to analogical mapping is right. *J. Exp. Theor. Artif. Intell.* **1998**, *10*, 231–257. [CrossRef]
46. Thrun, S.; Pratt, L. Learning to learn: Introduction and overview. In *Learning to Learn*; Springer: New York, NY, USA, 1998; pp. 3–17.
47. Pan, S.J.; Yang, Q. A survey on transfer learning. *IEEE Trans. Knowl. Data Eng.* **2009**, *22*, 1345–1359. [CrossRef]
48. Mohammadi, S.M.; Enshaeifar, S.; Hilton, A.; Dijk, D.J.; Wells, K. Transfer Learning for Clinical Sleep Pose Detection Using a Single 2D IR Camera. *IEEE Trans. Neural Syst. Rehabil. Eng.* **2021**, *29*, 290–299. [CrossRef]
49. Alotaibi, A. Automated and Intelligent System for Monitoring Swimming Pool Safety Based on the IoT and Transfer Learning. *Electronics* **2020**, *9*, 2082. [CrossRef]
50. Tan, C.; Sun, F.; Kong, T.; Zhang, W.; Yang, C.; Liu, C. A Survey on Deep Transfer Learning. In *Artificial Neural Networks and Machine Learning—ICANN 2018*; Kůrková, V., Manolopoulos, Y., Hammer, B., Iliadis, L., Maglogiannis, I., Eds.; Lecture Notes in Computer Science; Springer International Publishing: Cham, Switzerland, 2018; pp. 270–279. [CrossRef]
51. Wan, C.; Pan, R.; Li, J. Bi-weighting domain adaptation for cross-language text classification. In Proceedings of the Twenty-Second International Joint Conference on Artificial Intelligence Bi-Weighting Domain Adaptation for Cross-Language Text Classification, Barcelona, Spain, 16–22 July 2011.
52. Matasci, G.; Volpi, M.; Kanevski, M.; Bruzzone, L.; Tuia, D. Semisupervised Transfer Component Analysis for Domain Adaptation in Remote Sensing Image Classification. *IEEE Trans. Geosci. Remote Sens.* **2015**, *53*, 3550–3564. [CrossRef]
53. Li, W.; Gu, S.; Zhang, X.; Chen, T. Transfer learning for process fault diagnosis: Knowledge transfer from simulation to physical processes. *Comput. Chem. Eng.* **2020**, *139*, 106904. [CrossRef]
54. Hedengren, J.D. Level Regulation with MPC. 2018. Available online: https://apmonitor.com/do/index.php/Main/LevelControl (accessed on 15 November 2021).
55. Burchell, J.; le Roux, J.; Craig, I. Nonlinear model predictive control for improved water recovery and throughput stability for tailings reprocessing. *Control Eng. Pract.* **2023**, *131*, 105385. [CrossRef]
56. Beal, L.; Hill, D.; Martin, R.; Hedengren, J. GEKKO Optimization Suite. *Processes* **2018**, *6*, 106. [CrossRef]

57. Hedengren, J.D.; Shishavan, R.A.; Powell, K.M.; Edgar, T.F. Nonlinear modeling, estimation and predictive control in APMonitor. *Comput. Chem. Eng.* **2014**, *70*, 133–148. [CrossRef]
58. Oliveira, P.M.; Hedengren, J.D. An APMonitor Temperature Lab PID Control Experiment for Undergraduate Students. In Proceedings of the 2019 24th IEEE International Conference on Emerging Technologies and Factory Automation (ETFA), Zaragoza, Spain, 10–13 September 2019. [CrossRef]
59. Park, J.; Martin, R.A.; Kelly, J.D.; Hedengren, J.D. Benchmark temperature microcontroller for process dynamics and control. *Comput. Chem. Eng.* **2020**, *135*, 106736. [CrossRef]
60. Mejía, C.; Salazar, E.; Camacho, O. A comparative experimental evaluation of various Smith predictor approaches for a thermal process with large dead time. *Alex. Eng. J.* **2022**, *61*, 9377–9394. [CrossRef]
61. Yerolla, R.; Besta, C.S. Development of tuning free SISO PID controllers for First Order Plus Time Delay (FOPTD) and First Order Lag Plus Integral Plus Time Delay model (FOLIPD) systems based on partial model matching and experimental verification. *Results Control Optim.* **2021**, *5*, 100070. [CrossRef]
62. Sharma, S.; Padhy, P.K. An indirect approach for online identification of continuous time-delay systems. *Int. J. Numer. Model. Electron. Netw. Devices Fields* **2022**, *35*, e2947. [CrossRef]
63. de Moura Oliveira, P.B.; Hedengren, J.D.; Solteiro Pires, E.J. Swarm-Based Design of Proportional Integral and Derivative Controllers Using a Compromise Cost Function: An Arduino Temperature Laboratory Case Study. *Algorithms* **2020**, *13*, 315. [CrossRef]
64. de Moura Oliveira, P.; Hedengren, J.D.; Rossiter, J. Introducing Digital Controllers to Undergraduate Students using the TCLab Arduino Kit. *IFAC-PapersOnLine* **2020**, *53*, 17524–17529. [CrossRef]
65. Hedengren, J.D. Temperature Control Lab. 2021. Available online: http://apmonitor.com/pdc/index.php/Main/ArduinoTemperatureControl (accessed on 15 November 2021).
66. Abadi, M.; Agarwal, A.; Barham, P.; Brevdo, E.; Chen, Z.; Citro, C.; Corrado, G.S.; Davis, A.; Dean, J.; Devin, M.; et al. TensorFlow: Large-Scale Machine Learning on Heterogeneous Systems. 2015. Available online: https://www.tensorflow.org/ (accessed on 13 November 2022).
67. Hedengren, J.D.; Eaton, A.N. Overview of estimation methods for industrial dynamic systems. *Optim. Eng.* **2017**, *18*, 155–178. [CrossRef]
68. Raffel, C.; Shazeer, N.; Roberts, A.; Lee, K.; Narang, S.; Matena, M.; Zhou, Y.; Li, W.; Liu, P.J. Exploring the limits of transfer learning with a unified text-to-text transformer. *J. Mach. Learn. Res.* **2020**, *21*, 1–67.
69. Park, J. Hybrid Machine Learning and Physics-Based Modeling Approaches for Process Control and Optimization. Ph.D. Thesis, Brigham Young University, Provo, UT, USA, 2022.

MDPI AG
Grosspeteranlage 5
4052 Basel
Switzerland
Tel.: +41 61 683 77 34

Processes Editorial Office
E-mail: processes@mdpi.com
www.mdpi.com/journal/processes

www.ingramcontent.com/pod-product-compliance
Lightning Source LLC
Chambersburg PA
CBHW040834250926
51850CB00007BA/1308
9783725849031